T0192227

Responsible and Smart Land Management Interventions

Responsible and Smart Land Management Interventions

An African Context

Edited by
Walter Timo de Vries, John Tiah Bugri,
and Fatima Mandhu

CRC Press is an imprint of the
Taylor & Francis Group, an **informa** business

First edition published 2021
by CRC Press
6000 Broken Sound Parkway NW, Suite 300, Boca Raton, FL 33487-2742

and by CRC Press
2 Park Square, Milton Park, Abingdon, Oxon, OX14 4RN

CRC Press is an imprint of Taylor & Francis Group, LLC

ISBN: 978-0-367-33158-0 (hbk)
ISBN: 978-0-367-50094-8 (pbk)
ISBN: 978-1-003-02163-6 (ebk)

Typeset in Times
by Deanta Global Publishing Services, Chennai, India

Contents

SECTION IV Norms and Goals in Land Management Practice

SECTION V Land Management Methods and Tools

SECTION VI Future of Land Management

Preface

This book is made in the context of, and as a spin-off of, the ADLAND Project, "Advancing collaborative research in responsible and smart land management in and for Africa." This project was supported by Deutsche Gesellschaft für Internationale Zusammenarbeit (GIZ) on behalf of the German Federal Ministry for Economic Cooperation and Development (BMZ) and the German Federal Foreign Office (AA). The African Land Policy Centre (ALPC), the German government (through BMZ and GIZ), and the World Bank have jointly agreed to strengthen human and institutional capacities for implementing a sustainable and development-oriented land policy under the program "Strengthening Advisory Capacities for Land Governance in Africa (SLGA)," which has established the Network of Excellence on Land Governance in Africa (NELGA). The ADLAND project is part of the NELGA. It aims to advance the concept and praxis of responsible and smart land management, in the context of, and for the purpose of, being able to address the African land policy needs. This advancement is taking place through the networking of African and European Land Management research centers, implemented through a set of practical collaborative research and development activities. Researchers who are studying land management, land governance, and land polices in Africa have on various occasions for research development and research writing workshops in 2018 and 2019 collectively developed the content of this book.

The book aims to showcase new empirical findings on the conceptualization, design, and evaluation of land management interventions. It addresses two crucial aspects in this: how, and under which conditions, such interventions are responsible, and how such interventions can be supported by smart technologies. In this way, the book connects technical spatial design sciences with insights on socio-economic spatial relations. Conventional approaches in land management tend to be either too technocratic or lack a comprehensive or integrated view, and as a result do not work for most of the world's population. The approaches addressed in this book innovate land management, either by the enablement of smart technologies, and/or by properly linking societal needs to analytically driven solutions. Therefore, this book seeks to give a new kind of land management paradigm a theoretical and empirical basis. The chapters provide examples from different parts of Africa and from different study areas within the land management domain. The book is therefore useful for both academics, teachers, and land management practitioners and professionals. It can act as a new teaching resource for land management and land administration courses, and land related disciplines (e.g. geodesy, human geography, development studies, environmental planning), provide the basis for a new land management theory to be set against conventional land administration approaches and books, and deliver empirical data and cases for use in future research activities, taking place in both less developing and more developed contexts, and it offers, in a straightforward manner, a range of new tools for advocacy, design, and evaluation of land administration interventions. For the practitioner, the book delivers a contemporary update for the field land management sector, offers an overview of African developments, the current focus for innovative land administration design, provides a collection of cutting-edge tools from practice and for practice, with enough support data and methodological underpinnings to be readily utilized within the field.

The Editors

Prof. dr. ir. Walter Timo de Vries has been Chair of Land Management at the Technical University of Munich (TUM) since 2015. He is Study Dean for Geodesy and Geoinformation and Director of the Masters and PhD programs in Land Management and Geospatial Science. He has worked in numerous international projects in Asia, Africa, and South America, dealing with land information and land reform, geospatial data infrastructures, and professional training and education in land issues, cadastres, and information management. His current research interests at TUM include smart and responsible land management, urban and rural development, and land consolidation. Recent publications include the CRC/Taylor and Francis books *Advances in Responsible Land Administration*, *Responsible Land Management: Concept and Application in a Territorial Rural Context*, *Human Geodesy: Shaping a New Science and Profession for the World of Tomorrow*, and *Economic Versus Social Values in Land and Property Management: Two Sides of the Same Coin?*, amongst others.

Prof. John Tiah Bugri is an Associate Professor of Land Economy at the Department of Land Economy, Kwame Nkrumah University of Science and Technology (KNUST), Kumasi, Ghana. He is currently the Dean of the Faculty of Built Environment and formerly Head of the Department of Land Economy. He holds a PhD from the University of Greenwich, an MPhil from the University of Cambridge, and a BSc (Hons) in Land Economy from KNUST. He is particularly interested in people and land relations and focuses his research on the implications of human–land relations for sustainable land use and development, the dynamics of changes in political ecology on gender and livelihood sustainability, and the broader role of land governance and policy issues in poverty reduction in both agrarian and urban communities in developing countries. He has published widely in both local and international journals of repute and serves as a reviewer for several journals. He was the Country Coordinator of the World Bank Land Governance Assessment Framework (LGAF) for Ghana between 2011 and 2012 and is currently the Node Coordinator for Anglophone West Africa under the Network of Excellence for Land Governance in Africa (NELGA), a GIZ-funded project on Strengthening Advisory Capacities for Land Governance in Africa.

Dr. Fatima Mandhu is a post-doctoral fellow (since 2018) and lecturer at the University of Zambia (since 2011). The post-doctoral research title is "Oppression of women and nature in artisanal and small scale mining (ASM) in Zambia: A case study approach." She has served as Head of the Private Law Department at the University of Zambia in the School of Law. Her current interest in terms of research are mining law, land law, and gender, and she lectures in land law and property relations, as well as medical law. She is the project leader of the "Mineral Law in Africa: Zambia/South Collaborative Project" from the Zambian side. She is a participating member of the Southern African Node under the Network of Excellence for Land Governance in Africa (NELGA/NUST). She is the team leader for the scoping study on land governance and has submitted the draft chapter for Zambia. She is also involved in teaching on the LLM program and supervises both undergraduate as well as postgraduate students researching in land law. She has published a number of articles and has written book chapters. Recent publications include "Land Reform Policy Supporting Individual Land Rights and Diversification of the Zambian Economy: Realities on the Ground" in the *Zanzibar Yearbook of Law* (ZYBL) and "Land Titles or Deeds Registration as it relates to the new model of Land Registration system for Zambia" in the *Zambia Law Journal* (ZLJ).

Contributors

Bamiji Adeleye
Urban and Regional Planning Department
Federal University of Technology Minna
Minna, Nigeria

Ekundayo A. Adesina
Surveying and Geoinformatics Department
Federal University of Technology Minna
Minna, Nigeria

A.R. Ahmed
Department of Planning
University for Development Studies
Wa, Ghana

Kwasi Gyau Baffour Awuah
School of the Built Environment
University of Salford
Manchester, United Kingdom

Ninminga-Beka
Land Use and Spatial Planning Authority
Regional Office
Wa, Ghana

Tobias Bendzko
Chair of Land Management
Technical University of Munich
Munich, Germany

Charles Chavunduka
Department of Rural and Urban Planning
University of Zimbabwe
Harare, Zimbabwe

Uchendu E. Chigbu
Chair of Land Management
Technical University of Munich
Munich, Germany

Walter Dachaga
Chair of Land Management
Technical University of Munich
Munich, Germany

Pamela Durán-Díaz
Chair of Land Management
Technical University of Munich
Munich, Germany

Uwayezu Ernest
College of Science and Technology
Centre for Geographic Information Systems and Remote Sensing
University of Rwanda
Kigali, Rwanda

Andreas Hendricks
Bundeswehr University Munich
Neubiberg, Germany

Bupe Kabigi
University of Dodoma
Dodoma, Tanzania

Haule Kelvin
University of Dodoma
Dodoma, Tanzania

E.A. Kosoe
Department of Environment and Resource Studies
University for Development Studies
Wa, Ghana

E.D. Kuusaana
Department of Real Estate and Land Management
University for Development Studies
Tamale, Ghana

Monica Lengoiboni
Faculty of Geo-information Sciences and Earth Observation
University of Twente
Enschede, Netherlands

Ebelechukwu Maduekwe
Chair of Land Management
Technical University of Munich
Munich, Germany

Gianluca Miscione
University College Dublin
Dublin, Ireland

Gbenga Morenikeji
Estate Management and Valuation Department
Federal University of Technology Minna
Minna, Nigeria

Germain Muvunyi
INES Ruhengeri
Ruhengeri, Rwanda

Baslyd B. Nara
Faculty of Geo-information Sciences and Earth Observation
University of Twente
Enschede, Netherlands

Anthony Ntiador
Centre for Geospatial Intelligence Service
Institute of Local Government Studies
Accra, Ghana

Luke Obala
Department of Real Estate and Construction Management
University of Nairobi
Nairobi, Kenya

Joseph O. Odumosu
Surveying and Geoinformatics Department
Federal University of Technology Minna
Minna, Nigeria

Stellamaris Ogutu
The National Land Commission
Nairobi, Kenya

Lilian Mono Wabineno-Oryema
Department of Geomatics and Land Management
School of Built Environment
College of Engineering Design Art and Technology
Makerere University
Kampala, Uganda

Gloria Owona
Department of Land Administration and Management
INES-Ruhengeri
Musanze, Rwanda

Jossam Potel
Department of Land Administration and Management
INES-Ruhengeri
Musanze, Rwanda

Christine Richter
University of Twente
Enschede, Netherlands

Peterina Sakaria
Ministry of Land Reform
Windhoek, Namibia

Justin Tata
University of Juba
Juba, South Sudan

Wilson Tumsherure
Department of Land Administration and Management
INES-Ruhengeri
Musanze, Rwanda

Priscilla Tusiime
Technical University Munich
Munich, Germany

Frank Gyamfi-Yeboah
Department of Land Economy
Faculty of Built Environment
Kwame Nkrumah University of Science and Technology
Kumasi, Ghana

J.A. Zevenbergen
Faculty of Geo-information Sciences and Earth Observation
University of Twente
Enschede, Netherlands

Rafael Ziolkowski
University of Zurich
Zurich, Switzerland

Section I

Introducing Responsible and Smart Land Management

1 Framework of Responsible and Smart Land Management

Walter Timo de Vries, John Bugri, and Fathima Mandhu

CONTENTS

1.1 INTRODUCTION

Globally, a number of problems challenge the social-regulatory fabric of society with regard to land. Climate change, energy shortages, droughts, sudden outbreaks of new infectious diseases, migration, fake news, big data, and increasing privacy intrusion require new institutional forms of people-to-land relations to ensure sustainable and acceptable development. Problems occurring at local levels, such as boundary conflicts, disputed claims on land, international land grabbing of communal land, and natural disasters increasingly have a global effect. They lead to global land pressures and increasing vacant and unused land at the same time. They also lead to demographic and spatial changes, such as urbanization and increasing urban and rural divides. At the local level in Africa the global pressures manifest in different types of problems affecting the person-to-land relationships, including access, use, and ownership. The dualistic divide between urban and rural centering on tenure security has led to serious and complex land disputes. In Africa the interplay between the tenure systems that are regulated by statutory laws and customary tenures that are regulated by the customs and norms of society remains the focal point of the arguments. Obeng-Odoom (2012) argues that land reform in Africa is based on two theories: those who argue that land policies should be rooted in a theory of social capital, especially the African traditional land tenure system, and on the other hand, there are those who are convinced that individualized tenure systems are more effective and desirable. Resolving these conflicts requires a framework of responsible and smart land management.

1.2 LAND MANAGEMENT

To handle these problems properly one needs land management. The traditional way of presenting land management is a science of practice, i.e. a science that carefully examines the problems that professionals face in order to come up with solutions. The focus of such a science is to compare the practical solutions with the range of instruments available to a practitioner. Subsequently, the most effective, efficient, and appropriate combinations of alternatives or instruments are sought. This is not a problem in itself, as it helps practitioners to progressively expand their methodologies. Conversely, it helps the scientists to complete their insights on the practical work and the associated challenges.

On the other hand, however, this approach does not allow for a completely different view of the problem itself. The theory of "governance of problems" (Hoppe, 2011) states that problem-solving is essentially related to the epistemic community that shapes and defines problems. In other words, problems are not value-neutral. They reflect current political and ideological views and priorities. Therefore, a selection of issues also reflects the priorities, values, and contemporary beliefs in which the decisions are made. Therefore, as a scientist, one must understand the operant thoughts, which are largely normative, and the alternative thoughts that could provide new frameworks for the problems. In other words, the standards and techniques for solving problems are dynamic. Research on land management should therefore incorporate this dynamic.

Given this, de Vries and Chigbu (2017) describe land management as a combination of interventions in governance, law, socio-spatial relationships, economic opportunities, perceptions, and behavior. One can conceptualize this as an equation connecting how land interventions relate to and influence changes in multiple aspects relevant for land. If an intervention in land management can be quantified as ΔLM, then ΔLM is a function of (or otherwise put, it depends on, or yields to) the respective changes in governance, law, social-spatial relations, economic opportunities and dependencies, perceptions and beliefs, and behavior. In short form:

$$\Delta LM\ (\text{Land Management}) = f(\Delta G, \Delta L, \Delta S, \Delta E, \Delta P, \Delta B)$$

The idea that land interventions are both changes in themselves, and also derive changes, relates to Kötter et al. (2015) who argue that land management (and real estate management) is primarily action-oriented, namely towards components of spatial development and by corrective measures for the use of land and buildings. This does not happen in isolation of context and regulations. That's why Magel et al. (2016) emphasize the holistic nature of land management, from organizing engagements to changing spatial relationships to implementation at all levels of government. This holistic nature makes it so that while it is impossible to clearly distinguish land management from land administration or land policy and land governance, land management is specific in the sense that it concentrates on the use of land resources under current policy guidance and the legal framework for a given land area (Mattsson and Mansberger 2017). In order to describe land management, one needs, on the one hand, new ways to describe the relationship between action and changes in action and one needs new meta-concepts or theories, such as "human geodesy" (de Vries 2018, 2017). Part of the new meta-concepts is a more concrete understanding and application of what is "responsible" and "smart" land management.

1.3 RESPONSIBLE LAND MANAGEMENT

Discussions on responsible innovations refer to the question of who takes responsibility for the advancements and the effects of the innovations and whether the responsibility lies with a single actor or with a collective of actors. In this light, we call an innovation "responsible" when public authorities and private actors fundamentally aim to collaborate and are mutually accountable and responsive to each other in relation to the innovation. de Vries et al. (2015) describe the historical and epistemological roots of what can be considered "responsible." Rooted in public administration and political literature, the term "responsible" relates both to shared norms (in the sense of what is considered right or wrong in a particular socio-cultural context), and to the processes to which interventions and related actions are being conducted (e.g. whether people are being involved in the process and also take "responsibility" for certain actions), and to the effects which are being generated (related to the concept of accountability; people take responsibility if an adverse, unanticipated effect is emerging). This generates five specific aspects of "responsible," namely: multi-stakeholder focus, multidisciplinary, pro-active, international, and relevant/usable.

Building on these findings, de Vries and Chigbu (2017) developed a so-called 8R framework to plan and assess responsible land management interventions. This framework connects structures, processes, and impacts as key aspects of land interventions against eight indicators of responsible land management: responsiveness, resilience, robustness, reliability, respectedness, reflexivity, retraceability, and recognizability. Together they form an assessment and design matrix for land interventions (Amekwa et al. 2018).

1.4 SMART LAND MANAGEMENT

Land management, or perhaps more concretely a land intervention, is "smart" if it relies on passive and active information sensors before, during, and after the decision-making processes with regard to land. These sensors are generated both by technological means and based on voluntary and structured information contributions by people. Smartness, in other words, is the combination of both smart citizens, who are able to use information and communication technologies to advocate and pursue their interests, and on smart information-processing, i.e. facilities which can fuse data from all types of sources and platforms. The assumption is that smart(er) citizens are more capable of collectively shaping and deciding their own future, and as a result produce more intelligent, sustainable, and participative cities and societies. As this collective process of decision-making and the abundance of information as a basis for making decisions is changing, governance of cities changes. Whether smart land management is directly leading to responsible land management is not a given. Or, in other words, to what extent is smartness of citizens and of technologies changing the scale and type of governance, and increasing or decreasing the disparity between cities and the regions? And what can smart land management contribute to increasing responsibility?

With the definition that smartness is the combination of smart citizens and smart information-processing, the idea of smart land management is not only limited to employing geo-localization and wikification of places or massive use of ICT (information and communication technology) to address land management challenges. It is also about the accessibility and use of ICT for citizens and (state, non-state, and private) actors influencing land-related decisions; it's about the human and social capital accumulated through the experiences and activities of those actors; it's about the education level of the population to handle "smart" technology to foster and advocate certain land-related solutions and thereby generate localized land-related knowledge; and it's about the ability and capacity to derive a smart land economy, smart land mobility, and smart land-related environmental strategy.

Smart citizens and smart land-related actors produce smart land management, and vice versa. In other words, smart land management is a type of land management based on land-related agreements and change processes derived from and built on smart technologies. It will develop through learning and adaptive paradigms. Citizens and land-related actors build it together. Therefore, it is also an accumulation of intelligence. A smart city or region is therefore a city or region where its citizens are able to participate in shaping and deciding their future together with active and passive technology.

Despite this view that "smartness" lies in the combination of smart technologies and smart cities, in most of the smart city discourses, the definition of smartness is still not clear, and most authors confirm that it is multifaceted. The point of agreement for most writers is that cities must respond to changes in the context in which they operate. What amounts to intelligence depends on a number of underlying conditions such as the political system, geography, and dissemination of technology. A clear understanding is that smart solutions simply cannot be copied and pasted into new or different situation, therefore the value for each field needs to be evaluated differently to suit the circumstances into which it is being fitted(Pourahmad et al., 2018).

1.5 JUSTIFICATION AND EXPLANATION OF BOOK SECTIONS

Following this introduction, four distinct sections present a particular aspect of both responsible and smart land management, and a fifth section deals with future opportunities and further advances in developing responsible and smart land management. The five sections are:

1. Theory and new concepts of land management
2. Context and external drivers of land management
3. Norms and goals in land management practice
4. Land management methods and tools
5. Future of land management

Each of these sections are further introduced.

The theoretical section introduces a number of new concepts and approaches which are currently being investigated. Such matters include data quality and data veracity in the property market, spatial justice and its relationship to land tenure, the poverty paradox in relation to land, food security in the connection with land management and land interventions, and the concept of human recognition as a way to measure the extent of women's access to land.

The context and external drivers of land management involve the socio-institutional changes as well as the technical changes which are affecting both conceptual thinking about how land management interventions should be carried out, and which rules should guide land management interventions. Although regulatory changes related to land management keep on occurring in Africa, the direct effects of these are not always visible. However, they affect the planning practices and responsibilities. This chapter looks at this regulatory issue from various angles, including how it affects perceptions on progress in land reform, how it affects urban and rural land rights, how it changes the identity of the city, and how it affects tenure security. From a technological angle, the opportunity of blockchain is addressed, and how such an opportunity is institutionalized or affects institutional arrangements regarding land rights.

The next section deals with how practitioners shape the norms and goals of land management through their practices. It starts with a section on what land use planning and land governance ought to be, as compared to what it is in practice, followed by how the processes of congestion and urban expansion are managed. The systematic recording of land rights is additionally addressed, with a focus on what processes and systematic instruments are used and how these ultimately affect tenure security. Within this section there are also chapters dealing with the practice of conflict resolution and the practice of inter-agency collaboration. It is shown that the actual practices derive from the current land management intervention paradigms.

The subsequent section takes a closer look at the tools and instruments of land management. The ones which are discussed are: public value capture, the social tenure domain model, and the use of geospatial technologies. In all of the respective chapters there are innovative insights into how, where, and when the application of such tools are relevant and effective.

The final section of this book elaborates on future developments and how to advance responsible and smart land management. It focuses on persuasive and disruptive developments which are currently shaping the space within which land management can progress and identifies a number of key developments and opportunities which can make land management more responsible and smarter.

REFERENCES

Amekwa, Prince D., Walter Dachaga, Uchendu E. Chigbu, and Walter T. de Vries. "Responsible Land Management: The Basis for Evaluating Customary Land Management in Dormaa Ahenkro, in Ghana." 2018.

de Vries, Walter T., Rohan M. Bennett, and Jaap Zevenbergen. "Toward Responsible Land Administration." In: *Advances in Responsible Land Administration*, 3–14. CRC Press, 2015.

de Vries, Walter T., and Uchendu Eugene Chigbu. "Responsible Land Management - Concept and Application in a Territorial Rural Context." *fub. Flächenmanagement und Bodenordnung* 79(2) (April 2017): 65–73.

de Vries, Walter Timo. "Human Geodesy - Shaping a New Science and Profession for the World of Tomorrow." *FIG Working week 2017*, Helsinki, Finland, 2017.

———. "Suche Nach Neuen Grenzen Des Landmanagements." *zfv* 143(6) (2018): 373–383. doi:10.12902/zfv-0228-2018; https://geodaesie.info/zfv/zfv-62018/7946.

Hoppe, R. *Governance of Problems: Puzzling, Powering and Participation*. Portland, OR: Policy Press, 2011.

Kötter, Theo, Luz Berend, Andreas Drees, Sebastian Kropp, Hans Joachim Linke, Axel Lorig, Franz Reuter, et al. "Land-Und Immobilienmanagement–Begriffe, Handlungsfelder Und Strategien." *ZfV* 140(3) (2015): 136–146. doi:10.12902/zfv-0064-2015. http://geodaesie.info/system/files/privat/zfv_2015_3_Koetter_et-al.pdf.

Magel, H., F. Thiel, and J. Espinoza. "Bodenpolitik und Landmanagement – Eine Internationale Perspektive." In: Rummel, R. Et Al. (Hrsg.): *Handbuch Der Geodäsie, Band Bodenordnung und Landmanagement*. Springer Verlag *(Veröffentlichung Vorgesehen in 2016)*, 2016.

Mattsson, Hans, and Reinfried Mansberger. "Land Governance/Management Systems." *Land Ownership and Land Use Development: The Integration of Past, Present, and Future in Spatial Planning and Land Management Policies* 13 (2017).

Obeng-Odoom, Franklin. "Land Reforms in Africa: Theory, Practice, and Outcome." *Habitat International* 36(1) (2012): 161–170.

Pourahmad, Ahmad, Keramattolah Ziari, and Hossein Hataminejad. "Explanation of Concept and Features of a Smart City." *Bagh-e Nazar* 15(58) (2018): 5–26.

von Schomberg, René. "Why Responsible Innovation?" In: *International Handbook on Responsible Innovation*, Edward Elgar Publishing, 2019.

Section II

Theory of Responsible and Smart Land Management

2 Variation in Property Valuation and Market Data Sources in Ghana

Frank Gyamfi-Yeboah and Kwasi Gyau Baffour Awuah

CONTENTS

2.1 INTRODUCTION

Property valuation is fundamental to the healthy development of financial and property markets across the world. The effective and efficient operation of Africa's real estate markets depends largely on property valuation services. In fact, reliable valuation is seen as an important tool for good governance and transparent business activities. Currently, one of the major obstacles to real estate investment in Africa is market transparency challenges (Jones Lang Lasalle, 2013). However, real estate valuations carried out by surveyors in developing markets such as those of African countries are noted to be fraught with errors—inaccuracies, variations, and bias (see Obeng-Odoom and Ameyaw, 2011; Ayedun, Ogunba, and Oloyede, 2011; Babawale and Ajayi, 2011). This could potentially derail the current investment drive in Africa and ultimately threaten its socio-economic progress. It is, thus, necessary to streamline the process of valuation and improve availability of reliable market data to support ongoing efforts at promoting responsible land management. It must be stressed that heterogeneity in values produced by valuers for the same property injects uncertainties in decision-making and creates barriers in the implementation of smart land management interventions.

Valuations are essential for several reasons and are quite central to decision-making by market participants including financial institutions, investors, and governments. For instance, valuation may be required in lending decisions, insurance transactions, government divestiture and privatization programmes, taxation, and compensation in cases where properties are compulsorily acquired, among others. Regardless of the purpose, valuations do not only play an advisory role, but also serve as decision-making tools. Tretton (2007, p.482), for example, reckons that capital and rental valuations of real estate inform a major proportion of financial decisions in mature economies and the valuation of real estate, whether for sale, letting, or taxation purposes, is essential for all businesses. Consequently, inaccurate valuations are a recipe for wrong decision-making and could result in serious adverse repercussions, such as financial loss to investors and financial institutions, unhealthy operation of the property and financial markets, and undermining investor confidence. The collapse of the financial and property markets and the role inappropriate valuations played in the collapse during the 1990s in the UK as well as in that of the 2008 Asian financial crisis (see Ayuthaya and Swierczek, 2014) are instructive.

Several factors have been highlighted in the literature as possible causes of valuation errors including the choice of valuation methods and assumptions, quantity and quality of market data, and heuristic behavior of valuers or behavioral factors. The factor that is quite peculiar to developing markets is the lack of accurate and reliable market data. Property valuations, by their very nature, require the extensive use of data on market transactions to allow for meaningful comparisons between the subject of a valuation and market transactions in similar properties. Not surprisingly, all the methods of valuations require the element of comparison in their application. Consequently, the lack of accurate and reliable data is a major constraint on valuers' attempt to produce accurate and reliable valuations. Unlike advanced markets, most developing markets lack formal and organized databases of property transactions. Market transactions are often kept secret and where information is revealed, critical details are frequently omitted, making interpretation and analysis very difficult.

This chapter examines two related issues. Firstly, it provides empirical evidence on the extent of variation in values among valuers in Ghana. Variation, which measures the ability of two or more valuations to produce the same outcome, is one of three dimensions of valuation errors. The others are valuation accuracy and bias. It is pertinent to note that establishing variation requires that one measures the difference(s) between two or more valuations on the same property and under the same set of assumptions (Crosby, 2000). In this regard, the study requested professional valuers to undertake a market valuation of a hypothetical residential property as at a particular date based on the same set of instructions. Secondly, the study examines the sources and reliability of data used by valuers in the absence of formal organized databases in many developing markets such as Ghana. Even though there are no formal databases of property transactions, valuers often use methods that require that they rely on data sources for evidence of past transactions. It is, however, unclear which sources of data valuers use, and the extent to which such sources are reliable remains an empirical question. This study, therefore, seeks to investigate property market information sources for valuation practice in Ghana and to analyze the extent of reliability of the information they provide.

This study finds quite a high level of variation in valuation, consistent with anecdotal evidence. The level of variation found is substantially higher than the evidence reported for developed markets. The study also finds that valuers in Ghana most often rely on professional colleagues for data followed by their own database. It appears valuers have little faith in other data sources such as estate agents and the media. Since one of the likely causes of the high level of variation is the lack of accurate and reliable data, it is important that urgent steps are taken to create a database that would allow for the gathering of transaction data in a systematic way and in line with the requirements for a reliable market data.

The rest of the chapter is organized as follows: the next section reviews the literature on property valuation errors, followed by description of the data and methodology. These are followed by the presentation and discussion of the results after which a conclusion to the chapter is provided.

2.2 DETERMINANTS OF PROPERTY VALUATION ERRORS

The literature identifies several determinants of property valuation errors. However, in the main, these can be categorized into errors occasioned by the state of the property market, valuation models and assumptions, quantity and quality of market data, and heuristic behavior of valuers or behavioral factors (Iroham, Ogunba, and Oloyede, 2014). A core issue in property valuation is valuers' ability to interpret the property market correctly in their quest to assign value (Brown, 1992). Consequently, when there are doubts or uncertainties associated with the property market or misreading of the market by valuers, there is the likelihood that errors will occur (Matysiak and Wang, 1995). In their studies of the correspondence between valuations and subsequent prices in the UK, for example, Matysiak and Wang (1995) came to the conclusion that valuations tend to lag behind market prices during periods of rising markets, while they are usually above market prices in times of falling markets.

As stated previously, valuation is not a precise science. However, they rely on scientific processes and methods and are expected to be carried out in a methodical manner using the appropriate basis, set of assumptions, and methodology (Crosby, Lavers, and Murdoch, 1998).This requires understanding on the part of valuers as to clarity of instruction and purpose of valuation. It also requires valuers to undertake meticulous referencing of the subject matter of valuation—undertaking an inspection of the property to take inventory of its details, including its nature and complexity, checking the root of title and other encumbrances among other things, and employing the right basis of valuation, assumptions, and methodology. Without such a methodical approach, valuations are more likely to be engulfed in errors (Olawore et al., 2011). Indeed, Bretten and Wyatt (2001) in their study in the UK established that valuation models or methodologies have influence on valuation errors, while valuation of complex properties comparatively is more prone to errors.

Closely aligned with the state of the property market is availability of market data, such as evidence of sales and lettings of similar properties. Not only should data be adequate, it also needs to be of good quality (Olawore et al., 2011). Matysiak and Wang (1995) point out that inadequate market transactions or data are a recipe for property valuation errors. Brown (1992) also opines that valuations are a function of information. The better the information set, the better the valuation. Ratcliff (1968),, cited in Brown (1992, p.200), further states that:

> The prediction of value is no different than any other economic forecast. It is a prediction of human behaviour conditioned by a combination of dynamic known and unknown factors. It can never be predicted as a certainty and in some cases, the degree of uncertainty is high. Certainty of prediction is a function of the adequacy of the information on which the prediction is based and the skill and competence of the analyst.

Clearly, adequacy of information is paramount to good valuations and without it, errors are bound to occur. Peto (1997) acknowledges this view and makes the point that availability of market information should be seen from two standpoints: provision of accurate and timely data on transactions; and availability of data series and valuations to allow forecasting and modeling. To this end, several studies such as Peto (1997), Wyatt (1997), and Dunse et al. (1998) have been conducted in the developed world to improve property market data access and management for valuation purposes.

Behavioral factors—heuristic behaviors of valuers, including anchoring—have been identified as one of the causes of property valuation errors. Although research into this genre of causes of property valuation errors is fairly recent, it continues to receive attention and has become one of the topical areas in property research (Bretten and Wyatt, 2001; Iroham, Ogunba, and Oloyede, 2014). These heuristic behaviors encompass the use of simplifying short-cut means or rules of thumb developed over time to make value judgments and client pressures that influence value judgments. Iroham, Ogunba, and Oloyede (2014) identify four main heuristic behaviors. These are availability, representative, anchoring and adjustment, and positive heuristics. The availability heuristics are based on the experience decision-makers—in this instance, valuers—have had in the past with regard to solving the situation or problem at hand. This means that the strategy or solution that worked in the past is bound to be followed. The representative heuristics are more in the category of stereotyping, while anchoring and adjustment heuristics have to do with valuers making a priori judgments as to an estimate of value and then beginning to adjust the value as new information trickles in. The positive heuristics relate to valuers finding sets of data or information or design strategies to support their estimate of values rather than falsifying them.

2.3 DATA AND METHODOLOGY

Data for the study was obtained through the administration of a questionnaire instrument to professional real estate valuers who are members of the Ghana Institution of Surveyors (GhIS) in Accra, Ghana. The data instrument requested the valuers to provide an estimate of the market value for a hypothetical residential property at a particular date based on the same set of instructions. In addition, the respondents were asked to indicate the main sources of property market data they use and the reliability of each source.

The subject matter of the valuation was a leasehold estate in a single story three-bedroom house with an unexpired term of ten years sited on 0.093 ha of land located within the SHC neighborhood along the Liberation Road, Airport Residential Area, Accra, a commonly known area. This property was used because the design is well-known by valuers in the study site and market data for properties in the area where it was located is comparatively easy to come by. Thus, it was to ensure that a lot more valuers could participate in the research. All the property details were given to the respondents. These included the specific location of the property, its construction details, fixtures and fittings, external works, access to services, neighborhood characteristics, total floor area, sketch ground floor and location plans, title and planning/building permit status, and ground rent reserved. Also, information on the method employed by the respondents to undertake the valuation and why they employed the method was obtained.

Prior to the administration, a pre-testing of the questionnaire was undertaken to ensure that it passed the face and content validities test. This process among other things requested four experienced valuers to evaluate the questionnaire in terms of whether or not it covers what it was envisaged to address, and the effectiveness of how the research variables were to be measured. A total of 110 questionnaires were administered to the respondents. The respondents were selected based on purposive and snowball sampling techniques. These sampling techniques were employed due to a lack of a reliable sample frame. Although the GhIS provides a yearly list of valuers in good standing in Ghana, there is no such list specifically for valuers in Accra. Also, the lists do not often have the address and location of valuers. Accordingly, probability sampling could not be undertaken. A response rate of 63.64% was obtained.

The extent of variation in the market valuations obtained from the questionnaire survey was assessed by estimating the coefficient of variation and percentage median deviation of the valuations. Thus, the coefficient of variation of all the reported market values was estimated. The percentage median deviation was used because evaluation of the distribution of the reported market values was not normal based on both Levene and Simonov Kolmogorov tests. The median was, therefore, more representative of the distribution of the values than the mean (Field, 2005). The use of the percentage median variation to report the extent of variation is similar to the study by Adair et al. (1996), which used the percentage mean deviation to do so.

Using absolute values and ignoring negative signs, the percentage median deviation of each reported market value from the median market value was assessed as follows:

$$PMD_i = \left| \frac{x_i - MD}{MD} \right| \times 100 \tag{2.1}$$

where:

PMD_i = Percentage median deviation of each reported market value from the median market value;

x_i = Each reported market value; and

MD = Median market value.

The second part of the questionnaire instrument requested respondents to indicate how frequently they use different sources as well as to rate the reliability of these sources. A Likert scale was used, and the respondents were requested to rate the reliability of the identified property market data sources on a scale of 1–5 (1 = very unreliable, 2 = unreliable, 3 = quite reliable, 4 = reliable, and 5 = very reliable).

Generally, evaluation of an issue on a Likert scale is often undertaken based on consensus around the mean score. Tastle and Wierman (2007) intimate that such consensus could be assessed as follows:

$$\mathrm{Cns}\left(X \mid \mu_X\right) = 1 + \sum_{i=1}^{n} pi \log_2\left(1 - \frac{\left|X_i - \mu_X\right|}{d_X}\right) \tag{2.2}$$

where:

$\mathrm{Cns}\left(X \mid \mu_X\right)$ = Consensus around the mean score from the responses (the scores given the mean score);
X = The scores;
μ_X = The mean score;
X_i = Each score; and
d_X = The range of $X\left(d_X = X_{\max} - X_{\min}\right)$.

However, Tastle, Boasson, and Wierman (2009) on the basis of Equation (2.2) suggest that the evaluation could be undertaken using a preferred target (reference). Thus, μ_X could be replaced by the preferred target and in such circumstances, d_X should be multiplied by 2.

The research sought to evaluate the information obtained on the Likert scale in relative terms, and 5 was the highest score on all the scales. Thus, 5 was used as the preferred target, since a score on an attribute closer to or further away from 5 compared to the other attributes studied determined how highly that particular attribute was evaluated compared to the others. Thus, the following equation was used:

$$\mathrm{Agr}\left(X \mid 5\right) = 1 + \sum_{i=1}^{n} pi \log_2\left(1 - \frac{\left|X_i - 5\right|}{2d_X}\right) \tag{2.3}$$

where:

Agr = The level of agreement on evaluation of an attribute, and all other variables are as previously defined.

2.4 RESULTS

Seventy (70) out of a total of 110 questionnaires administered to the real estate valuers were received. This constituted 63.64% of the questionnaires that were administered. This is more than the response rate recorded from comparable studies such as Adair et al. (1996) (56%). However, for a few of the received questionnaires, some of the questions were not answered. These were taken into account in the data analyses.

Most of the respondents had less than 15 years of professional experience (30% were below 5 years of experience, 24.29% within 5–9 years, 15.71% within 10–14 years, and 18.57% above 25 years) (Figure 2.1). This result could be attributed to the increasing numbers of student intake on both the B.Sc. (Hons) Land Economy and Real Estate programmes at the KNUST over the last ten years or more compared to the initial years when the Land Economy program was introduced. Since graduates from these programmes have been the main source of professional membership for the Valuation and Estate Surveying (VES) division of the GhIS, it can be inferred that any change in student intake on the programmes would affect the membership state of that division. However, regarding the nature of practice of the respondents, 45.71%, 25.71%, and 28.57% of

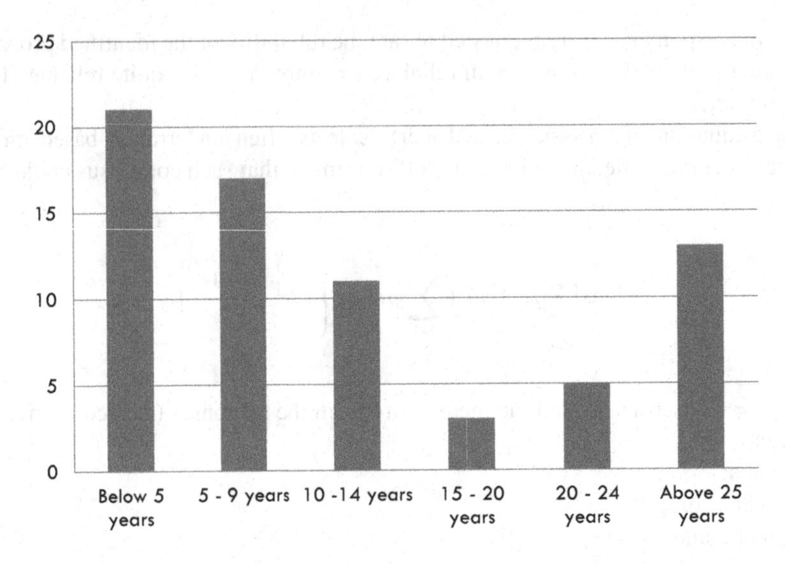

FIGURE 2.1 Respondents' years of experience.

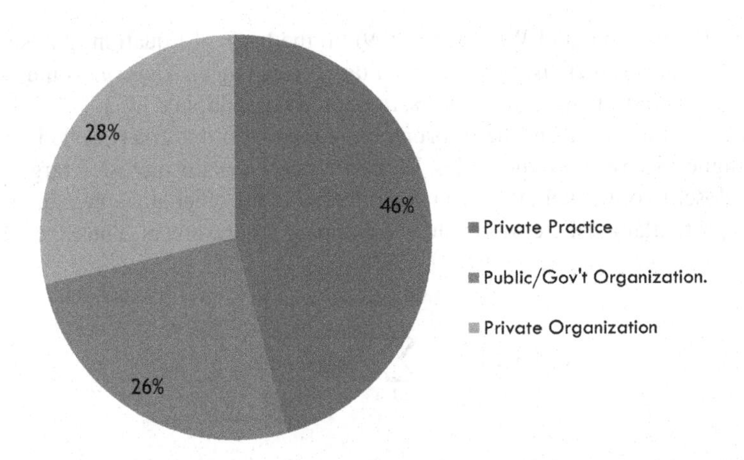

FIGURE 2.2 Nature of respondents' professional practice.

the respondents were in private practice, worked for government/public, and private organizations respectively (Figure 2.2).

2.4.1 Extent of Variation in Valuations

Tables 2.1 and 2.2 summarize the results on the extent of variation in the market value estimates received from the respondents on the three-bedroom residential property that constituted the subject matter of the valuations. Overall, the results indicate quite a level of variation, with a coefficient of variation of 0.63 (63%). The results also show that only 2.86%, 8.57%, and 14.29% of the valuations fell with ±5%, 10%, and 20% of the median for the reported value estimates respectively (Table 2.2). Further, these results show a much higher variation compared to Adair et al. (1996) in the UK, which established that the majority (80%) of the sampled valuations (n = 446) produced a variation from the mean of <20%, and Hansz and Diaz (2001) in the US, which produced a coefficient of variation of 0.098 (9.8%). This further emphasizes the extent of variation in the valuations.

A number of plausible reasons could be attributed to the high level of variability in the value estimates that were returned by the respondents. First, the subject matter of the valuations:

TABLE 2.1
Summary Statistics of the Reported Market Value Estimates

Statistic	Values
Minimum	90,700
Maximum	3,364,480
Mean	1,247,379
Median	1,134,650
Standard Deviation	784,428
Coefficient of Variation	0.63

TABLE 2.2
Percentage Mean and Median Variation in Valuations

	Percentage Variation	
	Mean	Median
<5	2.86%	2.86%
<10	7.14%	8.57%
<20	21.43%	14.29%
<30	34.29%	31.43%
<40	42.86%	42.86%
<50	57.14%	57.14%
<60	60.00%	60.00%
<70	72.86%	70.00%
<80	78.57%	71.43%
<90	85.71%	84.29%

a government leasehold interest in a residential property with an unexpired term of ten years may have appeared quite complex to the respondents. This is because in practice, valuations of such properties are often valued as a 50-year interest based on the assumption that government will renew the lease, although renewal is usually undertaken at a substantial cost, and it is not guaranteed. Secondly, the varying levels of experience of the sampled valuers (refer to the earlier discussion under the main subsection) may have resulted in different interpretation of the property market and the value estimates. Thirdly, the use of different methods of valuation to arrive at market value estimates could have accounted for the high variability in the valuations. The results show that seven different approaches (sales comparison (35.71%), replacement cost (38.57%), investment (15.71%), sales comparison and replacement cost (4.29%), sales comparison and replacement cost and investment (2.86%), sales comparison and investment (1.4%), and investment and replacement cost (1.4%)) were used by the respondents to produce the market value estimates. While this may reflect inadequate standardization in valuation practice, which could be corroborated by the numerous reasons reported by the respondents for the choice of a particular approach, it also mirrors the property market data challenge to valuation practitioners. This is particularly so, given that the valued hypothetical residential property was located in one of the prime and choicest areas of the case study country where the property market is very articulate, and that based on valuation theory, the market approach (sales comparison) should have been the preferred method of valuation.

2.4.2 Property Market Data Sources

Evaluation of the frequency of use of property market data sources by the respondents and their reliability was undertaken with Equation (2.2). A 5-point Likert scale was used to elicit the required responses. Results from the evaluation are reported in Tables 2.3 and 2.4.

In broad terms, the results demonstrate that valuers want to use property market data sources they perceive to be reliable. However, relying on professional colleagues for property market data was the most used property market data source (Agr|5 = 0.939) compared to the media, which is the least used (Agr|5 = 0.548). Valuation practitioners' own database was the second most used property market data source (Agr|5 = 0.826). This was followed by public institutions (Agr|5 = 0.797), estate developers (Agr|5 = 0.784), estate agents (Agr|5 = 0.717), and property owners (Agr|5 = 0.636), in that order. Ease of access to property market data, the need for valuation practitioners to check the reliability of their own databases, and the reliability of the property market data sources are possible reasons for the results. However, apart from the media as a property market data source whose usage corresponded to how the respondents rated its reliability, there were variations in how the respondents rated the reliability of the other property market data sources, compared with their frequency of use.

TABLE 2.3
Extent of Use of Property Market Data Sources

		Frequencies (%)										
Source	N	1	2	3	4	5	Min	Max	Mean	Median	Mode	Agr|5
Property Owner	70	2.86	21.43	37.14	18.57	20.00	1	5	3.31	3	3	0.636
Estate Agent	69	1.45	7.25	34.78	39.13	17.39	1	5	3.64	4	4	0.717
Professional Colleagues	69	0.00	0.00	8.70	13.04	78.26	3	5	4.70	5	5	0.939
Public Institutions	70	0.00	10.00	20.00	27.14	42.86	2	5	4.03	4	5	0.797
Estate Developers	69	1.45	10.14	14.49	37.68	36.23	1	5	3.97	4	4	0.784
Media	70	12.86	24.29	25.71	27.14	10.00	1	5	2.97	3	4	0.548
Own Database	66	3.03	6.06	10.61	30.30	50.00	1	5	4.18	4.5	5	0.826

1 = Do not use at all, 2 = Rarely, 3 = Quite often, 4 = Often, 5 = Very Often

TABLE 2.4
Reliability of Property Market Data Sources

		Frequencies (%)										
Source	N	1	2	3	4	5	Min	Max	Mean	Median	Mode	Agr|5
Property Owner	70	2.86	10.00	42.86	28.57	15.71	1	5	3.44	3	3	0.671
Estate Agent	69	1.45	11.59	42.03	39.13	5.80	1	5	3.36	3	3	0.657
Professional Colleagues	70	0.00	0.00	7.14	42.86	50.00	3	5	4.43	4.5	5	0.888
Public Institutions	70	0.00	5.71	25.71	31.43	37.14	2	5	4.00	4	5	0.794
Estate Developers	68	1.47	4.41	19.12	38.24	36.76	1	5	4.04	4	4	0.802
Media	67	7.46	22.39	50.75	17.91	1.49	1	5	2.84	3	3	0.528
Own Database	65	0.00	0.00	3.08	32.31	64.62	3	5	4.62	5	5	0.925

1 = Very unreliable, 2 = Unreliable, 3 = Quite reliable, 4 = Reliable, 5 = Very reliable

Although obtaining property market data from professional colleagues was the most often used data source, it was not perceived as the most reliable (Agr|5 = 0.888). On the contrary, practitioners' own database (Agr|5 = 0.925) was rated as the most reliable, even though it was not the most often used. Nevertheless, obtaining property market data from professional colleagues was perceived as more reliable than the other property market data sources. A possible reason for this finding could be the confidence that professionals have in the property market data collection and management capabilities of their colleagues. Further, obtaining property market data from real estate developers was rated as the next most reliable data source (Agr|5 = 0.802), followed by public institutions (Agr|5 = 0.794), property owners (Agr|5 = 0.671), and estate agents (Agr|5 = 0.657). The finding for real estate developers may be due to their ability to often provide current property market data. Inadequate record-keeping, and the tendency for provision of out-of-date data and the bureaucratic process associated with the provision of property market data may explain respondents' lack of confidence in the reliability of public institutions as a source of property market data. Also, the practice of not disclosing details of real estate transactions by property owners, and the poor data collection and documentation on real estate transactions by real estate agents are plausible explanations for the relatively low level of reliability respondents assigned to these sources.

2.5 DISCUSSIONS

Consistent with anecdotal evidence, the results indicate quite a high level of variation in the valuations produced by valuers in Ghana, as compared to the levels documented in the literature for developed markets. Obviously, the high level of variation should be of concern to all market participants as it can induce uncertainty into decision-making and impede the current investment drive into developing markets. It is important to stress that the interest held in the hypothetical property the respondents were asked to value may have played some part in accounting for the high level of variation reported. A leasehold interest with an unexpired term of ten years, though not uncommon, still presents challenges for valuers since there are currently no clear guidelines on the renewal of ground leases upon expiration. As a result, valuers often have to make assumptions regarding what is likely to occur when a ground lease expires. Moreover, there does not appear to be any consensus among valuers as to which assumption is reasonable or is likely, that is, whether an expired lease would be renewed and the price at which any extension will be granted. The fact that the legal and institutional framework governing ownership of land in Ghana since the coming into being of the 1992 constitution promotes leasehold interests, and in some cases, prohibits the disposition of the freehold interest, means that the valuation of leasehold interest will be predominant. It is therefore important that issues regarding lease renewals are clarified and coherent guidelines issued to allow for somewhat uniform assumptions to be made by valuers to help reduce excessive variations.

Apart from the apparently complex nature of the valuation, another plausible cause of the high levels of variation is property market data challenges faced by valuers. Paucity of property market data has been established as a challenge to valuation practice in Ghana. There is no formal organized database for keeping data on property transactions. As a result, valuers tend to use data from sources whose accuracy and reliability are difficult to verify. In fact, data from several of these sources may be questionable because of their lack of completeness. Issues such as lack of documentation of transactions, poor record-keeping and non-disclosure of transaction details are major problems that impede access to reliable property market data for valuation. Notwithstanding the foregoing, this work provides useful information not only to valuation practitioners, but to allied professionals such as land administrators, planners, lawyers, quantity surveyors and project managers, land/property taxation, and financial institutions and investors as to where to obtain property market data for decision-making on land/property transactions.

Real estate valuation practice should enhance real estate business and market operations. Thus, it is essential that good quality valuations with incidence of limited variations and other errors are produced. This may require continuous re-evaluation of standards for valuation practice at the professional body level in the face of rapidly changing local and international real estate markets. It may also require valuation practitioners to examine their approach to valuation practice, and in particular, the need to obtain reliable property market data. This includes working in harmony and collaboration with all the stakeholders in the valuation/real estate industry to promote standardization and access to good quality property market data to improve practice.

2.6 CONCLUSION

This chapter has provided empirical evidence on the extent of variation in the value estimates produced by valuers in a developing market characterized by a paucity of reliable and adequate market data. Analysis of data obtained through a questionnaire survey of professional real estate valuers in Ghana indicates quite a high level of variation in valuations. This is consistent with anecdotal evidence but substantially higher than the evidence reported for developed markets. The effects of this variation and heterogeneity among valuers are: an ineffective land market; wrong/inaccurate estimations in values when acquiring land for infrastructural projects; and problems/disputes when providing compensation for acquiring land. Reliable property market information is critical to the production of professional and ethical valuations. However, access to such information for valuation practice in Ghana remains a challenge. A shift in the current practices of how market information is collected, managed, provided, and accessed is required. This suggests a need to examine the existing property market data sources for valuation practice and the reliability of the data they produce to provide input into any initiative to address the market data challenges faced by valuers in the country.

The study established seven sources of property market information for valuation practice in Ghana. These are: public and quasi-public institutions; property owners; property valuation practitioners and professional property consultancy firms; lawyers who deal with property transactions (professional colleagues); real estate developers; practitioners' own databases; informal real estate agents; and the media/online transactions. In terms of reliability of the market data produced by the sources, valuers' own databases were regarded as the most reliable source followed by the databases of their professional colleagues, real estate developers, public institutions, property owners, estate agents, and the media in that order. Further, it was found that the access to reliable property market information problem is due to the incomplete and scattered nature of market data; non-disclosure of details of property transactions; data integrity concerns; and lack of requisite training and experience especially for estate agents in collecting and managing market data. The findings further shed light on the link between the degree of uncertainty associated with the determination of value and the adequacy of reliable and accurate data.

Given the central role reliable valuations and market data play in responsible land management, it is important that urgent steps are taken to create a database that would allow for the gathering of transaction data in a systematic way and in line with the requirements for reliable market data. This will not only reduce the level of variation in values among valuers but more importantly, it would improve decision-making and engender more confidence among players in the land market.

REFERENCES

Adair, Alastair, Norman Hutchison, Bryan MacGregor, Stanley McGreal, and Nanda Nanthakumaran. "An analysis of valuation variation in the UK commercial property market: Hager and Lord Revisited." *Journal of Property Valuation and Investment* 14(5) (1996): 34–47.

Ayuthaya, N., and F. W. Swierczek. "Factors influencing variation in value and investors confidence." *IOSR Journal of Business and Management* 16(5) (2014): 41–51.

Ayedun, C.A., O.A. Ogunba, and S.A. Oloyede. "Empirical verification of the accuracy of valuation estimates emanating from Nigerian valuers: A case study of Lagos Metropolis." *International Journal of Marketing Studies* 3(4) (2011): 117–129.

Babawale, G.K., and C.A. Ajayi. "Variance in residential property valuation in Lagos, Nigeria." *Property Management* 29(3) (2011): 222–237.

Bretten, James, and Peter Wyatt. "Variance in commercial property valuations for lending purposes: An empirical study." *Journal of Property Investment and Finance* 19(3) (2001): 267–282.

Brown, Gerald R. "Valuation accuracy: Developing the economic issues." *Journal of Property Research* 9(3) (1992): 199–207.

Crosby, Neil. "Valuation accuracy, variation and bias in the context of standards and expectations." *Journal of Property Investment and Finance* 18(2) (2000): 130–161.

Crosby, Neil, Anthony Lavers, and John Murdoch. "Property valuation variation and the 'margin of error' in the UK." *Journal of Property Research* 15(4) (1998): 305–330.

Dunse, Neil, Colin Jones, Allison Orr, and Heather Tarbet. "The extent and limitations of local commercial property market data." *Journal of Property Valuation and Investment* 16(5) (1998): 455–473.

Field, Andy. *Discovering Statistics Using SPSS* (2nd edn.). London: Sage Publications, (2005).

Hansz, J. Andrew, and Julian Diaz III. "Valuation bias in commercial appraisal: A transaction price feedback experiment." *Real Estate Economics* 29(4) (2001): 553–565.

Iroham, C.O., O.A. Ogunba, and S.A. Oloyede. "Effect of principal heuristics on accuracy of property valuation in Nigeria." *Journal of Land and Rural Studies* 2(1) (2014): 89–111.

Jones Lang Lasalle (JLL). "Twenty African cities emerge as next frontier for commercial real estate growth between now and 2020." http://www.joneslanglasalle.co.uk/unitedkingdom/engb/pages/NewsItem.aspx?ItemID=27948. [Accessed: September, 2013].

Matysiak, George, and Peijie Wang. "Commercial property market prices and valuations: Analysing the correspondence." *Journal of Property Research* 12(3) (1995): 181–202.

Obeng-Odoom, Franklin. "Real estate agents in Ghana: A suitable case for regulation?" *Regional Studies* 45(3) (2011): 403–416.

Obeng-Odoom, Franklin, and Stephen Ameyaw. "The state of the surveying profession in Africa: A Ghanaian perspective." *Property Management* 29(3) (2011): 262–284.

Olawore, Akin, Austin Otegbulu, and G.K. Babawale. "Valuers' perception of potential sources of inaccuracy in plant and machinery valuation in Nigeria." *Property Management* 29(3) (2011): 238–261.

Peto, Robert. "Market information management for better valuations: Part II-data availability and application." *Journal of Property Valuation and Investment* 15(5) (1997): 411–422.

Ratcliff, Richard U. "Capitalized income is not market value." *Appraisal Journal* 36(1) (1968): 33–40.

Tastle, William J., and Mark J. Wierman. "Consensus and dissention: A measure of ordinal dispersion." *International Journal of Approximate Reasoning* 45(3) (2007): 531–545.

Tastle, William J., Emil Boasson, and Mark J. Wierman. "Assessing team performance in information systems projects." *Information System Educational Journal* 7(90) (2009): 1545–679X.

Tretton, David. "Where is the world of property valuation for taxation purposes going?" *Journal of Property Investment and Finance* 25(5) (2007): 482–514.

Wyatt, Peter J. "The development of a GIS-based property information system for real estate valuation." *International Journal of Geographical Information Science* 11(5) (1997): 435–450.

3 Patterns of Spatial Justice and Land Tenure Security from Urban Land Management

Insights from Kigali City, Rwanda

Uwayezu Ernest and Walter Timo De Vries

CONTENTS

3.1 INTRODUCTION

Pursuing spatial justice in land management can remedy different forms of spatial injustices and social inequalities resulting from unjust social and political interactions which require the management of spatial resources (Rawls, 1971, p.214). In urban areas, spatial injustices and social inequalities derive from neoliberal processes of urban (re)development, which deprive some urban dwellers of access to land, housing, and basic urban amenities. Remedies for these injustices require embedding spatial justice in both the rules and the processes of urban management. Generally, spatial justice refers to the spatial dimension of social justice. It consists of conceptual apparatus of moral and social norms, rules, and practices seeking equality of rights and opportunities for all people to access and/or use various spatial resources, alongside the (re)organization of any geographic space. This aspect of social justice relates to Rawls' principles of equal rights and opportunities to the access and use of basic resources for all people within any society (Rawls, 1971, pp.213–214). For the purpose of this chapter, spatial justice is discussed from the perspective of advancing individuals' rights to land, housing, and urban amenities. These are key inputs to all people's wellbeing. Recognition and respect of these rights is sought in both rules and processes of urban land management (Chatterton, 2010), with much attention to the rights of poor and vulnerable groups so that they can improve their living conditions (Gibbs, 2009). This claim resonates with the aspirations of responsible land management. It advocates for land agencies and development options that deliver their interventions in a responsible way, aligned with the needs of owners and users of land resources (de Vries and Chigbu, 2017). In praxis, spatial justice and responsible land management share in common three dimensions, converging towards one aspiration of meeting (or being responsive to) basic needs and promoting tenure security for all users of land resource (de Vries and Chigbu, 2017; Uwayezu and de Vries, 2018). These dimensions are rules (for spatial justice) connected with structures (for responsible land management), processes, and outcomes (Zevenbergen, de Vries, and Bennett, 2015; Lefebvre, 1991; Marcuse, 2010). *Rules* include formal policies, laws, constitutions, legal directives, and informal social norms, codes of conduct, conventions, and political decisions affecting the management of land and other spatial resources. To this, responsible land management adds the technical aspects of interventions connected to these rules and make "structures" as a whole. *Processes* embrace the implementation procedures of these rules, interventions, and different activities pertaining to land management, in collaboration with local community. *Outcomes* imply the results expected in access and use of spatial resources by all citizens towards improving their living conditions. Land tenure security derives from these outcomes, especially from increased social and spatial integration of all people in spatial management processes and defiance to all practices depriving them of their real properties (Uwayezu and de Vries, 2019). This chapter deconstructs patterns of spatial justice in rules, processes, and outcomes of urban land management in Kigali city and its potential for land tenure security for the poor and vulnerable urban dwellers. These patterns are explored in the broad contours of promoting inclusive and sustainable urban development in Rwanda (Ministry of Finance and Economic Planning, 2013), which includes the relocation of these people from high risk zones (including wetlands and steep slope areas) to serviced sites. Their relocation is also framed in the general goals of the Kigali city master plan of ensuring access to basic urban facilities for all urban dwellers, and sustainable use of urban land (City of Kigali, 2016). Following the introduction, spatial justice is firstly conceptualized and connected to land tenure security. Secondly, trends of spatial justice emerging from the resettlement process of poor and vulnerable informal urban dwellers in Kigali city are identified. Thirdly, the implications of this process for tenure security are ascertained. A general conclusion is drawn at the end.

3.2 SETTING SPATIAL JUSTICE INTO URBAN LAND MANAGEMENT

Spatial justice discourse gained its momentum in urban management literature in the 2000s (Harvey, 2010). Spatial justice proponents started to decry uneven processes of urbanization engendering spatial injustices through social-spatial segregation depriving poor urban dwellers of access to social and material goods (Rawls, 1999). They argued for spatial justice, consisting of rational distribution of urban resources among all urban inhabitants, based on the principles of equality of rights and opportunities so that they can improve their living conditions and participate equally in the socio-economic development processes. To achieve this, four forms of spatial justice consisting of procedural, recognitional, redistributive, and intra- and inter-generational justice have to be pursued in the management of any geographic space (Fainstein, 2014; Fraser and Honneth, 2003; Rawls, 1999).

Procedural justice means appropriate urban (re)development and/or land management rules crafted in a participatory and consultative manner. These rules must recognize the rights to access and use land resources for all people, regardless of their socio-economic statuses (poor, low or middle-income, or rich) and forms of tenure. Community participation and dialogue in making these rules and related structures are also normative values making land management responsive to all people's needs. This renders land management processes and outcomes more accountable, reliable, transparent, and acceptable (de Vries and Chigbu, 2017). From a procedural justice perspective, these outcomes include remedies to material resources deprivation through compensation or their restitution when land management projects deprive urban dwellers of their properties. *Recognitional justice* embraces the respect of all people's rights to land and other resources, regardless of the types of land tenure (formal or informal), when implementing the rules. It requires devising land development options which permit all people, especially poor and other disadvantaged people to access or use land resources in order to meet their basic needs (Young, 1990). Similarly, responsible land management advocates must recognize individual socio-economic status and tenure aspects when designing and implementing land management systems (de Vries and Chigbu, 2017). *Redistributive justice* requires fair distribution of land and other material resources and services proportionally to the needs of all categories of people. Like the responsiveness pattern of responsible land management, *redistributive justice* helps overcome material resources deprivation that undermines human wellbeing (Magel, 2016). *Intra- and inter-generational justice* appeals for effective rules and processes of land management, providing different categories of people of the current and future generations with equal opportunities to access and use material resources (Magel, 2016). Similar to the *responsibility* aspect of responsible land management, pursuing *intra-generational justice* results in equal chances to access and use land and other resources for all people of the same generation. While the *intra-generational justice* is focused on prioritizing access to these resources for the least advantaged in the society, *inter-generational justice* inclines to equality and opportunities for access and use of spatial resources for all people of present and future generations. Embedding these claims in land management rules and processes advances the recognition and respect of all people's land rights (access, ownership, use, etc.) and their transfer to future generations as a desideratum for their welfare (Picard et al., 2015).

Drawing from the above, procedural justice is required for recognitional and redistributive justice to advance equality of rights and opportunities for all people in accessing material resources and meeting their basic needs (Rawls, 1999). Combining procedural, recognitional, redistributive, and intra- and inter-generational justice is required for redressing resources deprivation for the worst-off (Magel, 2016). Therefore, they are pursued in combination to attain their universal aspirations. These aspirations are also claimed by responsible land management. They include participatory and transparent resources management, rendered by just and inclusive land management rules, applied for improving the quality of life and economic wellbeing of all people, and enhanced tenure security (Uwayezu and de Vries, 2019; Zevenbergen, de Vries, and Bennett, 2015). In practice, tenure security is enhanced through a framework of urban land management, grounded on a spatial justice claim. Therefore, we can grasp the connection between spatial justice and land tenure security based on these aspirations.

3.3 CONNECTING SPATIAL JUSTICE TO LAND TENURE SECURITY

Spatial justice aspirations include remedies for deprivation of land and other material resources for all people and their socio-spatial integration into spatial development processes. Achieving them also enhances tenure security. Land tenure security is defined as the landowners' perception of their rights to land and attached properties on a constant basis, free from eviction, or interference from outside sources (Simbizi, Bennett, and Zevenbergen, 2014). Secure tenure protects property owners against arbitrary removal from their lands or residences and provides them with a certain degree of confidence that they will not lose these properties within some future time period. Incontestably, the loss can occur, following legal procedures which must be objectively applied and which can include just compensation, either in cash or in-kind, through relocation processes (UN-Habitat 1996).

The contemporary literature discusses land tenure security as a composite concept of three elements: *de jure* (or legal), *de facto*, and perceived security (Payne, 2001). The *de jure* tenure security is connected to legal registration of land rights through land titling (Simbizi, Bennett, and Zevenbergen, 2014). The *de facto* tenure security emerges from recognition and enforced respect of individuals' rights to land, by social and political-administrative institutions, even when these rights are not recorded (Payne, 2001). The perceived tenure security is connected to stability and durability of property ownership within a socio-political environment which recognizes property rights for all people and prevents their displacement from their properties (van Gelder, 2009). However, in urban areas, any of the three elements of tenure security can flare and converge into tenure insecurity through eviction or displacement of property owners, due to urban (re)development schemes which do not tolerate different forms of tenure. Displacement, being one pattern of tenure insecurity, can also result from land use regulations which are not aligned with financial capacity of property owners to use their lands, even when they possess land titles (Payne, 2001).

One way to counteract this insecurity of tenure would be to reinvigorate spatial justice in the urban space management, because it advances the social value of land and integration of all urban dwellers in the urban space, from which emerges tenure security. Alternatively, restoration of property rights through just compensation constitutes a remedial option to displacement and alternative approach to promote land tenure security from a spatial justice lens (Uwayezu and de Vries, 2019). These logical frames echo the potential of spatial justice to land tenure security, which is among the key outcomes of responsible land management. We probe this using the urban (re)development experience in Kigali city.

3.4 THE CASE STUDY

Our case study is Kigali, the capital city of Rwanda, in Eastern Africa. This city has experienced a rapid population growth during the last 20 years. In the 1960s, it was a small center whose main function was the political administration (City of Kigali, 2013). The built-up area increased from 3 km^2 in 1962 to 15 km^2 in 1984 and reached 93 km^2 in 2012 (Rwanda Environment Management Authority, 2017). Its population increased from 6,000 inhabitants in 1962 to 235,664 inhabitants in 1991, and reached 1,135,428 inhabitants in 2012. This population growth results from an increasing annual birth rate (of 2.6%) combined with a high (10%) rural-urban migration net (Ministry of Finance and Economic Planning and National Institute of Statistics of Rwanda, 2014). However, this demographic growth of Kigali city has not been coupled with the provision of basic infrastructure and serviced land for housing development (Manirakiza and Ansoms, 2014). This resulted in an escalating development of informal settlements, deprived of access to basic urban amenities and environmental degradation (Rwanda Environment Management Authority, 2017). These settlements cover 62% of residential areas and host 79% of Kigali city inhabitants, mainly composed of poor and low-income groups (National Institute of Statistics of Rwanda, 2015). To curb the growth of informal settlements, the Government of Rwanda and administration of Kigali city have been implementing detailed master plans, which re-organize the urban space and extend the provision of urban amenities (City of Kigali, 2013). Unfortunately, the initial implementation of these master plans had inadvertently been fueling land tenure insecurity. This was reflected in the displacement of poor and low-income dwellers from Kigali city, through slum clearance operations carried out until 2014 (Finn, 2018; Goodfellow, 2014).

From 2015, various approaches to implementing Kigali city master plans in spatially just ways, which promote social protection of the poor and low-income dwellers of informal settlements have been adopted (City of Kigali, 2016). These approaches include the resettlement of poor people and vulnerable groups, whose houses are in high risk zones, in serviced sites. This process reflects various aspects of spatial justice which are required in redressing deprivation of rights to land, housing and urban services for the disadvantaged urban dwellers (Uwayezu and de Vries, 2018). This chapter therefore deconstructs the potential patterns of spatial justice and land tenure security emerging from this process.

3.5 DATA SOURCES AND METHODS

Primary data for this study were collected from January to March 2018 and January to March 2019 in three districts of Kigali city. We used interviews and Focus Group Discussions (FGD) with heads of households in three villages, where poor and vulnerable informal settlements dwellers have been resettled (see Figure 3.3). Our key informants range between 26 and 28 people, randomly selected from 104 households resettled in each village. Interviews were also conducted with land managers, urban planners, heads of social units in each district, and local leaders at each of the three villages. These interviews and FGD covered the following topics: selection of resettled households, status of land and housing ownership before the resettlement, perceptions of the resettled people on their land tenure security before and after their resettlement, access to basic amenities and employment in the new settlements, and types of support provided to them for the reconstitution of their livelihoods. Field observations consisted of collecting data on the quality of housing and access to basic amenities. Secondary data came from the review of laws, policies, and regulations related to urban (re) development in Rwanda, Kigali city development plans, and related activities reports. Qualitative content analysis of the provisions of the rules in use and pattern matching between these provisions, their implementation practices, and their outcomes was used to derive findings. We combined this analytical approach with indicators evaluating whether Kigali city (re)development advances spatial justice and land tenure security. The evaluative framework is presented in the next section.

3.6 EVALUATIVE FRAMEWORK

We used an analytical framework of spatial justice comprising a series of indicators connected to four forms of spatial justice and its three dimensions: rules, processes, and outcomes. "Rules" stands for law, policies, zoning regulations, master and physical development plans, political decisions applied in the (re)development of Kigali city, and resettlement of displaced people. "Processes" relates to implementation of these rules. "Outcomes" covers the aspects of peoples' relations to space and includes access to housing and urban amenities, and tenure security (Uwayezu and de Vries, 2018). The framework is illustrated in Figure 3.1.

As Figure 3.1 shows, we applied the evaluative indicators connected to the three dimensions of spatial justice (discussed in the introduction) and its four forms, to derive insights on how Kigali

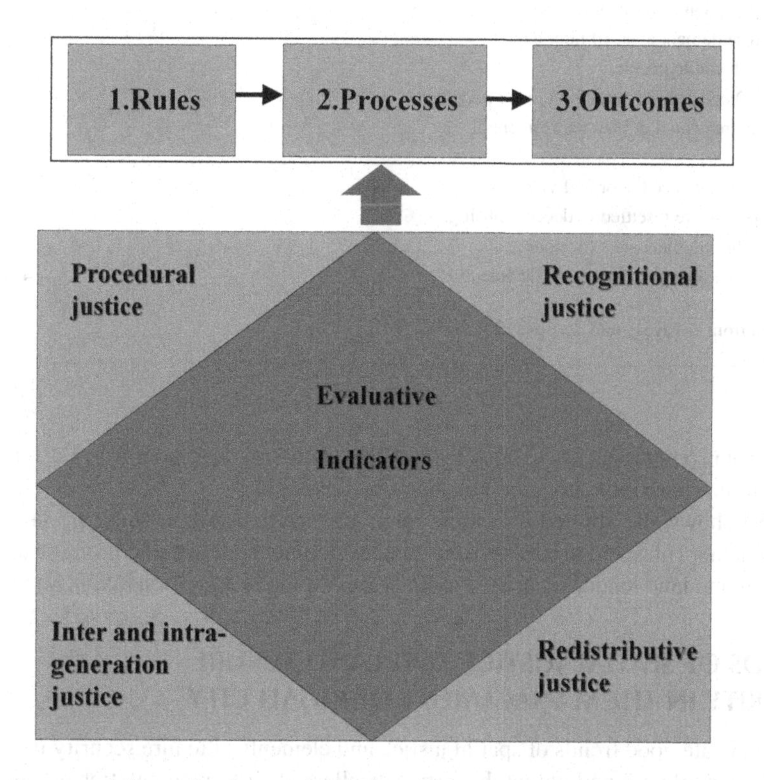

FIGURE 3.1 Framework for evaluating spatial justice and land tenure security in land resources management. (Source: Adapted from Uwayezu and de Vries, 2018).

TABLE 3.1
Indicators Evaluating Trends of Spatial Justice and Land Tenure Security in the Resettlement Processes of Poor and Vulnerable Urban Dwellers in Kigali City

		Related form of spatial justice			
Dimension of spatial justice	**Evaluative indicators**	**Procedural**	**Recognitional**	**Redistributive**	**Inter- and intra-generational**
Rules	1. Urban development and zoning rules promote the development of housing for poor and vulnerable groups.	✓	✓	✓	✓
	2. Zoning rules promote the provision of basic urban amenities in settlements of poor and vulnerable groups.	✓	✓	✓	✓
	3. Urban development rules promote the allocation of land and/or housing to poor and low-income groups.	✓	✓	✓	✓
Processes	4. Spatial development plans include resettlement sites for poor and vulnerable people living in slums.	✓	✓	✓	
	5. Urban development budgets include funds for housing development for poor and vulnerable groups.	✓	✓	✓	
	6. Relocation plans include funds for the provision of basic amenities in resettlement areas of poor and vulnerable people.		✓	✓	
Outcomes	7. Displaced poor and vulnerable people are relocated in planned residential areas.		✓	✓	✓
	8. All displaced poor and vulnerable people are resettled in decent housing.		✓	✓	✓
	9. Resettlement sites for poor and vulnerable groups have basic amenities.		✓	✓	✓

Sources: Adapted from Uwayezu and de Vries, 2018.

city management promotes access to housing and urban amenities for poor urban dwellers and their security of tenure. These indicators are presented in Table 3.1.

As Table 3.1 shows, the applied indicators relate to specific aspects which are largely observed in assessing whether rules and processes of urban development are just and promote access to basic urban resources and land tenure security. Findings are presented and discussed in Section 3.7 below.

3.7 TRENDS OF SPATIAL JUSTICE AND LAND TENURE SECURITY IN THE MANAGEMENT OF KIGALI CITY

Our findings indicate good trends of spatial justice and elements of tenure security in rules underlying the resettlement of poor and vulnerable urban dwellers, their implementation practices, and their outcomes. These findings are summarized in Figure 3.2.

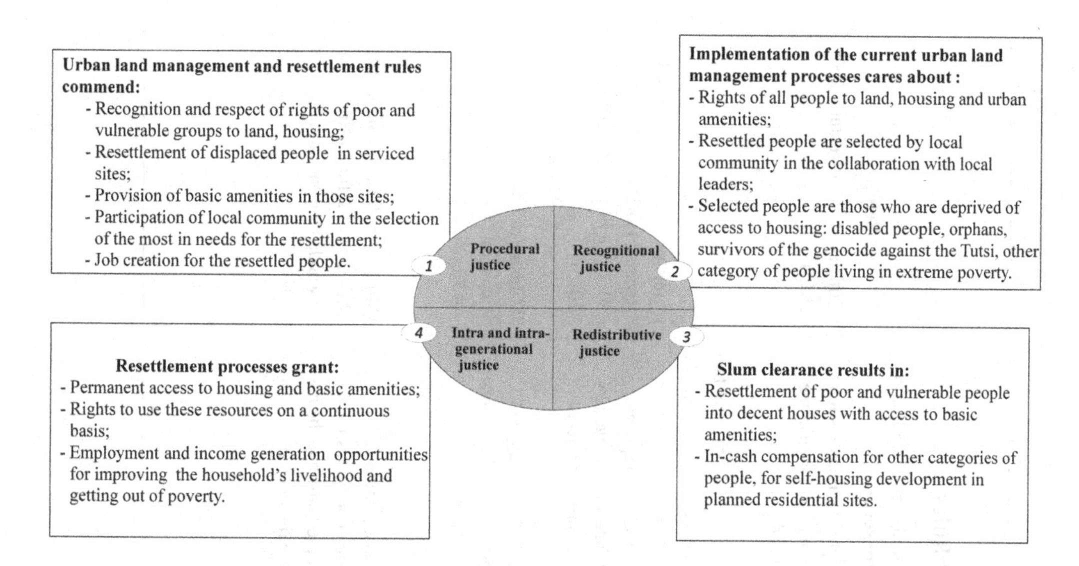

FIGURE 3.2 Patterns of spatial justice and land tenure derived from Kigali city (re)development schemes. (Data source: Review of rules in use, and field observations (January–March, 2018 and January–March, 2019)).

As Figure 3.2 shows, procedural justice (1) is embedded in urban land management rules binding Kigali city authorities, urban planners, and other decision-makers to promote inclusive and participatory urban (re)development, with due respect of rights to land, housing, and basic urban amenities for poor and vulnerable urban dwellers. Recognitional justice (2) is reflected in the increased recognition of these rights alongside the implementation practices of the rules in use. Redistributive justice (3) is reflected in the resettlement of these people in decent houses, developed in planned urban neighborhoods. It is also embedded in the cash compensation for people who cannot be resettled by the government due to limited financial resources. This has increased the perceptions of dwellers of informal settlements on their inclusion in the urban space and the recognition of their property rights. The integration of poor and vulnerable people in the city has also spurred the inter- and intra-generation justice (4), because it resulted in the creation of a good social and spatial environment in which they enjoy their rights to basic goods and commodities. Through increased feelings of their inclusion in the city and the financial support they receive, they invest in income generation activities to get out of poverty and improve their livelihoods. The next subsection discusses in detail how these outcomes have been attained.

3.7.1 Patterns of Spatial Justice into Urban (Re)development Rules

Findings on the potential of rules underlying the resettlement of the poor and vulnerable urban dwellers to spatial justice and land tenure security alongside the development of Kigali city are summarized in Table 3.2. It shows the identified patterns of spatial justice and land tenure security within these rules and which are reflected in the inclusion of the resettled people in the urban space and their access to quality housing and basic urban resources.

Table 3.2 shows that the current rules underlying the Kigali city management exhibit all forms of spatial justice, converging towards two main aspirations: (1) advancing the rights of all urbanites to inhabit and use the urban space through decreased exclusion, and (2) their support in improving their livelihoods. In the next subsection, we discuss how these aspirations are attained in practice.

3.7.2 The Resettlement of Poor and Vulnerable People and Access to Basic Amenities

The main tangible pattern of spatial justice deriving from the resettlement of poor and vurnerable urban dwellers in Kigali city is their access to decent houses. These houses have been developed

TABLE 3.2
Features of Spatial Justice and Land Tenure Security Identified in Kigali City (Re)development Rules

Form of spatial justice	Identified actions as stated in rules	Reviewed legal framework	Features of spatial justice and land tenure security	References
Procedural	Promotion of access to adequate and safe housing, basic infrastructure, and services for all people.	Urbanization policy; human settlement policy; Kigali city development plans.	• Equality of rights to urban resources. • Inclusion of poor and low-income people in urban development processes. • Redressing resources deprivation for the least advantaged.	(Ministry of Infrastructure, 2009, 2015)
Recognitional	• Relocating people living in high risk zones to serviced sites; • Provision of basic amenities and job creation in those sites.	Human settlement, urban housing, and construction policy; housing development policy; Kigali city development plans.	• Compensation and reparation to resources deprivation for people whose land rights are infringed by urban (re)development projects.	(Ministry of Infrastructure and Rwanda Housing Authority, 2011; City of Kigali, 2013, 2016)
Redistributive	• Allocation of public funds for land acquisition and housing development for the worst-off. • Financial support to resettled people for meeting basic life needs.	Housing development policy; economic development and poverty reduction strategy; Kigali city development plans.	• Restitution of rights to land and housing by prioritizing the poorest and vulnerable people.	(Ministry of Finance and Economic Planning, 2013, City of Kigali, 2013)
Inter- and intra-generational justice	• Development of inclusive, safe, resilient, and sustainable city. • Supporting job creation and income generation activities for all human settlements to improve living conditions of their inhabitants.	Urbanization and housing policies. Kigali city development plans.	• Resources redistribution and improved living conditions for the worst-off people constitute the base for their future livelihoods.	(Ministry of Infrastructure, 2015, Ministry of Infrastructure and Rwanda Housing Authority, 2011)

Sources: Interviews, review, and content analysis of urban (re)development rules in use.

FIGURE 3.3 Housing quality before and after the resettlement process of poor and vulnerable urban dwellers in Kigali city. (Data source: Field observations (January–March, 2018 and January–March, 2019)).

since the financial year 2016–2017 in each of the three districts of Kigali city. As shown in Figure 3.3, they are in the forms of detached houses and apartments, within a configuration of urban village.

People who are resettled in these villages include the survivors of the Rwandan genocide against the Tutsi, historically marginalized people, very poor widows, and old people. Their resettlement embraces some patterns of procedural justice, since it reflects the political zeal for improving the living conditions of poor and marginalized groups, as stated in the national urbanization policy (Ministry of Infrastructure, 2015). Procedural justice is also embedded in the participation of the local community in selecting the beneficiaries of this resettlement process. Priority is given to households meeting the criteria defined by the Ministry of Local Government (MINALOC), which is the leading agency in the development of social housing. Selected households are very poor people whose annual consumption expenditure is less than US$180. They also have to fulfill one of the following criteria: they neither have a house nor can build it themselves, have a disability and cannot build their own houses, live in houses rented by the government or charitable organizations, or have self-developed poor-quality houses in high risk zone where their lives can be endangered. Resettling these people is also framed in the moral realm of recognitional and redistributive justice pledging to prioritize poor people when the intended actions are remediation of deprivation in housing and basic commodities. A redistributive pattern of their resettlement is also reflected in its funding schemes. This activity is funded by Kigali city and its three constituent districts in the framework of the developing social housing for poor and vulnerable people. By March 2019, the budget allocated to each developed village was estimated at US$3.5 million.* Instigating features of spatial justice in this resettlement process offers other interesting insights, if we refer to what the unambiguous meaning of spatial justice conceals, from the lens of intra- and inter-generation justice. It is the restoration of rights to spatial resources or redistribution of sources for wellbeing, in accordance with the needs of people who have been historically deprived of access to these resources. Figure 3.4

* Financial reports of Gasabo, Kicukiro and Nyarugenge districts.

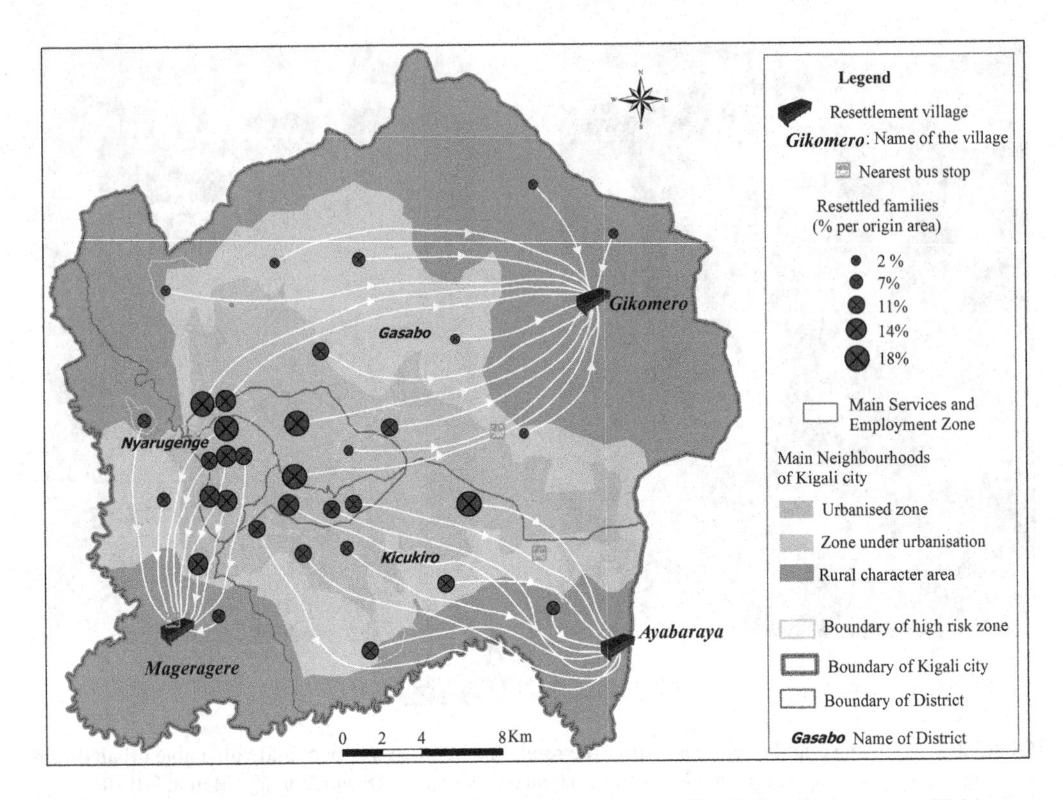

FIGURE 3.4 Resettlement sites for poor and vulnerable groups in Kigali city. (Data source: Field survey (January–March 2019)).

shows that most of the resettled people had been living in high-risk areas, where they were exposed to different calamities, such as landslides and flooding.

The wellbeing aspect also relates to the protection of their lives, in addition to the opportunities for owning shelter on a continuous basis, and accessing and using basic urban amenities. In addition, there is a kindergarten in each village, while primary schools and health posts and/or centers are located within 500 meters.

3.7.3 Access to Basic Goods and Employment

Claims of spatial justice include the recognition of all people's rights to other goods, required for exerting their basic capabilities and sustaining their livelihoods. The resettled people receive support for income generation and production of some food stuffs. As Figure 3.5 shows, this support consists of developing poultry and small-scale horticulture which are fully supported by Kigali city and its constituent districts.

Apart from these funded activities, there are other social funding schemes helping the resttled people to improve their living conditions. Each family receives a monthly subsistence allowance covering the cost for water, electricity, and other daily comodities. The amounts of money they receive vary according to the size of the household as follows:* 7,500 rwandan francs (Rfws†) for a single household, 12,000 Rfws for a household composed of two people, 15,000 Rfws for three people, 18,000 Rfws for four people, and 21,000 Rfws for five people or more. In addition, districts pay for their health insurance, while their children are exempted from the payment of school fees. These

* Interviews with resettled households and financial reports of social affairs units of the three Districts of Kigali city.

† The exchange rate was 884 Rwandan francs for 1 USD in March 2018.

FIGURE 3.5 Income generation activities in residential sites of resettled poor and vulnerable people. (Data source: Field observations (January–March 2019).)

forms of support are embedded in recognitional, redistributive, and inter-and intra-generational justice, and in universal principles of human rights requiring actors in spatial resources management to ensure that disadvantaged people have access to essential goods which sustain their survival.

It is worth noting that these people are resettled far from the core urban area. As Figure 3.4 shows, the resettlement sites for poor and vulnerable groups in Kicukiro and Gasabo Districts (the Ayabaraya and Gikomero sites) are located in remote rural areas, compared to their former neighborhoods which are within the core urban area. These sites are even very distant to the nearest bus stop connecting them to core urban area. Their dislocation from the core urban area creates difficulty for active people in keeping their small jobs, on which they had financially relied on before their resettlement. This is the main problem that they face because they will hardly find other employment opportunities in their current neighborhoods.

3.7.4 Featuring Land Tenure Security from the Resettlement Processes

The resettlement of poor and vulnerable people in serviced sites also relates to social and spatial integration. From this emerges land tenure security through remediation of the evictions that would affect the resettled people alongside the implementation of the new master plan of Kigali city. In fact, these people developed their dwellings in the 1980–1990s, due to the inability of Kigali city administration to provide them with the options to develop their dwellings in planned areas (Rwanda Environment Management Authority, 2015). They had therefore illegally developed houses on their lands located in high slope areas (over 30 degrees: Rwanda Environment Management Authority, 2015). Others had illegally developed their houses on public lands, including wetlands and forest areas. Over time, these settlements were tolerated by Kigali city authorities. This political tolerance resulted in *de facto* tenure security for these inhabitants. With this tolerance followed the provision of social facilities such as water, electricity, transportation network, education, and health facilities, in accordance with the recognition of rights of dwellers of these areas to basic urban amenities.

From this recognition emerged the perceived tenure security. Land registration undertaken in 2009–2012 resulted in the provision of certificates of property ownership, except for urban dwellers whose houses had been developed on public lands. The non-registration of their property rights has subverted the existing perceived and *de facto* tenure security. Tenure insecurity increased in 2013 following the passage of detailed master plans of Kigali city, which recomended the clearance of all settlements located on steep slope areas and wetlands which were unsuitable zones for residential use. However, the insecurity of tenure that could result from the implementaion of the master plans has been counteracted through devising resettlement and compensation options for the inhabitants of these areas, so that they can have access to other dwelling units.

This political decision emanated from rules related to housing development and the national urbanization policy which bolstered the resettlement of people who are displaced alongside the implementation of various urban (re)development schemes (Ministry of Infrastructure, 2009). They explicitly uphold the key claim of procedural justice. This rests on people's relations to land and the restoration of their property rights if they are infringed by spatial (re)development programs. Pursuing procedural justice through compliance with these rules has therefore re-established *de facto* and perceived tenure security as reported by the resettled inhabitants and actors involved in their resettlement process. They argue that: "The resettlement has resulted from the political recognition of rights to housing for the resettled people. This recognition is affirmed through their relocation in residential villages, delineated in the current master plans."* Inherent to the meaning of land tenure security, the perceptions of the resettled people reflect the *de facto* tenure security, connected to their formal integration in the urban space. Another element of land tenure security is the perceived, emerging from this resettlement process, and the increased feelings of resettled people on the permanency of their current dwellings. This element of tenure security was also carved from the pursuit of recognitional, redistributive, and inter- and intra-generational justice which has been instilled by the political decisions† to improve living conditions of people living in high risk areas and to protect their lives through their resettlement in serviced sites. However, these resettled people do not currently have rights to sell their houses or use them in the form of collateral for bank loans. They will be granted the full ownership documents after five years of occupation.‡ This may result in *de jure* tenure security. These documents can also be used for seeking loans, even though the restriction on sale remains valid. Despite this condition, their resettlement process has generally enhanced their tenure security.

3.8 CONCLUSION

This chapter has dicussed how political leaders and urban planners and managers in Kigali city have been phasing out the exclusive urban (re)development towards integrating all categories of urban dwellers in the urban space. This was explored through the process of resettling poor and vulnerable urban dwellers in decent houses. Their resettlement and the associated suppport for income generation are features of different aspects of spatial justice and tenure security. Procedural justice is echoed in the rules underlying this process. These rules explicitly convey the state's rhetoric on inclusive urban (re)development advancing access to basic resources such as land, housing, and urban amenities for the poor. The implementation of these rules features some patterns of recognitional and redistributive justice, reflected in the increased recognition of rights of the resettled people to these resources, and the significant public funds allocated to their resettlement and the financial support they receive to get out of poverty. There are also features of recognitional, redistributive, and intra- and inter-generational justice embedded in the selection of people who are most in need of housing. The establishment of their permanent dwellings along their resettlement processes features some aspects of tenure security through increased recognition of the rights to land and housing. These outcomes explicitly relate to the general aspiration of responsible land management, which is the responsiveness to these basic needs and tenure security for users of land resource. This means that pursuing spatial justice in urban management contributes to responsible land management. Similarly, responsible land management can advance spatial justice since their aspirations are intertwined, as discussed in our introduction. Despite good trends of spatial justice emanating from the resettlement of poor and other homeless urban dwellers in Kigali city, it is worth noting that they were resettled far from the urbanised areas which offer various employment opportunities. We suggest increased recognition of their rights to jobs by situating further resettlement sites in the vicinity of the core urban area.This would also decrease the feelings of resettled people on their exclusion from the urbanized area.

* Source: Interviews and household survey.
† Source: Interviews with land managers, urban planners, head of social affairs units.
‡ Source: Government regulations related to allocation of social housing to poor and vulnerable people.

REFERENCES

Chatterton, P. 2010. "Seeking the urban common: Furthering the debate on spatial justice". *City* 14(6):625–628. doi:10.1080/13604813.2010.525304.

City of Kigali. 2013. *The City of Kigali Development Plan (2013–2018)*. Kigali, Rwanda.

City of Kigali. 2016. *Kigali: A Rising Star in Africa. Five Year Achievements (2011–2015)*. Kigali, Rwanda.

de Vries, T.W., and Chigbu, E.U. 2017. "Responsible land management-Concept and application in a territorial rural context". *fub. Flächenmanagement und Bodenordnung* 2: 65–73.

Fainstein, S.S. 2014. "The just city". *International Journal of Urban Sciences* 18(1):1–18. doi:10.1080/12265934.2013.834643.

Finn, B. 2018. "Quietly chasing Kigali: Young men and the intolerance of informality in Rwanda's capital city". *Urban Forum* 29(2):205–218. doi:10.1007/s12132-017-9327-y.

Fraser, N., and Honneth, A. 2003. *Redistribution or Recognition? A Political-Philosophical Exchange*. London, UK, and New York: Verso.

Gibbs, M. 2009. "Using restorative justice to resolve historical injustices of Indigenous peoples". *Contemporary Justice Review* 12(1):45–57. doi:10.1080/10282580802681725.

Goodfellow, T. 2014. "Rwanda's political settlement and the urban transition: Expropriation, construction and taxation in Kigali". *Journal of Eastern African Studies* 8(2):311–329. doi:10.1080/17531055.2014.891714.

Harvey, D. 2010. *Social Justice and the City*, vol. 1, Geographies of Justice and Social Transformation Series. Athens, GA: University of Georgia Press.

Lefebvre, H. 1991. *The Production of Space*. Oxford, UK: Basil Blackwell.

Magel, H. 2016. "Räumliche Gerechtigkeit-Ein Thema für Landentwickler und sonstige Geodäten?!" *Zeitschrift für Geodäsie, Geoinformation und Landmanagement* 141:377–383. doi:10.12902/zfv-0144-2016.

Manirakiza, V., and Ansoms, A. 2014. "Modernizing Kigali: The struggle for space in the Rwandan urban context". In: *Losing Your Land Dispossession in the Great Lakes*, edited by A. Ansoms and Hilhorst, T., 218. Woodbridge, NJ: Boydell and Brewer.

Marcuse, P. 2010. "Spatial justice: Derivative but causal of social justice in: Justice et injustices spatiales". In: *Justices et Injustices Spatiales*, edited by B. Bret, Gervais-Lambony, P., Hancock, C. and Landy, F., 75–94. Paris: Presses Univeritaires de Paris Ouest.

Ministry of Finance and Economic Planning. 2013. *Economic Development and Poverty Reduction Strategy II, 2013–2018*. Kigali, Rwanda: Ministry of Finance and Economic Planning.

Ministry of Finance and Economic Planning, and National Institute of Statistics of Rwanda. 2014. "Fourth population and housing census, Rwanda, 2012". Kigali, Rwanda.

Ministry of Infrastructure. 2009. *Updated Version of the National Human Settlement Policy in Rwanda*. Kigali, Rwanda.

Ministry of Infrastructure. 2015. *National Urbanization Policy Kigali*. Rwanda.

Ministry of Infrastructure and Rwanda Housing Authority. 2011. *Compilation of Policies Concerning Human Settlement, Urban Housing and Construction in Rwanda*. Kigali, Republic of Rwanda.

National Institute of Statistics of Rwanda. 2015. *Rwanda Poverty Profile Report 2013/2014: Results of Integrated Household Living Conditions Survey (EICV 4) Kigali, Rwanda*.

Payne, G. 2001. "Urban land tenure options: Titles or rights?" *Habitat International* 25(3):415–429. doi:10.1016/S0197-3975(01)00014-5.

Picard, L.A., Buss, T.F., Seybolt, T.B., and Lelei, M.C. 2015. *Sustainable Development and Human Security in Africa: Governance as the Missing Link*, edited by D.H. Rosenbloom and Rabin, J., vol. 196. London and New York: CRC Press.

Rawls, J. 1971. *A Theory of Justice*. Cambridge, MA: Harvard University Press.

Rawls, J. 1999. *A Theory of Justice*. Revised edition. Cambridge, MA: University Press Cambridge.

Rwanda Environment Management Authority (REMA). 2015. *Kigali: State of Environment and Outlook Report* (p. 244). Kigali, Rwanda.

Rwanda Environment Management Authority. 2017. *State of the Environment and Outlook Report Kigali*. Rwanda.

Simbizi, M.C.D., Bennett, R.M., and Zevenbergen, J. 2014. "Land tenure security: Revisiting and refining the concept for Sub-Saharan Africa's rural poor". *Land Use Policy* 36:231–238. doi:10.1016/j.landusepol.2013.08.006.

UN-Habitat. 1996. *The Habitat Agenda: Goals and Principles, Commitments and Global Plan for Action*. Istanbul, Turkey.

Uwayezu, E., and de Vries, T.W. 2018. "Indicators for measuring spatial justice and land tenure security for poor and low income urban dwellers". *Land* 7(3):34. doi:10.3390/land7030084.

Uwayezu, E., and de Vries, T.W. 2019. "Scoping land tenure security for the poor and low-income urban dwellers from a spatial justice lens". *Habitat International* 91:10. doi:10.1016/j.habitatint.2019.102016.

van Gelder, J.L. 2009. "Legal tenure security, perceived tenure security and housing improvement in Buenos Aires: An attempt towards integration". *International Journal of Urban and Regional Research* 33(2):126–146. doi:10.1111/j.1468-2427.2009.00833.x.

Young, I.M. 1990. *Justice and the Politics of Difference*. Princeton, NJ: Princeton University Press.

Zevenbergen, J., de Vries, T.W., and Bennett, R.M. 2015. *Advances in Responsible Land Administration*. New York: CRC Press.

4 The Paradox of Poverty amidst Potentially Plentiful Natural Resources of Land in Rwanda

Jossam Potel, Gloria Owona, and Wilson Tumsherure

CONTENTS

4.1 INTRODUCTION

According to UNDP (2016), the core concept today is that poverty has been related to income. However, "income" is itself no less problematic a concept than "poverty," and it too has to be carefully and precisely elaborated. Other resources such as assets, income in kind, and subsidies to public services and employment should be imputed to arrive at a comprehensive but accurate measure of income. People can be said to be in poverty when they are deprived of income and other resources needed to obtain the conditions of life, for example, the diets, material goods, amenities, standards, and services that enable them to play the roles, meet the obligations and participate in the relationships and customs of their society (UNDP, 2016).

The determination of a poverty line cannot be based on a random selection of a low income level. Only scientific criteria independent of income can justify where the poverty line should be drawn. The key is therefore to define a threshold of income below which people are found to be poor. The measure of poverty must be decided on the basis of evidence about each and every sphere of the range of social and individual activities people perform in fulfillment of individual and family needs, and social obligations (GoR, 2014).

The comparison of poverty to income is the basis for ascertaining the threshold amount of income ordinarily required by households of different compositions to surmount poverty. The application of this method permits analysis of trends in poverty in and across different countries. According to the

World Bank, poverty is more than just the amount of money a person has. It is a multidimensional issue that concerns one's level of access to health services, availability of educational opportunities, and quality of life (WB, 2018).

The post-genocide Rwanda has seen a formidable change in the economic life of the country. From 2001–2014, Rwanda's GDP growth rate was about 8% each year. However, almost 17% of Rwandans are still living below the nation's poverty rate (NISR, 2017). The reasons why Rwanda is currently considered a poor country is because of the following: the history of the prolonged conflicts which date back to the late 1950s that culminated in genocide against the Tutsi in 1994, where over a million Tutsi were massacred; lack of valuable natural resources; the fact that most people in Rwanda practice subsistence farming; and the country is geographically land-locked (GOR, 2008).

This study contributes to responsible land management, as in order to fight poverty, especially in the poor regions of Africa, land is a central engine for this effort; therefore the paradox of poverty amidst potentially plentiful natural resources of land should be given due attention in land management and administration.

4.1.1 What is Poverty?

There is no single definition of poverty; different authors have different definitions of the word. However, they all agree on one thing, which is a lack or limited capacity to afford basic social needs. Poverty is a complex phenomenon that manifests in different ways and can be studied in many different ways (Richard, 2009).

According to the United States government, poverty is a state of deprivation, lacking the usual or the socially acceptable amount of money or material possessions. Some authors define poverty as a lack of material wellbeing considered the minimum acceptable in the society where they live, or as a deprivation of human needs (Misra, 2005).

The human rights approach underlines the multidimensional nature of poverty, describing it in terms of range of interrelated and mutually reinforcing deprivations, and drawing attention to the stigma, discrimination, insecurity, and social exclusion associated with poverty (UNDP, 2016).

According to the National Institute of Statistics of Rwanda (NISR), the measurement of poverty can be done in different ways. The three principal approaches of measurement are monetary, non-monetary, and subjective measurements (GOR, 2012).

The extremes of rural poverty in the developing nations are outrageous. Rurality and rural poverty are like two sides of the same coin. In Rwanda, poverty levels tend to be a descending trend, but in the rural areas, the trend of decline is not statistically significant. According to NISR (2017), rural poverty is at 43.1% nationally, of which 18.1% of the rural poor are in extreme poverty. The most puzzling issue is that most rural areas have plenty of fertile land, but the people living amidst these natural resources tend to be poor.

Although there are economic resources like land, forests, rivers, lakes, minerals, and mountains that are supposed to help the poor to move out of poverty, many of these people have not been able to successfully utilize them, and hence they have remained poor. The Rwandan Government has taken a number of actions to spur the transition from a poor, agricultural state to a middle income and service economy. The country has spent a significant amount of money in improving infrastructure, even though the state of the current infrastructure still remains a challenge for the economic growth (GOR, 2008).

It is therefore paradoxical to find poverty in rural areas but amidst richness in natural resources and the will of the government to invest richly in rural people to curb poverty. This is the resource curse hypothesis that highlights the view that areas that have plenty of natural resources tend to perform poorer than those areas with scarce natural resources (Nsaikila, 2015). Poverty in rural Rwanda is manifested through low standards of living, low incomes, poor health, and other anthropometric indicators like low food ratio, poor education attainment, illiteracy, high disease prevalence, etc.

4.1.2 Problem Statement

Land as a natural resource is supposed to be a key engine in enabling people to eliminate and eradicate poverty, especially in the context of rural areas. Poverty in rural Rwanda is manifested through low standards of living, low incomes, poor health, and other anthropometric indicators like low food ratio, poor education attainment, illiteracy, high disease prevalence, etc. This is as a result of underutilization of land, hence yielding no optimum output. Even where output is realized, it is not sufficient enough to meet the domestic food needs of families and later for market to earn some income, hence leaving many people extremely poor (UNDP/UNEP, 2006).

Ideally, when rural areas have plenty of natural resources, they are supposed to be better off in socio-economic terms. Rural poverty is best understood in terms of the clusters of disadvantage that Robert Chambers (1984) described as the deprivation trap in terms of poverty proper, physical weakness, isolation, vulnerability, powerlessness. Therefore, the aim of this research is to examine the paradox of poverty amidst potentially plentiful of natural resources of land in Rwanda.

4.1.3 Purpose and Objectives of the Study

The purpose of this study was to analyze the relationship between land as a natural resource and poverty levels in rural Rwanda.

In order to attain that purpose, the following objectives were formulated:

a) To assess the level of poverty in rural Rwanda
b) To examine people's perception in regards to poverty in rural Rwanda
c) To analyze the level of poverty amidst potentially plentiful of natural resources of land in Rwanda.

4.2 METHODOLOGY

This section provides a comprehensive description of the methodology adopted for the research. It includes the research design, describes the population and sample selected, identifies the instruments that were used, establishes the validity and reliability of the instruments, and discusses the methods of data analysis.

4.2.1 Research Design

In order to achieve the objectives of the study, a cross-sectional survey study was carried out. The research design was both quantitative and qualitative. The choice of the cross-sectional survey design was made because the researchers dealt with a small representative fraction of the population in the study and an inference was drawn about the population based on results of the findings (Richard, 2006). Data was gathered from the rural poor who have access to land and other natural resources. Cross-sectional research design was deemed to be the most appropriate for this study which sought to establish the relationship between rural poverty and natural resource potential in Rwanda.

4.2.2 Study Area

The study areas are the four provinces of Rwanda, namely the northern, eastern, western, and southern provinces. From these provinces, 20 rural districts were chosen, and these districts were chosen basing on the fact that they have the highest poverty rates (NISR, 2017). These districts include the following: Northern province districts include Rulindo, Gakenke, Musanze, Burera, and Gicumbi; Eastern province districts include Ngoma, Nyagatare, Gatsibo, Kirehe, and Bugesera;

Western province districts include Rutsiro, Karongi, Nyabihu, Ngororero, and Nyamasheke; and Southern province districts include Nyanza, Gisagara, Nyaruguru, Huye, and Nyamagabe. The map in Figure 4.1 shows the districts in Rwanda and mainly shows the study area.

4.2.3 Study Population, Sample Size, and Selection Techniques

The sample of 30 respondents per rural district was selected. Rwanda has 30 districts of which 20 are categorized as rural districts, or to be precise, districts with rural characteristics. Therefore, a total of 600 respondents were selected for this study. The sample size was determined using Roscoe's (1975) rule of thumb. Roscoe contends that any sample size between 30 and 500 is sufficient to give credible results. This sample size was considered because it adequately cuts across the social, economic, and political spectrum in the study area and thus sufficient data was collected

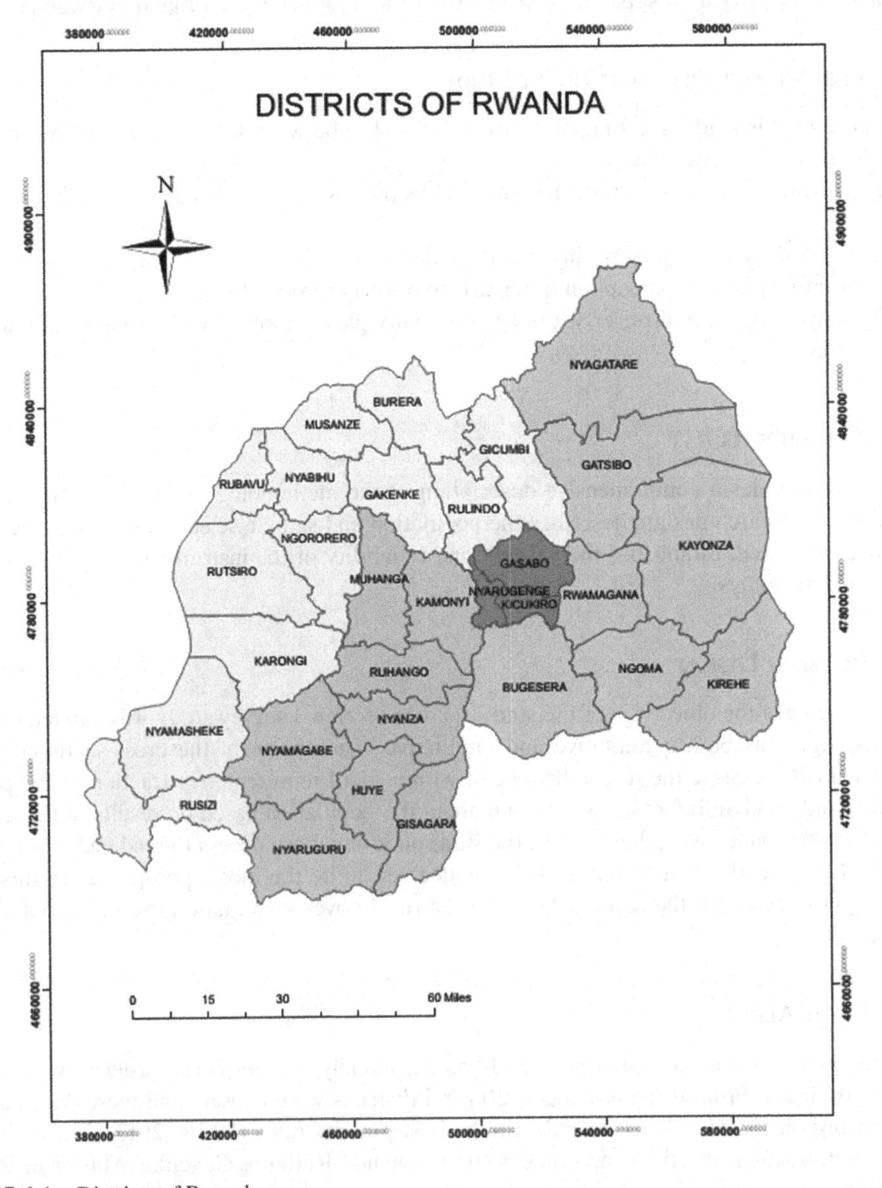

FIGURE 4.1 Districts of Rwanda.

to fulfill the objectives of the study. Accordingly, the researcher used a simple random sampling technique. This is a sampling technique where every item in the population has an even chance and likelihood of being selected in the sample. The selection of items completely depends on chance or probability. The unit of analysis for this study was the poor rural dwellers who have access to land. The researcher used it to select a representative sample without any bias from the study population. This was done to ensure that each member within the study population had an equal chance of being selected in the random sample.

4.2.4 Data Collection Instruments

These methods involved using the primary and secondary data collection methods. In this study, two types of instruments of data collection were used, namely questionnaires and document reviews.

Structured questionnaires were administered to respondents. These questionnaires consisted of open- and closed-ended questions. This helped in achieving the stated objectives. The questionnaires assessed the level of poverty in rural Rwanda, examined people's perception in regards to poverty in rural Rwanda, and analyzed the level of poverty amidst the potentially plentiful natural resources of land in Rwanda.

The criteria used to design the questionnaires were as follows: the language in the questionnaires had to be clear, the questions had to be specific, and the questions had to have linguistic completeness and grammatical consistency; and the criteria used to distribute the questionnaires were as follows: the questions created had to be tested by random individuals within the sample to check whether the questions were fully understood, and the reactions of the participants to the different questions were analyzed so as to avoid confusion and errors while answering the questions, after which the questionnaires were handed directly to the respondents in the chosen sample frame (ECRM, 2012).

Documentary review involved viewing and analyzing various documents to complement the research and generate supportive information obtained from textbooks, newspapers, the world wide web, and journals that were deemed relevant to the research.

4.2.5 Data Analysis

Descriptive statistical methods were employed in data analysis based on data and information collected from primary and secondary sources on the existence of poverty amidst the availability of significant amounts of natural resources and the findings were integrated into the report using percentages, the highest percentages representing the majority of the opinion of the study which was the basis upon which the conclusions were drawn. During the research, the SPSS 17.0 software program and the statistical tools such as tallying, averages, percentages, and tables were used to summarize the data collected from the questionnaires. The results were then presented in different forms of tables corresponding to the items in the questionnaires.

4.2.6 Validity and Reliability

Validity refers to the extent to which the data collection instruments employed (to measure variables) in a study actually measures what they claim (or are intended) to measure. In this regard, attention was paid to the content and construct validity of the data collection instruments which dealt with the degree to which the instrument(s) relevantly and adequately explore(s) the variable(s) under investigation and how well the constituent elements of the instrument(s) are arranged for the purpose of accurately eliciting the data required for the study (Golafshani, 2003).

Reliability is the description of the degree of precision, consistency, repeatability, and dependability with which the instruments elicited the data that was required for the study. It was established through a pilot test of the questionnaire.

4.2.7 Theoretical Framework

As UNDP (2016) outlines, there are a number of issues involved in defining and measuring poverty. Is it confined to material aspects of life, or does it also include social, cultural, and political aspects? Is it about what may be achieved, given the resources available and the prevailing environment, or what is actually achieved? Should definitions and measurement methods be applied in the same way in all countries and used for comparisons? Are there "objective" methods, or are value judgments involved? What is the rationale for defining a poverty line? Should it be absolute, as in the Millennium Development Goals and in most developing countries, or relative, as in the rich OECD countries?

Poverty defined as a lack of wellbeing is clearly multidimensional. The MDG and World Bank $1.2-a-day poverty line, being a solely monetary measure, has therefore rightly been criticized as inadequate and mono-dimensional. Yet, it is the most widely recognized poverty estimate in use today. The objective of this chapter is to demonstrate that, even if we accept a monetary measure of poverty, the $1.2-a-day poverty line is unrealistically low, thus misleading policy-makers and the public on the extent of global poverty and the scale of redistribution needed to remove it (WB, 2015).

The basis for the $1.2-a-day poverty line is simply that it is the median of ten of the lowest national poverty lines in the world. It is not derived from any consideration of wellbeing or basic needs. The World Bank economists most involved in this area recently called it "frugal," stating that it "must be deemed a conservative estimate whereby aggregate poverty in the developing world is defined by the perceptions of poverty found in the poorest countries." The implication is that the $1.2-a-day poverty line is unreasonably low. Almost certainly it is lower than developed world populations would consider morally justifiable. Indeed, the World Bank does increasingly quote poverty indices for a $2.4-a-day line that seems to be based simply on a doubling of the $1.2-a-day poverty line (Kakwani, 2016).

Consider a country where the incidence of poverty is falling by 1 percentage point per annum but population is rising by 2 percentage points per annum. Poverty as normally measured—the number of poor people as a proportion of total population—has fallen. But the absolute number of the poor has risen. This is not just a statistical curiosum, but it can occur, and has occurred in countries, like those in Africa, where population growth is relatively high (Kakwani, 2016).

In most sub-Saharan African countries, land is the prime means of making livelihood and a core source for investment, accumulating wealth, and transferring it between generations. Land is also a key element of household wealth. In such countries, land constitutes over 70% of household assets (Deininger, 2003). Since land comprises a large share of the asset portfolio of the poor in many developing countries, giving secure property rights to land they already possess can greatly increase the net wealth of poor people (Misra, 2005). Many questions have been raised on why Africa still remains in extreme poverty irrespective of the abundance of large and fertile land for agriculture (Deininger, 2003).

Because land embraces a large share of the asset portfolio of the poor in many developing countries, giving secure property rights to land they already possess can greatly increase the net wealth of poor people. By allowing them to make productive use of their labor, land ownership makes them less reliant on wage labor, thereby reducing their vulnerability (Deininger, 2003). If land governance institutions are not working well or are poorly coordinated, inefficient, or corrupt, transaction costs will be high, thereby reducing the level of transactions below what would be socially optimal and, in many cases,, excluding the poor completely.

Tables 4.1 and 4.2 are used to depict the rate of poverty in Rwanda. From Table 4.1 it can be seen that the poverty rate is high in rural Rwanda (43.1%), compared to urban poverty which stands at 15.2%. In terms of the poverty levels, the rural provinces also still manifest high levels of poverty of which people in the western (47%), southern (41.4%), and northern (42.3%) provinces are still poorest. Furthermore, 16% of people nationally are still in extreme poverty, of which 18.1% are people in extreme poverty in rural areas.

TABLE 4.1
Rate of Poverty in Rwanda

Regions	Total poverty (%)	Extreme poverty (%)
Rwanda (National)	38.2	16.0
Urban	15.8	5.9
Rural	43.1	18.1
Provinces		
Kigali city	13.9	4.2
Southern	41.4	16.9
Western	47.1	21.6
Northern	42.3	17.4
Eastern	37.4	15.3

Source: NISR 2017.

TABLE 4.2
Rate of Poverty in the Districts of Rwanda

District	Poverty rate (%)
Nyamasheke	69
Gisagara	56
Rulindo	53
Karongi	52
Nyaruguru	51
Burera	49
Rutsiro	48
Nyamagabe	47
Ngororero	46
Nyabihu	45
Nyanza	44
Nyagatare	43.9
Kirehe	43
Gatsibo	42.9
Musanze	42
Bugesera	41
Huye	40
Ruhango	38
Ngoma	37
Rubavu	36
Gicumbi	35
Gakenke	35
Rusizi	34
Muhanga	33
Kayonza	28
Kamonyi	23
Rwamagana	18
Gasabo	16
Nyarugenge	13
Kicukiro	12

Source: NISR, 2017.

4.2.8 Conclusion to this Section

From the trend analysis based on the tables indicated in the above section, poverty rates in Rwanda are declining, but not at an impressive rate, and land as a resource is still a problem in terms of ownership, especially in rural areas. The majority of Rwandans still don't have enough land to carry out agriculture and animal husbandry. This is partly attributed to high population growth rates especially in the rural areas.

4.3 RESULTS AND INTERPRETATIONS

This section presents the results after the analysis of the data collected from the different questionnaires. The results are present in form of frequency tables developed from SPSS 17.0 software.

a) Level of Poverty in Rwanda

Table 4.3 was used to depict the number of the sampled population that own land in Rwanda. It was observed that 95% people own land, while 5% people do not own land. From the above data, the majority of the respondents own land, which implies that land is available.

Table 4.4 is used to depict the size of land that is owned by the sampled population compared to the number of the people in the household. It was observed that 41% have less than 1 hectare,

TABLE 4.3
Level of Land Ownership

Do you have land?	Frequency	Percentage
Yes	567	95
No	33	5
Total	600	100

Source: Primary data.

TABLE 4.4
Size of Land versus Family Size

Size (hectares)	Frequency	Percentage
<1	233	41
1 to 2	132	23
2 to 3	92	16
3 to 5	73	13
>5	37	7
Total	567	100

Size (hectares)	Number of people living in the house	Frequency	Percentage
<1	4–5	269	47
1 to 2	3–4	127	23
2 to 3	2–3	80	14
3 to 5	1–2	64	11
>5	1	27	5
Total		567	100

Source: Primary data.

23% own between 1 and 2 hectares, 16% own between 2 and 3hectares, 13% own between 3 and 5 hectares and 7% own more than 5 hectares. However, 47% have 4–5 people living in less than 1 hectare, 23% have 3–4 people living in 1 to 2 hectares, 14% have 2 to 3 people living in 3 to 5 hectares, and 5% have one person living on less than 5 hectares of land. Since the majority of rural people depend on less than 1 hectare it implies that the land which is the major source of livelihood is inadequate to minimize their poverty levels. This also implies that the higher the number of the people in the household, the lower the amount land owned, and the greater the likelihood of higher poverty level in the family.

Table 4.5 was used to depict the degree of land use satisfaction. It was observed that 44% indicated that the land is not enough to satisfy their needs, 13% indicated that the land use is good, but it is not enough to satisfy their needs, 14% get just enough agricultural produce to satisfy their families, and 29% get enough agricultural produce to both satisfy their families and to sell at the markets. This implies that the degree of land ownership is not satisfactory to most rural people in Rwanda.

From Table 4.6, 40% of Rwandans have food security, 40% have marginal food security since agricultural land is optimally used, 17% have moderate food security, and 3% have severe food insecurity. However, food consumption is not necessarily for farmers in rural areas. In any case, farmers are the ones who turn to be among the food insecure people, because they sell food for income to urban dwellers (Table 4.7).

Table 4.8 above is used to depict the different categories of household wealth in rural Rwanda. It was observed that 50% are well-off, and this is because: they possess stable houses made of bricks and covered with iron sheets which they construct by themselves; have over five domestic animals, for example, cows, goats, and pig; have means of transportation in their families, for example, bicycles and motorbikes; have land with gardens and grow food for sale in the market; own a radio and a television set; have electricity or use solar; have access to clean water; and eat thrice a day. 31%

TABLE 4.5
Degree of Land Use Satisfaction

Degree of land use satisfaction	Frequency	Percentage
Not enough	247	44
Good, but not enough to satisfy my family	75	13
Enough to satisfy my family	79	14
Enough to reserve and sell at the market	166	29
Total	567	100

Source: Primary data.

TABLE 4.6
National Household Food Security Status

Food security	Percentage
Food secure	40
Marginally food secure	40
Moderately food insecure	17
Severely food Insecure	3
Total	100

Source: Comprehensive Food Security and Vulnerability Analysis, 2018.

TABLE 4.7
Welfare Levels in Rural Rwanda

Category of household	Frequency	Percentage
Relatively well-off	297	50
Medium	186	31
Poor	92	15
Very poor	25	4
Total	600	100

Source: Primary data.

TABLE 4.8
Level of Poverty

Are you poor?	Frequency	Percentage
Yes	47	8
No	553	92
Total	600	100

Source: Primary data.

have medium well-off households because: they possess stable houses made of wood and covered by iron sheets constructed by themselves; have land and can grow food crops for all seasons; have less than five domestic animals, for example, cows, goats, and pigs; own a radio, can pay for health insurance for every family member; and eat thrice a day. 15% have poor households because: they live in unstable houses made of mud and grass, or stable houses but donated/constructed for them by the government; have no savings; have no domestic animals; use government support to take their children to schools; get health insurance from the government; and eat twice a day. 4% have very poor households because: they have no houses or unstable houses made out of mud and grass; have no land to grow food crops; have no domestic animals; get food through working for others or from donations from neighbors; get health insurance from the government; and eat once or rarely eat in a day.

b) Examination of People's Perception in Regard to Poverty in Rural Rwanda

The sampled population has different definitions of what poverty is. They said that: poverty is when one has no money, it is when one lacks land, it is when one cannot buy him/herself his or her social needs, it is when one falls sick and is unable to go to hospital due to lack of money, it is when a man cannot feed his family, and it is when one has no land and no job.

Table 4.8 above is used to depict whether one is poor. It was observed that 92% said that they are not poor because of the following reasons: they have money, fertile land, can buy the basic needs, and are able to go to hospital once sick, and they can feed their families, while 8% said that they are poor because of the following: lack of land, infertile land, lack of market for agricultural products, old age, lack of employment, diseases, and many dependents who happen to be family members.

c) Analysis of the Level of Poverty amidst Potentially Plentiful Natural Resources of Land in Rwanda

TABLE 4.9
Relationship between Land Size, Number of People Living in a Household, and Poverty Level

Size (hectares)	Number of people living in the house	Poverty level at household level	Percentage
<1	4–5	210	37
1 to 2	3–4	117	21
2 to 3	2–3	97	17
3 to 5	1–2	89	15
>5	1	54	10
Total		567	100

Source: Primary data.

The third objective of this research was to analyze the level of poverty amidst plentiful natural resources of land in Rwanda. In most parts of rural Rwanda, people have land which is fertile and able to support their agriculture life. They grow crops and rear domestic animals to support their subsistence way of living. When asked whether they consider themselves poor, only 8% agree to this, with the rest claiming not to be poor since they can feed themselves, and then they claim lack of transport infrastructure to access towns and markets for their agricultural products prevent them getting liquid cash.

When observing the level of poverty in Rwanda, it varies depending on who is analyzing or classifying the poor people. Classification of a person as poor or not should not be based on universal conditions/criteria, but should be based on individual analysis of personal situations. People in many rural villages may be lacking liquid cash/money, they may be without bank accounts, without monthly or daily paying jobs, but they have land where they grow their food and construct their houses to meet all the social primary needs.

While observing the living conditions of rural people in Rwanda, one wonders whether these people should be called poor. But referring to the commonly agreed definition of poverty, definitely these people would be called poor. However, referring to the conditions that one may meet to be classified as a poor person, in most cases this does not apply to these rural people.

Table 4.9 depicts the relationship between land size, the number of people living in a household and poverty level of the sampled population. It was observed that a big household of 4–5 people have less than one hectare and poverty is such households is high at 37%. Households of 3–4 people have between 1–2 hectares of land and the poverty level is at 21%, while households of 2–3 people have 2–3 hectares of land, and poverty is at 17%. Households of 1–2 people have 3–5hectare of land and poverty is at 15%. Finally, a household of one person who has land of more 5 hectares has a poverty rate of 10%. When asked whether he is poor, one respondent reacted “No, that is an insult, how can you call me poor, I have my land, I have my cows, how can you call me a poor person?” He stated this after mentioning that he only gets money once he sells either a cow or a goat. While conducting this research, the researcher wanted to explore whether the rural people’s way of living reflect the general definition of poverty.

4.3.1 Conclusion to the Section

From the above section, it was observed that the paradox of poverty tends to be manifested in households with many family members where the land share tends to be small.

Furthermore, land ownership and land distribution among the rural poor is increasingly becoming a big problem due to population explosion.

4.4 CONCLUSION AND RECOMMENDATIONS

Based on the findings of the study, the following observations can be made:

567 respondents (95%) of the rural people sampled own land, while 33 respondents (which is 5%), have no land, implying that majority of rural people own land. This concludes that land ownership in rural Rwanda is high.

On the size of land owned, 41% have less than 1 hectare, 23% own between 1 and 2 hectares, 16% own between 2 and 3 hectares, 13% own between 3 and 5 hectares, and 7% own more than 5 hectares. The conclusion from the data above is that majority of respondents revealed that they have limited land which is below 1 hectare, implying that the possibility of being poor is high.

Furthermore, the research revealed that the level of satisfaction over land ownership also varies. 44% indicated that the land is not enough to satisfy their needs, 13% indicated that the land use is good, but it is not enough to satisfy their needs, 14% get just enough agricultural produce to satisfy their families, and 29% get enough agricultural produce to both satisfy their families and to sell at the markets. Therefore, one can conclude that land distribution in rural Rwanda is not enough to satisfy local needs.

Finally, the study established the relationship between land ownership, family size, and poverty level. It was established that big households of 4–5 people have less than 1 hectare, and poverty in such households is high, at 37%. Households of 3–4 people have between 1–2 hectares of land and have poverty levels at 21%, while households of 2–3 people have 2–3 hectares of land and have poverty levels at 17%. A household of 1–2 people has land of 3–5 hectares and has poverty levels at 15%. Finally, a household of one person has land of more than 5 hectares and has a poverty rate of 10%. Therefore, the study concludes that the more people in the households, the smaller the land size owned and the higher the poverty, and the reverse is also true.

REFERENCES

Chambers, R (1984). Beyond the Green Revolution: A Selective Essay. In: *Understanding Green Revolutions*. Cambridge, UK: Cambridge University Press, 362–79.

Deininger K (2003). *Land Policies for Growth and Poverty Reduction*. Washington DC: Oxford University Press.

Golafshani N (2003). Understanding Reliability and Validity in Qualitative Research. *Journal of Technology Education*, 8(4): 597–606.

GOR (2008). *Rwanda: Poverty Reduction Strategy Paper.* Washington D.C: International Monetary Fund.

GOR (2012). *The Evolution of Poverty in Rwanda from 2000 to 2011.*

GoR (2014). *Rwanda Poverty Profile Report*. Kigali: NISR.

International, F. H. (2012). Qualitative Research Methods Overview. *ECRM*. Academic publishing.

Kakwani N (2016). Poverty and Well Being. In UNDP, *Poverty in Focus*. Brasilia.

NISR (2017). *Comprehensive Food Security and Vulnerability Analysis.*

Nsaikila M (2015). *Poverty, Resource Endowments and Conflicts in Subsaharan Africa: A Reexamination of the Resource Curse Hypothesis.* MA Thesis. Department of Economics and Decision Sciences Western Illinois University.

Misra S N (2005). *Poverty and Its Alleviation.* New Delhi: Deep & Deep publications Pvt. Ltd.

UNDP (2016). *Poverty in Focus*. Brasilia, Brazil: International Poverty Centre.

UNDP/UNEP (2006). *Environment and Poverty Reduction in Rwanda.*

WB (2015). *The World Bank Group Twin Goals and the Millennium Development Goals 2014/2015.*

WB (2018). *The Innovation Paradox*. Tokyo.

Richard (2006). Qualitative research design. *Book Thinking Research.* Sage, 73–104.

Roscoe J T (1975). *Fundamental Research Statistics for the Behavior Sciences*. 2nd edition. New York: Holt, Rinehart and Winston.

5 Influences of Community Land Rights and Tenure Security Intervention Processes on Food Security in Northwest Ghana

Baslyd B. Nara, Monica Lengoiboni, and J.A. Zevenbergen

CONTENTS

5.1 INTRODUCTION

A joint FAO, IFAD, UNICEF, and WFP (2018) report claims that globally, many people lack land and related rights to produce food for themselves. They may therefore lack agricultural revenue for food accessibility too, thus worsening food insecurity (FAO, 2019). For instance, WFP says "on any given day, it has 5,000 trucks, 20 ships and 70 planes delivering food to those in most need," yet global food insecurity still persists and affects one in nine people (FAO et al., 2018). Fertile agricultural land is a major source of livelihood for many people in Asia and Africa, but it is scarce, thereby threatening food production (Duncan and Brants, 2004; Lawry et al., 2014; Ruerd, 2011). This is partly due to demographic growth changing customary laws and practices where state legislations and implementation are either weak or absent, or where there is low farming input (Bugri, 2008; Naab, Dinye, and Kasanga, 2013). Little attention is paid to smallholder subsistence farming, even though such farms generate majority of food consumed in the world (Lawry et al., 2014; Mwesigye, Matsumoto, and Otsuka, 2017). Also, little research covers "unnoticed" tenure insecure groups like settlers (i.e. permanently resident migrants) in customary areas in Ghana who possess fewer secure secondary land rights, which tends to hinder their farming. This research seeks to suggest responsible land management interventions from local practices to first secure land rights and tenure in order to enhance subsistence farming for food security in line with de Vries and Chigbu (2017) and Zevenbergen; de Vries, and Bennet (2016, pp.6–7). Land management is responsible if it is resilient, robust, reliable, respected, reflexive, retraceable, and recognizable (de Vries and Chigbu, 2017). Such interventions are important, as settlers in some communities in Ghana constitute up to 80% of the population as compared to 30% in Ivory Coast (Cotula, Toulmin, and Hesse, 2004). In both countries, farmers often suffer land rights and tenure challenges such as evictions, rights variations, reduced farm sizes, etc. which have consequences for food security (Cotula and Toulmin, 2004; Lawry et al., 2014).

Food security is when "all people, at all times, have physical, social and economic access to sufficient, safe and nutritious food that meets their dietary needs and food preferences for an active and healthy life" (Pinstrup-Andersen, 2009). This chapter first focuses on the food availability component from their own farming, upon which access, nutrition, and stability dimensions revolve. Land rights and tenure security, often based on customary tenure and service arrangements, remain a major means for food supply and income for most people in sub-Saharan Africa (Lawry et al., 2014; Ruerd, 2011; Simbizi, 2016). These service arrangements are basically non-monetary tributes which landholders are required to seasonally render to landowners to acknowledge landowners' ownership rights and renew landholders' access, holding, use, and other rights. Therefore, when these arrangements hinder farming, agricultural revenue is reduced, thereby affecting farmers' food accessibility since subsistence farmers sometimes sell some of their produce to buy foods they lack.

Since customary tenure is socio-culturally unique, responsible land management interventions require addressing contextual user requirements to secure land rights. This, however, may necessitate formal state facilitation to locally re-negotiate and alter service arrangements to be compatible with changing trends.

5.1.1 Land Tenure Security, Challenges, and Mechanisms to Address Them

A landowner in northwest Ghana is *tendana* in Kunfabiala and Sing, *tortina* in Nimoro and Piina 1, or *tengansob* in Fielmua and Piina 2. He controls family/clan lands by ensuring that rights of both members and non-members are secure based on custom and within the formal legal framework. Land tenure security involves the protection that landholders have against involuntary removal from the land they hold (Almeida and Wassel, 2016; Boudreaux and Sacks, 2009; Simbizi, 2016). It is the perception that tenure is secure and influences land use or farming decisions (Bugri, 2008; Simbizi, 2016). Secure and large land are a precondition for "profitable" farming for most people in sub-Saharan Africa (Payne, 2004; Ruerd, 2011). Customary tenure and service arrangements can provide adequate security because land rights, once allocated, are rarely revoked under customary law and practices (Kasanga and Kotey, 2001; Place, 2009; Platteau, 1996). The rare circumstances for revocation are: absence of heir, gross misbehavior (denying landowner's ownership); and abandoning the land.

A complex tenure system exists in most developing countries, so its degree of security and influence on livelihoods is crucial (Chauveau et al., 2007). Tenure security, in the form of formal legal, customary, or religious land rights, can provide some predictability and access to fundamental rights, including to food and housing (Wickeri and Kalhan, 2010). Cotula and Toulmin (2004) suggest that the state can legitimize land rights by validating (documenting) local practices. But Zevenbergen et al. (2013) caution that registration alone does not secure tenure. De Soto equates land tenure security to recognition of existing rights by means of formalization (Brasselle et al., 2002). However, Lawry et al. (2014) question the impact of formalizing customary land rights on investments, especially in sub-Saharan Africa. Cotula and Mathieu (2008) found that in four sub-Saharan African countries, neither titled tenure nor land transfer rights affected farm productivity. Also, Almeida and Wassel (2016) found that the current land law in Timor-Leste does not provide legal rights for those without any documentation, yet most of the respondents without documentation still consider their tenure to be secure. That is why for Lund (2000), land tenure security exists when an individual perceives that s/he has land rights on a continuous basis, free from imposition or interference from outside sources, and the ability to reap the benefits of investments in it, either in use or upon transfer. Bugri found that 80% of his sample farmers in neighboring northeastern Ghana with no registered title felt that their land rights and tenure were secure. Also Obeng-Odoom (2012) clarified that it is the perception of secure tenure that matters, not necessarily a formal legal mechanism.

Customary land law is noted to historically offer the best security of tenure to individuals, families, and local communities (Akrofi and Whittal, 2011), but levels of tenure security differ on gender grounds (Duncan and Brants, 2004). Yet, tenure remaining relatively secure for different categories of people, irrespective of age, disability, and status, is still challenging. On the whole, secure land tenure, whether legal, *de facto*, or perceived, is the recognition of one's bundle of rights for a given period which is long enough to support investment and recouping the benefits (Lambrecht and Asare, 2016; Nguyen, 2014; Van Gelder and Luciano, 2015). For Boudreaux and Sacks (2009), forcible eviction and deprivation of land rights places responsibility on formal authorities and customary custodians. Therefore, a critical investigation is worthwhile to assess local interventions to strengthen land rights and secure tenure. It is also important to ascertain how adequate these local procedures are in addressing land rights and tenure security for farming and food security.

5.2 METHODS

This explorative study was conducted from June–December, 2018. This duration witnessed the commencement of farming activities through to harvesting, facilitating assessment of how changing customary tenure and service arrangements affect land rights and influence farming. Direct narrations of respondents' personal experiences on avenues used in resolving their tenure challenges in Figure 5.1 below were obtained, analyzed, presented, and discussed. Key issues they responded to included tenure security and farming and stakeholders' role in securing land rights, as well as community perception, challenges, and indicators of food security.

The research used focus group discussions (FGDs), key informant interviews, institutional interviews, and satellite image interpretation. Focus groups ranged between eight and 12 people (Hancock, Ockleford, and Windridge, 2009; Kothari, 2014). Separate FGDs of male and female elders, aged 60+, and youths aged between 18 and 59 (Republic of Ghana, 2012) were conducted. The disabled FGDs were categorized into males and females irrespective of age. A total of 36 FGDs were conducted in six communities, involving about 400 individuals. Additionally, six key informant interviews were conducted with three each of indigenous landowners and heads of settler groups. Two institutional interviews were conducted with the regional Lands Commission (LC) and Wa Central Customary Lands Secretariat (WCCLS). The categorization of the focus groups facilitated collection of data specific to each group. It became clear that age and disability do not influence land rights and tenure in the communities, but rather status (landowners or settlers) and gender. This guided the rest of the analysis.

5.2.1 Data Analysis

The data comprised audio-recordings, hand-written proceedings of FGDs, interviews, and satellite images. The results have been presented in maps, graphs, and charts. The emerging issues were juxtaposed with the literature in descriptions and discussions, based on which an informed opinion is expressed in the conclusion and recommendations.

5.2.2 Study Area

The study area is the Upper West region (also referred to as the northwest) of Ghana (Figure 5.1), where the nature of customary land rights and tenure system (explained later) may have attracted

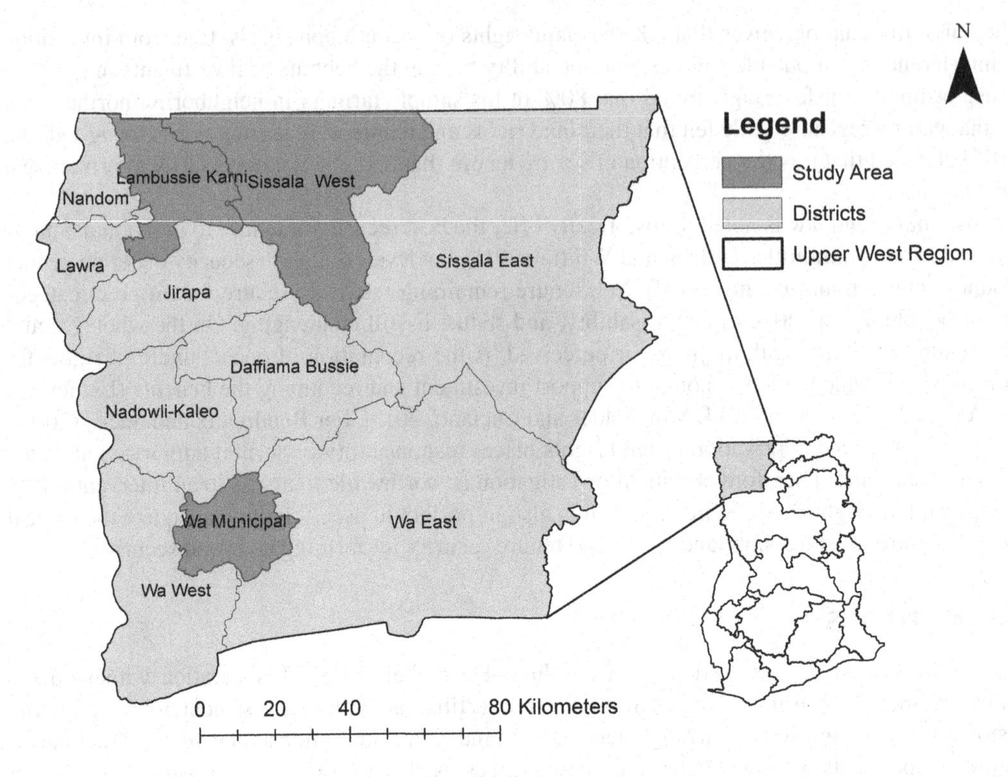

FIGURE 5.1 Study area – northwest Ghana. (Adapted from Free Spatial Data by Country by DIVA-GIS, 2017: http://www.diva-gis.org/gdata.)

people from Ghana and Burkina Faso as settlers since time immemorial. The communities together cover a total land area of 3,641.74 km^2. They were purposively selected due to the prevalent dominance of subsistence and settler farming (Republic of Ghana, 2012). Also, its land tenure system is unique in Ghana, i.e. initially without land sale or sharecropping. There are intermittent subtle, and sometimes escalated, land rights disagreements between landowners and settlers affecting farming and food security. Additionally, there is reported out-migration of youth claiming they can no longer secure food from their own farming activities, due to weakening land rights and increasing tenure insecurity.

The communities visited included Kunfabiala and Sing in Wa Municipality with approximately 234.74 km^2 of landmass and the most urbanized place in the region. The other two largely rural districts are Lambusie-Karni and Sissala West, having 1,356.6 km^2 and 2,050.4 km^2 of land area respectively. They are located over 200 km north, sharing a boundary with each other and southern Burkina Faso. The choice of rural and urban areas was to determine the extent to which their urbanity or rurality affects land rights and tenure security, and influences farming and food security.

5.3 RESULTS

5.3.1 Local Processes of Resolving Land Rights and Tenure Challenges

The regular "payment" by settlers of the "token gift" called *kagyin* or *kaju* to landlords in accordance with custom was indicated by landowners as the settlers' duty. Meanwhile, settlers admitted their failure to meet this obligation due to smaller infertile lands and continuously poor harvests, which settlers claim is obvious to landowners. Settlers in Fielmua claim that for them, "token gift-giving" tenure service ended centuries ago, marking their assumption of landowner status which their original landowners in Nimoro are contesting in court. All the FGDs indicated that such disagreements weaken rights, as landlords are currently selling lands that settlers claim to have become infertile, and reclaiming lands that settlers hold without rendering the required customary services. In addition to the above, and to offer interventions in land rights and tenure challenges, other avenues exist. These are specific local "offices," institutions, and organizations who are direct or indirect stakeholders in customary land rights and tenure security in northwest Ghana. From Figure 5.1 below, a landholder having issues goes to the group head to mediate it. If it fails, it goes to the *tendana* involving the CLS for spiritual or administrative intervention. When that fails, it goes to the chief to arbitrate and also moves to the Lands Commission for notification (verification), and if that fails too, then it goes to court for a final adjudication.

The people reminisced about the peace that accompanied farming, food availability, and food security when this order was strictly respected. The FGDs indicated that seeming non-adherence to the procedure in Figure 5.2 is because there are virtually no punitive measures meted out for non-compliance. A land management intervention that is responsible based on user requirements to secure land rights and tenure is desired by communities to curb the growing disregard for people's rights and foster inclusiveness. FGD participants mentioned that stipulated re-negotiation and subsequent documentation can curtail both landowners' and especially settlers' weak land rights. FGD participants also complained that the laid down structure to secure land rights in Figure 5.2 would be effective with strict legal backing and penalties for offenders in the constantly changing socio-economic environment.

FGD participants further stated that land rights and tenure insecurity stem from subtle unilateral re-allocation of settler lands for sand, stone, and gravel winning rendering the land uncultivable and facilitating its conversion to non-agricultural uses. Consequently, resistance from settlers then worsens the situation, affecting farming and food availability with lingering mistrust and fear of possible attacks. In the FGDs, it came out that to resolve the current tenure insecurity, there

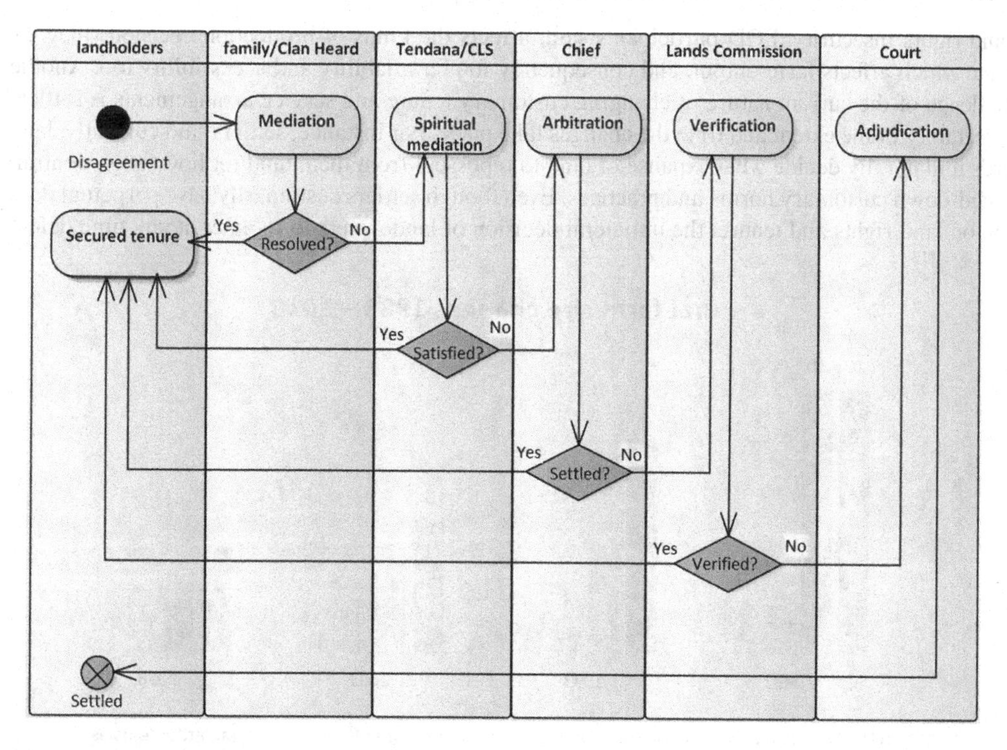

FIGURE 5.2 Procedure for settling customary tenure challenges. (Author's construct, 2019, using Enterprise Architect.)

have been calls for renegotiation of customary terms involving all stakeholders with state/outsider facilitation and state endorsement as a measure of responsible land management in line with the community's continuum of land rights. Also, the FGDs indicated that special tenure packages for specific groups, agreed upon and applicable in each area is feasible to minimize frequent and often contested land rights changes/challenges. Settler FGDs specifically proposed that landowners allow settlers to permanently own land not exceeding 2 ha. Otherwise, these settlers cannot envisage any end to their tenure challenges, suspicions, and consequent food insecurity in the following intervention measure:

> we appeal to be spared some minimum farmland of about 2 hectares as our reward for 'protecting the land for landowners' against encroachers, other claimants and potential attacks over the centuries. We have never let them down and they will also not let us down at this crucial moment in spite of current monetary motivations for land sale. This can be an effective land rights and tenure security intervention to enable us continue to farm for our food needs. Otherwise, the thought of complete future landlessness rather emboldens us to resist tenure changes being introduced by landowners but this worsens tenure insecurity and consequently affects food security.

The settlers hinted that such purely customary agreements should then be documented by the Lands Commission to prevent any unilateral variations to tenure in the future, as they are currently witnessing reducing farmlands.

The case of women is different because women's access and control of land remains almost entirely dependent on their relationship as daughters or wives of men. It emerged from the FGDs that wives farm mainly "for soup" crops like vegetables and groundnuts on smaller farms, evident in Figures 5.3 and 5.4, to complement husbands' or household harvests.

5.3.2 Minimizing the Challenges of Changing Customary Tenure Security

Land rights insecurity, FGD participants said, affects the kinds of production decisions they can make which affects farm output, and consequently food availability and accessibility too. Another challenge of the current nature of changing customary tenure and service arrangements is settlers' uncertainty of the extent and time the changes take place. For instance, settlers said currently, landlords unilaterally decide what expanse of land to repossess from them (and for how long), contrary to laid-down customary norms and practices. Even though settlers customarily have perpetual duration on land rights and tenure, the unilateral decision of landowners to re-enter at any time defeats

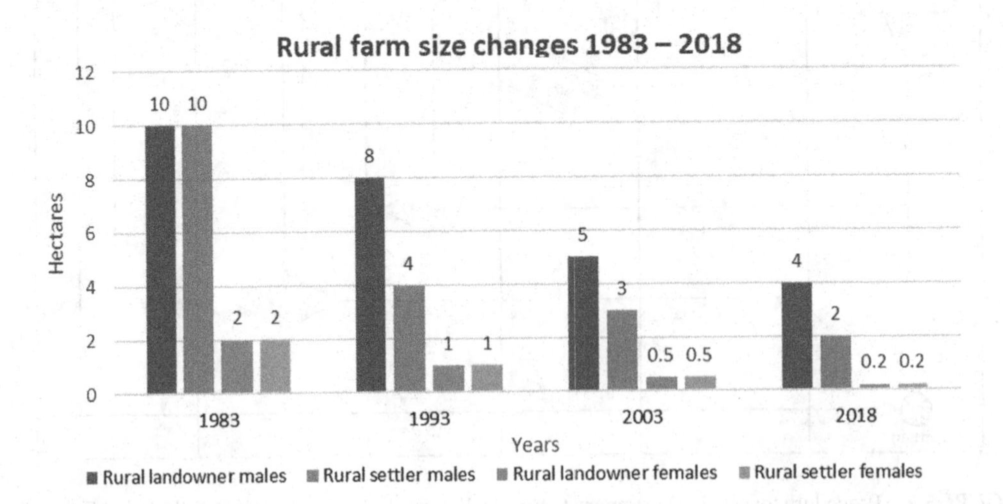

FIGURE 5.3 Trend of rural farm size changes. (Author's construct, 2019.)

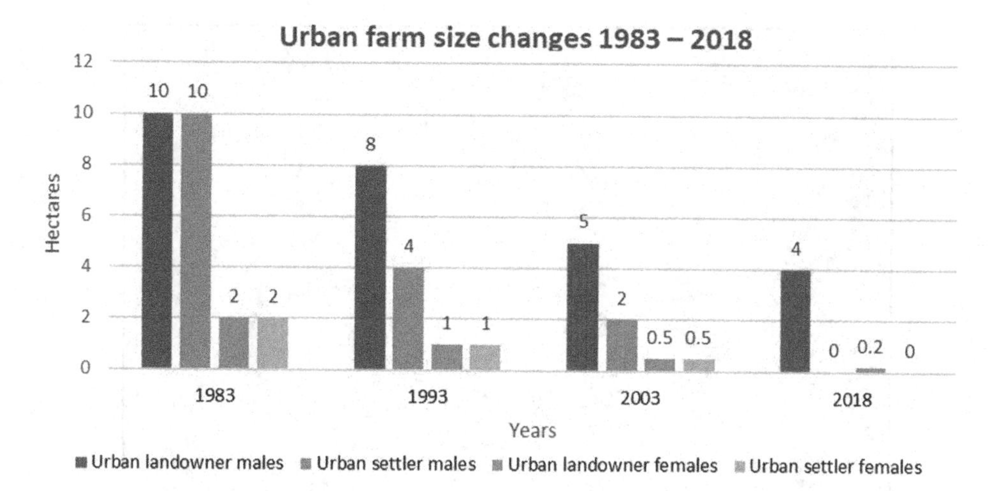

FIGURE 5.4 Trend of urban farm size changes. (Authors construct, 2019.)

that right. Settlers believe this can be minimized with state/outsider intervention. Also, the absence of settler rights to trees in Piina, for instance, was discussed in the focus groups. Settlers said this partly weakens their farming decision-making because it is nearly impossible for settlers to plant trees as an investment to help re-fertilize farms. These together leave settlers with "nothing to live for," because they claim some do not know their roots, and so are referred to as settlers, but not visitors.

Settlers said they quietly, but strongly, resisted attempts to dispossess them of the lands they occupy. So, when those in Kunfabiala, for instance, heard of eviction intentions, they responded by making more permanent structures to secure their settlements at least. Due to the commercial forces driving land rights changes, according to Chauveau et al. (2007), many landlords do not feel obliged to respect the customary norms guiding tenure and service arrangements. To minimize tenure insecurity, the enactment and enforcement of a well-publicized legal framework rooted in customarily re-negotiated land rights was constantly re-echoed, especially by settlers and women. The reason they advocate this intervention measure is because settlers claim "the ancestors are dead," an indication of loss of trust in the "spiritual verdict" regarding disagreements on tenure. This "death" makes landowners abuse land rights of marginalized groups without caution. All FGD participants explained that in times past, major contrary tenure decisions automatically attracted "invitation to the ancestry," i.e. death. This deterred people from engaging in arbitrariness on land, but urbanization and foreign religions have partly contributed to land rights changes and consequent tenure insecurity. FGD participants pointed out that rights and land tenure can further be made secure by government-led facilitation to protect the vulnerable. They suggested that government should 1) make or streamline laws, 2) ensure strict law enforcement, 3) promote affordable and socio-culturally acceptable payment, and 4) promote transparent land documentation by collaborating with customary people and experienced private land documentation agencies. Furthermore, settlers said that to overcome the challenge of continuous farm size reduction caused by these transformations, the state must also legislate lower and upper landholding limits beyond which defined categories of settlers cannot hold. This will prevent settlers from claiming too much land at the expense of landlords and vice versa. By this, they all believe land rights security, peaceful co-existence, and food security can be promoted.

The satellite images in Figures 5.5 and 5.6 show that housing development in Kunfabiala between 2006 and 2018 seems minimal. This may defeat the direct influence of urbanization solely changing customary tenure in terms of land availability and rights. FGDs revealed that the expectation for urbanization characterized by physical development has prompted land demarcation. The result,

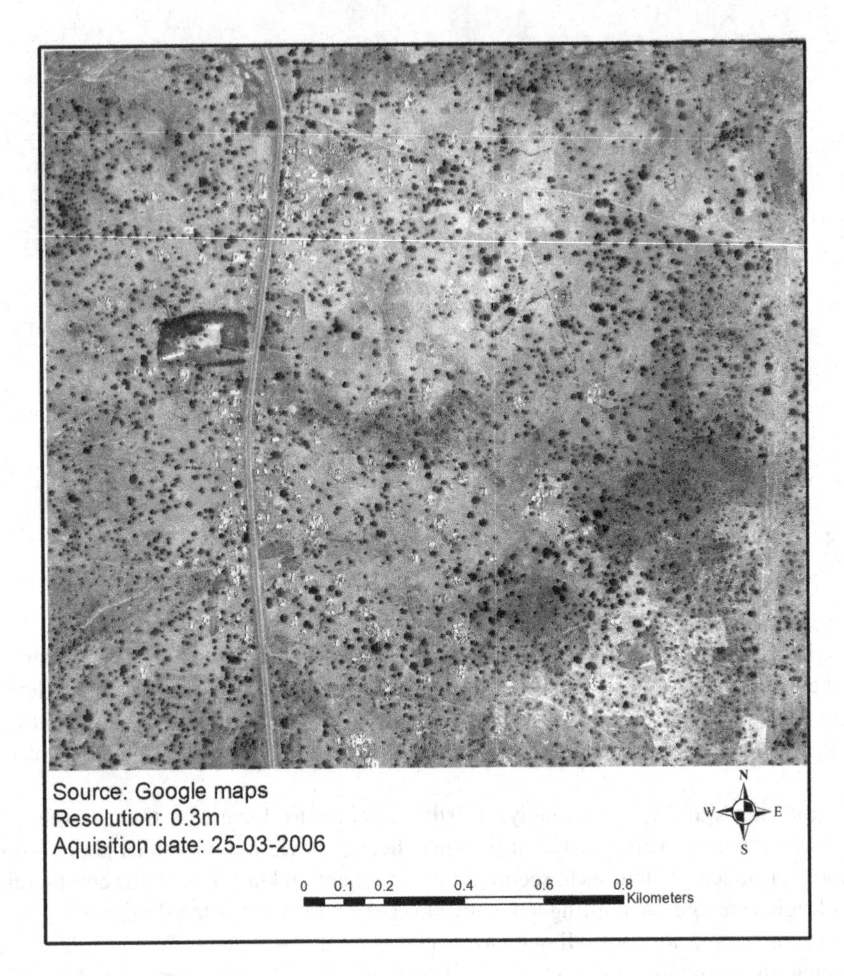

FIGURE 5.5 Building development and land availability, 2006. (Adapted from Google Earth Pro.)

they confirm, is that settler farmers now resort to land encroachment to farm for their household food supplies. To settlers, these point to some need for adopting a combination of socio-cultural, legal, and administrative participatory land management interventions that can be referred to as responsible in the short- to medium-term to address land rights and tenure insecurity. FGDs also raised that the government should prepare and implement development plans that cater for subsistence farming even in urban areas to minimize the threat of food insecurity. The medium to long term was emphasized, since development will definitely catch up in all these communities (whether urban or rural) sometime in the future.

5.3.3 Local Perception on Food Security and Land Rights Changes

Pinstrup-Andersen (2009) identifies transitory and permanent food insecurity. The former is periodic while the latter is long-term food insufficiency. Food may be secured through production, borrowing, exchange, purchases, or food aid properly processed and stored. Meanwhile for the FGDs, food security is fundamentally the production and availability of food in the household for their consumption from one harvest to the next. They believe that food availability through production or purchase provides food security. For the people, since physical, financial, and even social challenges can hinder access to food from the market, they rely largely on their own food production, and also

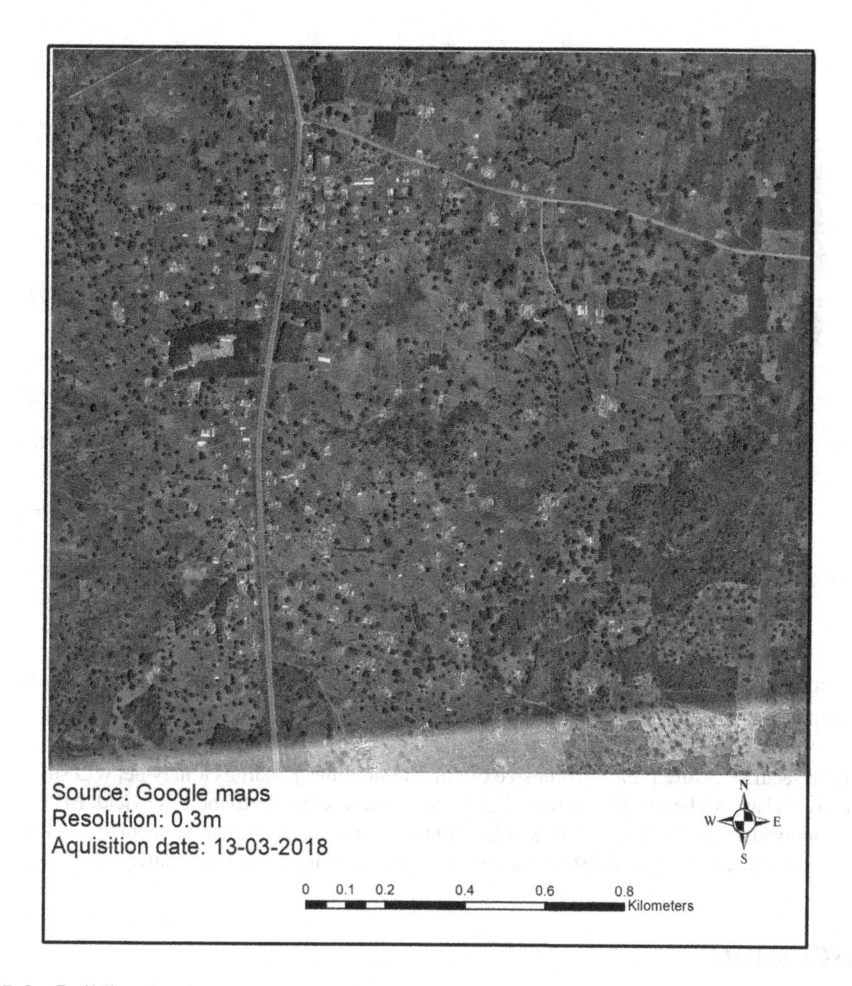

FIGURE 5.6 Building development and land availability, 2018. (Adapted from Google Earth Pro.)

because food can be secured premised on land rights and tenure security (Holden and Ghebru, 2016; Nguyen, 2014; Savenije et al., 2017). This was confirmed by respondents when one in Kunfabiala remarked rhetorically that: "if you lack land, on what will you farm to produce your own food?"

So, it came out during the FGDs that with secure tenure and other land-related opportunities like mortgage, transfer, and credit access, it may facilitate increased farm investments and food production, as shown in Nguyen (2014). Therefore, weakening land rights and tenure security can affect farming and food availability. Figure 5.7 illustrates the diminishing food production and reduction of monthly food availability trends as land rights weaken and tenure becomes insecure. This spanned over three decades since 1983 and the creation of the northwest as the Upper West region, which may have activated land transactions and showed the prospect of land value increases. This situation makes the drive towards responsible and fit-for-purpose land management a step in the right direction.

Figure 5.7 shows transitory food insecurity, since household food stocks last up to eight months. There are other challenges to food security which landowners mentioned as non-use of modern farming inputs, followed by an army worm epidemic, erratic rainfall, and post-harvest losses. The landowners asserted that in the past, they could manage their food stocks to last until the following season. They therefore had some surplus to sell and buy foods they do not grow themselves. So streamlined tenure, they hope, can assure people of reaping the benefits from their investments for

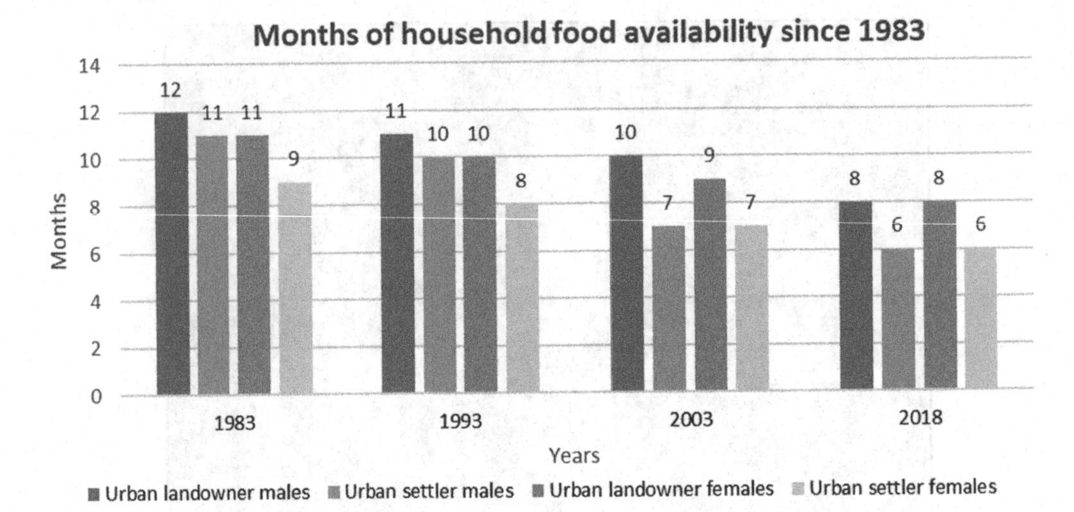

FIGURE 5.7 Monthly trend of food availability from own farming, 1983 and 2018. (Author's construct, 2019.)

which reason they will invest more by using modern inputs to increase production. But settlers in Piina number 2 stress that:

> with tenure security assured, the current size of our landholdings [small as it may be] will suffice since we will buy and apply chemicals – fertilizer, pesticides and weedicides to increase yield. In the absence of these chemicals, we need large parcels of land to produce more. Now that we seem to be under siege, the worst is feared and 'only bulldozers can move us out [an indication of resistance].

5.4 DISCUSSION

Holden and Ghebru, (2016) emphasized that secure access to sufficient land is an important means of promoting subsistence farming and achieving food security on customary lands. But Rao et al. (2016) indicated that there are currently weak laws to regulate various aspects of land, and either formal or local customary laws were not adequately used. Even though legislation can contribute to a high level of land tenure security in formal environments, it sometimes fails to protect some categories of local farmers (Barry and Danso, 2014; Deininger and Jin, 2006). This study contends that food insecurity occurs partly because of non-adherence to customary and formal land laws, and over-generalizing formal laws by disregarding context differences. For instance, the land tenure system in the study area operates on moral considerations rather than for profit or commerce. External influences have now introduced commercial-for-profit potential on land. Coupled with weak local land governance and poor implementation of national land laws, this threatens tenure security and consequently, food security of all subsistence farmers in varying degrees (Barry and Danso, 2014; Ho, 2001). This is supported by Barry and Danso's findings (2014) that existing formal land laws in Ghana were enacted without considering the unique context of users. For this reason, landholders, but especially owners, do not find current land laws very beneficial to them. Landowners therefore disregard these laws with almost no legal consequences, since tenure arrangements are not documented and scarcely enforced by formal laws. The available literature emphasizes that documentation alone cannot secure land tenure, especially in environments with a weak legal framework and institutional capacity and commitment (Zevenbergen et al., 2013). FGD respondents also insist that their perceptions alone no longer provide customary tenure security, except for landlords (in a few instances), because it is the courts that have finality on all land litigations in Ghana. Furthermore, women and settler respondents indicated that their land rights are no longer secure based on the customary practices, norms, and guarantees alone.

Even though local people recognize that other factors influence food security, they are convinced in line with Lawry et al. (2014), Ruerd (2011), and Savenije et al. (2017) that the primary factor is land rights and tenure security that enable them to farm their own food. It is especially true in this case, because the local people confess that they do not possess alternative skills with which to increase farm investment or diversify their food supply sources. Also, respondents concluded that if there is no means to farm their own food, it implies they particularly will have weak financial power to access food from the market. As a result, women, but especially male settlers said they welcome efforts to intervene in their tenure insecurity to facilitate continuous farming for their food supply and security. Even landowners agreed that perception alone is no longer sufficient to secure tenure, because there is a risk of unfair changes in future to the detriment of succeeding generations.

This therefore bridges the land rights and tenure security controversy in the literature regarding perception by Bugri (2008) and Simbizi, Bennett, and Zevenbergen (2014), documentation by Cotula and Toulmin (2004), and registration by Zevenbergen et al. (2013) to secure land rights. The reliance on either perception or documentation for land rights security in part depends on whether one's land rights are primary, in the case of landowners, or secondary in the case of settlers and women. This is clear on the ground, as the people of Fielmua are regarded as settlers, but insisted, based on perception and *de facto* possession, that their land tenure is secure. But this claim is immediately in question as it is being challenged in court by the original landowners; meanwhile, the court has freedom and finality to rule for or against. The current generation of both Fielmua and Nimoro have never witnessed the performance of any tenure services between the two peoples in their life. Yet the Nimoro people (original landlords) believe there is a need to reactivate the receipt of tenure services and "gifts" from the people of Fielmua (latecomer settler "landowners"). But the people of Fielmua rather find this to be the re-introduction of a practice that was ended mutually by both groups' ancestors long before any of the current settler and landowner generations were born. Based on the above discussions, further research is required to identify the most appropriate, effective, and efficient interventions which respond to responsible land management.

5.5 CONCLUSION AND POLICY RECOMMENDATIONS

The research objectives focused on addressing tenure challenges with locally based approaches to promote farming and food security. Agriculture is a main economic activity in northwest Ghana, and access to land is a fundamental means for food supplies. In the short- to medium-term, changes in land tenure systems may produce winners (urban landowners gaining money from land sales) and losers (women and settlers losing farmland). In the long run, all the people may lose out when these disagreements escalate. Both groups do not completely adhere to local intervention structures/avenues to address tenure insecurity issues due to discriminatory customary practices, weak formal land laws, and non-speedy adjudication in courts. This does not encourage compliance to either law or custom to address land rights and tenure insecurity. Recurring disagreements arising from weak institutional and legal system and non-documentation of customary arrangements governing land tenure raises inherent weaknesses in a transforming society.

In spite of its inherent relevance in making land available to many people in the area, the customary tenure system is challenged. And so, government intervention is still required in line with responsible land management through further research into blending formal legal recognition with local involvement. In this way, current land rights are ensured, but future needs for land are not compromised. Documentation with state facilitation and active participation of users can promote future land rights and tenure security, thus making land management more responsible. Furthermore, the involvement of private players, closely monitored by government, is vital for tenure documentation, thereby making land management more responsible (based on Meridia's successes in other parts of the world and in the western region of Ghana). Legal protection of previously obtained land rights based on customary arrangements may sanitize local land access and tenure security. It is especially so for marginalized groups like women and male settlers. Legal recognition of existing land rights and locally based institutions like the Customary Lands Secretariats (CLSs) using alternative dispute resolution (ADR) mechanisms to secure tenure are equally relevant for responsible land management interventions. Land management can also be responsible if it involves all stakeholders resolving land rights and tenure challenges. For instance, the formal laws can mandate CLSs to settle related disputes so that cases referred by them should be allowed to be heard in court in order to eliminate the numerous unresolved land cases in Ghana's courts for decades (Biitir and Nara, 2016). The dynamics influencing tenure security are unique. This calls for a process to design and implement context specific participatory responsible land management interventions for each spatio-cultural setting. The customary dimension of land access, land rights, and tenure security may promote farming for food supply and food security. This chapter advocates documentation of community-based land rights backed by law with strict legal enforcement, since local people now voluntarily accept that land documentation can strengthen their land rights and secure tenure for the future. The people have therefore suggested that the state through its representatives like the courts, quasi-judicial bodies, and Lands Commissions (LC) should collaborate with CLSs, customary leaders, and institutions to legislate and regulate an appropriate locally acceptable hierarchical land management structure. This makes land management responsible, since all stakeholders' needs may have been incorporated into it to meet the land needs of both current and future generations.

ACKNOWLEDGMENTS

I acknowledge financing from Nuffic (now OKP), the participation of interviewees, Lands Commission (LC), and Wa Central Customary Land Secretariat (WCCLS) in the Upper West Region, as well as Meridia Ghana, for responding to questions via email. Others are Benedict Akpem for transcribing FGDs and Williams Miller Appau, Tahiru Alhassan, Osman Mohammed Banyellibu, and Urbanus Wedaba for data collection support and critical technical inputs, not forgetting Maxwell Owusu, University of Twente, Netherlands.

REFERENCES

Akrofi, E. O., & Whittal, J. (2011). Traditional governance and customary peri-urban land delivery: A case study of Asokore-Mampong in Ghana. *AfricanGeo*, (October 2016). doi:10.13140/RG.2.1.4056.0480

Almeida, B., & Wassel, T. (2016). Survey on access to land, tenure security and land conflicts in Timor-Leste, 41. Retrieved from http://asiafoundation.org/publication/survey-access-land-tenure-security-land-conflicts-timor-leste/.

Barry, M., & Danso, E. K. (2014). Tenure security, land registration and customary tenure in a peri-urban Accra community. *Land Use Policy, 39*, 358–365. doi:10.1016/j.landusepol.2014.01.017

Biitir, S. B., & Nara, B. B. (2016). The role of Customary Land Secretariats in promoting good local land governance in Ghana. *Land Use Policy, 50*. doi:10.1016/j.landusepol.2015.10.024

Boudreaux, B. K., & Sacks, D. (2009). *Land Tenure Security and Agricultural Productivity*. Available at: https://www.mercatus.org/publications/development-economics/land-tenure-security-and-agricultural-productivity.

Brasselle, A. S., Gaspart, F., & Platteau, J. P. (2002). Land tenure security and investment incentives: Puzzling evidence from Burkina Faso. *Journal of Development Economics, 67*(2), 373–418. doi:10.1016/S0304-3878(01)00190-0

Bugri, J. T. (2008). The dynamics of tenure security, agricultural production and environmental degradation in Africa: Evidence from stakeholders in north-east Ghana. *Land Use Policy, 25*(2), 271–285. doi:10.1016/j.landusepol.2007.08.002

Chauveau, Jean-Pierre, Cissé, Salmana, Colin, Jean-Philippe, Cotula, Lorenzo, Delville, Philippe Lavigne, Neves, Bernardete, & Quan, J. T. (2007). *Changes in "Customary" Land Tenure Systems in Africa Edited by Lorenzo Cotula*. (Lorenzo Cotula, Ed.). IIED, UK Available online via : https://pubs.iied.org/pdfs/12537IIED.pdf.

Cotula, Lorenzo, & Mathieu, P. (2008). Legal empowerment in practice: Using legal tools to secure land rights in Africa. *Environmental Law.* IIED, UK. Available via: https://dlc.dlib.indiana.edu/dlc/bitstream/handle/10535/5812/12552IIED.pdf?sequence=1&isAllowed=y

Cotula, L., & Toulmin, C. (2004). Till to tiller: International migration, remittances and land rights in West Africa. *International Institute for Environment and Development, 132*(132), 1–92.

Cotula, L., Toulmin, C., & Hesse, C. (2004). *Land Tenure and Administration in Africa: Lessons of Experience and Emerging Issues.* IIED, UK.

de Vries, W. T., & Chigbu, E. (2017). Responsible land management-concept and application in a territorial rural context verantwortungsvolles landmanagement im Rahmen der Ländlichen Entwicklung.

Deininger, K., & Jin, S. (2006). Tenure security and land-related investment: Evidence from Ethiopia. *European Economic Review, 50*(5), 1245–1277. doi:10.1016/j.euroecorev.2005.02.001

Duncan, B. A., & Brants, C. (2004). *Access to and Control Over Land from a Gender Perspective: FAO Corporate Document Repository.* Retrieved from http://www.fao.org/docrep/007/ae501e/ae501e00.htm.

FAO. (2019). *Food Security and Nutrition in the World. IEEE Journal of Selected Topics in Applied Earth Observations and Remote Sensing.* doi:10.1109/JSTARS.2014.2300145

FAO, IFAD, UNICEF, WFP, W. (2018). *The State of food Security and Nutrition in the World 2018. Building Climate Resilience for Food Security and Nutrition.* doi:10.1093/cjres/rst006

Hancock, B., Windridge, K., & Ockleford, E. (2009). An introduction to qualitative research. *Research Development Society for East Midlands Yorkshire and Humber.* Nottingham, UK: University of Nottingham.

Ho, P. (2001). Who owns China ' s land ? Policies , property rights and deliberate institutional ambiguity. *The China Quarterly, 166*(166), 394–421.

Holden, S. T., & Ghebru, H. (2016). Land tenure reforms, tenure security and food security in poor agrarian economies: Causal linkages and research gaps. *Global Food Security, 10,* 21–28. doi:10.1016/j.gfs.2016.07.002

Kasanga, K., & Kotey, N. A. (2001). Land management in Ghana: Building on tradition and modernity. *Russell The Journal of the Bertrand Russell Archives, (February),* 1–42. doi:ISBN: 1-899825-69-X

Kothari. (2014). *Research Methodology Methods and Techniques.* New Delhi, India: New Age International.

Lambrecht, I., & Asare, S. (2016). The complexity of local tenure systems: A smallholders' perspective on tenure in Ghana. *Land Use Policy, 58,* 251–263. doi:10.1016/j.landusepol.2016.07.029

Lawry, S., Samii, C., Hall, R., Leopold, A., Hornby, D., Mtero, F., … Samii, C. (2014). The impact of land property rights interventions on investment and agricultural productivity in developing countries. doi:10.4073/csr.2014.1

Lund, C. (2000). African land tenure: Questioning basic assumptions. *Drylands Issue Papers, 28.*

Mwesigye, F., Matsumoto, T., & Otsuka, K. (2017). Population pressure, rural-to-rural migration and evolution of land tenure institutions: The case of Uganda. *Land Use Policy, 65,* 1–14. doi:10.1016/j.landusepol.2017.03.020

Naab, F. Z., Dinye, R. D., & Kasanga, R. K. (2013). Urbanisation and its impact on agricultural lands in growing cities in developing countries: A case study of Tamale in Ghana. *Modern Social Science Journal, 2*(2), 256–287. Retrieved from http://scik.org/index.php/mssj/article/view/993.

Nguyen, H. (2014). *Farmers' land tenure Security in Vietnam and China.* Retrieved from https://www.rug.nl/research/portal/files/14332380/Complete_dissertation.pdf.

Obeng-Odoom, F. (2012). Land reforms in Africa: Theory, practice, and outcome. *Habitat International, 36*(1), 161–170. doi:10.1016/j.habitatint.2011.07.001

Payne, G. (2004). Land tenure and property rights: An introduction. *Habitat International, 28*(2), 167–179. doi:10.1016/S0197-3975(03)00066-3

Pinstrup-Andersen, P. (2009). Food security: Definition and measurement. *Food Security, 1*(1), 5–7. doi:10.1007/s12571-008-0002-y

Place, F. (2009). Land tenure and agricultural productivity in Africa: A comparative analysis of the economics literature and recent policy strategies and reforms. *World Development, 37*(8), 1326–1336. doi:10.1016/j.worlddev.2008.08.020

Platteau, J. P. (1996). The evolutionary theory of land rights as applied to sub-Saharan Africa: A critical assessment. *Development and Change, 27*(1), 29–86. doi:10.1111/j.1467-7660.1996.tb00578.x

Rao, F., Spoor, M., Ma, X., & Shi, X. (2016). Land tenure (in)security and crop-tree intercropping in rural Xinjiang, China. *Land Use Policy, 50,* 102–114. doi:10.1016/j.landusepol.2015.09.001

Republic of Ghana. (2012). *Population and Housing Census Report, 2010.* Retrieved from http://www.statsghana.gov.gh/docfiles/2010phc/Census2010_Summary_report_of_final_results.pdf.

Ruerd, R. (2011). *Improving Food Security Report.* Netherlands.

Savenije, H., Baltissen, G., Van Ruijven, M., Verkuijl, H., Hazelzet, M., & Van Dijk, K. (2017). *Improving the Positive Impacts of Investments on Smallholder Livelihoods and the Landscapes They Live in Improving the Positive Impacts of Investments on Smallholder Livelihoods and the Landscapes They Live in 2 Disclaimer.*

Simbizi, D.M.C. (2016). *Measuring Land Tenure Security: A Pro-Poor Perspective.* doi:10.3990/1.9789036540544

Simbizi, M. C. D., Bennett, R. M., & Zevenbergen, J. (2014). Land tenure security: Revisiting and refining the concept for Sub-Saharan Africa's rural poor. *Land Use Policy, 36*, 231–238. doi:10.1016/j.landusepol.2013.08.006

Van Gelder, J.-L., & Luciano, E. C. (2015). Tenure security as a predictor of housing investment in low-income settlements: Testing a tripartite model. *Environment and Planning a Field United Nations Habitat, 47*(2), 485–500. doi:10.1068/a130151p

Wickeri, Elisabeth, & Kalhan, A. (2010). Land rights issues in international human rights law, *4*, 10.

Zevenbergen, J., Augustinus, C., Antonio, D., & Bennett, R. (2013). Pro-poor land administration: Principles for recording the land rights of the underrepresented. *Land Use Policy, 31*(March), 595–604. doi:10.1016/j.landusepol.2012.09.005

Zevenbergen, Jaap, De Vries, Walter, & Bennet, R. (2016). Advances in responsible land administration. *Journal of Spatial Science, 61*(1). doi:10.1080/14498596.2016.1145619

6 Human Recognition and Land Use Behavior

A Structural Equation Modeling Approach from Malawi

Ebelechukwu Maduekwe and Walter Timo de Vries

CONTENTS

6.1 INTRODUCTION

Land management interventions can only be responsible if the relations between people and land is properly understood, and if the underlying relationship and land decision dynamics reasons are effectively incorporated in the evaluation of land-related interventions (de Vries and Chigbu, 2017). Contemporary land administration literature models relations between people and land as either a relation of rights, a relation of use, or a relation of values (Lemmen, van Oosterom, and Bennett, 2015; Williamson et al., 2010; Henssen, 1995). In these models, people are captured by the concept "subject" or "party" and are usually considered a rather static entity. Within a group, it is assumed that certain interpersonal relations exist. However, the relationship dynamics which influence the decision to own or use a piece of land are usually not accounted for in land management models. Thus, it is not obvious during land allocation, registration, or adjudication processes how to properly assess land tenure relations on an individual level.

Additional discourses on responsible land management and land governance (Zevenbergen, de Vries, and Bennett, 2015 and FAO, 2012) emphasize that responsibilities are rooted in multidisciplinary design. It argues that insights and conventions within land administration are better connected and integrated with theoretical concepts from other disciplines, like sustainable resource management, economics, and development studies. In this chapter we introduce the concept of human recognition to improve the comprehension of interpersonal relations and their effects on land use behavior. Human recognition addresses the extent to which individuals are viewed and valued by others, as well as treated based on this value, with effects on recipient's wellbeing, but contribute to aggregate economic development.

Therefore, using structural equation modeling (SEM), we examine the influence of human recognition on land use behavior in selected communities in Malawi. Particularly focusing on women farmers, we analyze to what extent certain land use behavior and participation in community discourse are influenced by self/household and the institutional human recognition around them. Our study contributes to the literature on the role intangible concepts of development like human recognition play in responsible land use and community discourse. It also highlights the role of women farmers as self-agents of perception on barriers to responsible land use versus the role of community agents in understanding and meeting the needs of their community members.

The rest of the chapter is organized as follows: Section 6.2 looks at land use and tenure in Malawi. Methodology and descriptive results are presented in Section 6.3 and Section 6.4 presents the hypothesized relationships. The results are discussed and concluded with policy implications in Section 6.5.

6.2 LAND TENURE IN MALAWI

Malawi developed a national land policy aimed at improving customary landholders' tenure security (Djurfeldt et al., 2018, p.605; Peters, 2010). Since the new policy was passed towards the end of 2016 and the study data was collected five months after that, we argue that the effects of the new policy are yet to be reflected. We address the policy in place in the surveyed villages and districts, which is the customary land tenure. 66% of Malawi's land is held under customary law, and kinship identifies who has rights to customary land, varying across regions and ethnic groups (Kishindo, 2010, p.90). Two social systems define how land rights are passed on. They are the patrilineal system, predominant in the northern region where land rights are passed from father to son, and the matrilineal system, where land rights are passed on from mothers to daughters. For instance, Kishindo's (2010, pp.92–97) study of Yao households (matrilineal) residing in Kachenga village, Balaka district, Southern Malawi notes that decisions on selection of seed for the next cropping cycle and other minor decisions are carried out by the women. However, purchase of inputs, labor, and income disposal were done by the men. Linkages also exists between secure land tenure, decision-making, and welfare. Djurfeldt et al. (2018, p.607) observe the influence of female land rights on decision-making and household spending patterns in the matrilineal village of Khasu. Although Holden and Otsuka (2014, pp.92–93), Lovo (2016, p.219) and Holden and Ghebru (2016) note that increased land tenure security increases observable efforts to invest in long-term land development and soil conservation for fertility, these findings were not examined by gender. In a nutshell, findings on the role of women as equal partners in responsible land management are lacking, and one reason could be the gender roles established in the community that do not fully recognize women as land managers and thus, value-adders in overall land management.

6.2.1 LAND USE BEHAVIOR

Studies have explored the impact of challenges like climate (Asfaw et al., 2016, p.646; Kassie et al., 2008; Arslan et al., 2014), adoption cost (Sylwester, 2004), credit and market imperfections (Carter and Barrett, 2006), and tenure and community norms on land use (World Bank, 2006, pp.18–30; Asfaw et al., 2016, p.643; Meinzen-Dick et al., 2019, p.74) on responsible land use. For example, the adoption of improved agronomic practices such as cover crops and crop rotations, has been associated with better farm performance, improved income, and overall environmental sustainability by Knowler and Bradshaw (2007), Teklewold, Kassie, and Shiferaw (2013), Snapp et al. (2002, p.159), and Waldman et al. (2016, pp.1087–1088). Improved land use encompasses the responsible use of land resources, whose exploitation, enhancement, and investment are carried out such that both its current and future potential to meet human needs are advanced (deWrachien, 2010, p.472). Such land use behavior retains the land's fertility and supports the production of food and other renewable resources for long-term use.

There is increased awareness of the role gender plays in household decision dynamics in adopting certain land use behaviour. For example, Waldman et al. (2016, p.1094) observe that although households in Malawi are more likely to plant legume crops because of soil fertility and cultural context, female-headed households are less likely to do so if they think that it involves more labor input. These perceptions reflect the investment responsibilities within households where women are constrained in labor input. Meinzen-Dick et al. (1997, p.1312) argues that the gender differential in agricultural productivity may also be fueled by education, access, time, labor, or other forms of human development. As observed by Kodoth (2001, pp.291–92), the gendered pattern in land use is heavily influenced by women's position in the family, community, and ethnic group. That is, how women are viewed and valued in the society impacts women's power dynamics, with significant influence on land use behaviour.

Community participation for women is important to improve gender asymmetry. Removal of gendered institutions will support women farmers' short and long-term investments in land improvements. However, there are still gaps in addressing land use behavior through the lens of intangible forms of human development, like human recognition for women farmers.

Because of the patrilineal and matrilineal systems that exogenously defined land use and inheritance patterns, Malawi offers an excellent opportunity to examine self/household and institutional human recognition as an intangible influence on land use behavior for women farmers. Looking from a gender and human rights perspective, such analysis can help a better understanding of some of the constraints faced by women farmers. It will also help highlight how human recognition overall affects and promotes responsible land use behaviour.

6.3 HYPOTHESIS AND METHODOLOGICAL APPROACH

6.3.1 Structural Equation Modeling (SEM)

Structural equation modeling (SEM) is mostly used for estimating relationships between multiple endogenous and exogenous factors. SEM with latent variables (LV) further allow the ability to model more abstract concepts constructed as measurement models which cannot be directly captured in a linear model nor be measured by known variables (Denny et al., 2018, p.10).

SEM has been used extensively to model abstract relationships in agriculture (Jaijit, Paoprasert, and Pichitlamken, 2018; Denny et al., 2018, p.10; Najafabadi, 2014) and land use change (Wang, Li, and Yang, 2015). Wang, Li, and Yang (2015) employed SEM to assess the effect of land use change on regional climate in southern China and found that vegetation latent variables significantly reduce climate effect. Using SEM, Toma et al. (2018, p.864) analyzed the technological information transfer on uptake of innovative crop technologies and found that economic characteristics, including income and farm labor, have the strongest effects. Liu and Luo (2018, p.11) used SEM to analyze the driving factors between farmers' land protection perception and land use change in northeast China. The authors found that, among others, external factors like insecure land rights reduce farmers' willingness to engage in land protection. Najafabadi (2014) also used SEM to explore the motivating and challenging elements affecting organic farming adoption in Iran through a gender perspective. The author found that attitudes to organic agriculture adoption were slightly variable by gender for farmers.

We used the SEM to examine the relationship between self/household and community-reported institutional human recognition, favoring land use behavior for women farmers in Malawi. In particular, we ask:

1. To what extent is land use behaviour influenced by latent self/household and institutional human recognition?
2. Does latent self/household and human institutional recognition influence women farmers' community participation?

Our SEM consists of two components: the measurement component, showing the relationships between the latent variable and its related observed variables, and the structural component which describes the relationships between the latent variables. We specify our equation as follows:

The measurement model for self/household and institutional human recognition:

$$x_i = \alpha_i + \beta_i X + \varepsilon \cdot x_i \tag{6.1}$$

The measurement model for land use behaviour and community participation:

$$y_i = \alpha_i + \beta_i Y + \varepsilon \cdot y_i \tag{6.2}$$

where x_i and y_i are vectors of endogenous and exogenous predictor variables, $i = 1, 2, 3, \ldots, n$, respectively, and β measures the impact of the latent constructs, while $\varepsilon{\cdot}x_i$ and $\varepsilon{\cdot}y_i$ are vectors of measurement errors in x and y. In our specification, it encompasses both observed variables used for the latent exogenous and endogenous constructs.

The structural component which measures the impact of the exogenous latent variable on the endogenous variable of interest is given as follows:

$$Y = BY + \Gamma X + \zeta \tag{6.3}$$

Where Y measures the endogenous latent constructs of interest (land use behaviour and community participation), X is the exogenous latent constructs (self/household and institutional human recognition), B measures the relationship among the endogenous latent constructs, Γ measures the impact of the exogenous latent construct on the endogenous latent construct, and ζ is the residual of Y.

Figure 6.1 illustrates the relationship between the latent human recognition and land use behavior. The arrows show the directions of influence. Firstly, the latent construct of self/household human recognition contains three variables: an additive score for self-reported human recognition received in the household, an additive score of self-reported human recognition received in the community and a measure of human recognition generated using the Alkire-Foster method on indicators of violence, humiliation, power, and autonomy. Secondly, the latent construct of institutional recognition contains three observed variables. They are three additive scores of community agent-reported recognition measures on a) ease of access to rights of property and divorce, b) ease of participating autonomously in binding village-level decisions, and c) a measure of community recognition for general and female members. We constructed these institutional scores by taking the average response from the community agents by district surveyed. Finally, the latent endogenous construct for community participation contains two variables on public speaking capacities for a) comfortably speaking for redirection of public funds for developing community infrastructure, and b) comfortably speaking out against abuse of public office.

Land use behaviour (LUB) is a latent construct that includes three variables capturing the behavior towards land use in agriculture: two categorical measure of the final decisionmaker for agricultural land improvement undertakings and agricultural input use, and the sum total of responsible land use practices implemented in the last cropping cycle from the following list: crop rotation, use

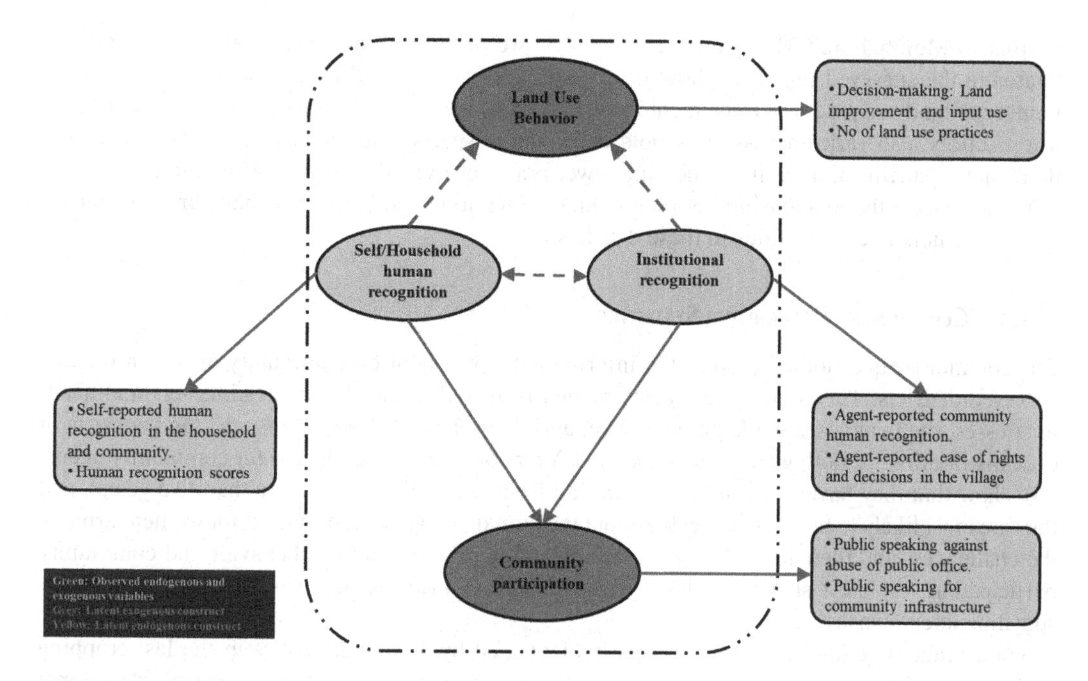

FIGURE 6.1 Proposed model of interaction between land use behavior, self/household human, and institutional recognition.

of improved seed, use of irrigation technology, and use of herbicides and pesticides, as well as other forms of farm enhancements. We estimate the model using maximum likelihood estimation (ML) and report model estimates. We also use modification indices to include covariance matrices which are relevant to our model (available on request). We use multiple fit indices to evaluate whether our model was a good fit reporting the comparative fit index (CFI), the root mean squared error of approximation (RMSEA), and chi-square goodness-of-fit statistics. A satisfactory fit is indicated by CFI values greater than 0.95, non-significant chi-square (Hu and Bentler, 1998; Hu and Bentler, 1999) and RMSEA values less than 0.05 (Steiger, 1990).

6.3.2 Study Area Statistics

We conducted a farm household surveys in two districts in the central region and three districts in the southern region of Malawi: Lilongwe (7 communities), Salima (7 communities), Mangochi (8 communities), Chiradzulu (7 communities), and Nsanje (10 communities). The fieldwork was done between May and July 2017. Data was collected from ca.190 respondents per district yielding a total of ca.950 respondents. We designed two questionnaires (community and household) to measure individual, household, and community human recognition.

The study area also encompasses several climatic and agro-ecological zones with cooler seasons found in Lilongwe, Salima, and Mangochi and hotter seasons in Chiradzulu and Nsanje. Rain-fed agriculture is predominant, and the rainy season runs between November and March. These districts present different patterns in economic development and cultural institutions for women that influence the breadth of human recognition. Lilongwe and Mangochi are both predominantly matrilineal districts, while Nsanje is a patrilineal district. Chiradzulu and Salima presents a mixture of both lineages. For the final analysis, we selected women farmers in households with agricultural land and who reported crop and land improvements for the last cropping cycle, giving a total of 619 respondents.

Agriculture very important in Malawi (Asfaw et al., 2016, p.645; Munthali and Murayama, 2013, p.158; Malley et al., 2009, p.176). On average, most smallholder farmers farm on less than 1 ha.

Similar to Munthali and Murayama (2013, p.160), we observed that maize is the main crop cultivated in the surveyed regions, followed by beans and peas as well as groundnuts. However, the number of land use practices implemented varied across first crops.* Similarly, the number of land use practices also varies across household heads and districts, with households in Nsanje, a predominantly patrilineal area, implementing fewer practices overall. Just like Waldman et al. (2016, p.1088), we treat the relationship between human recognition and land use behaviour as distinct to women farmers in communities in these districts.

6.3.3 Community Descriptive Statistics

The community questionnaire collected information from about 60 community agents in the five surveyed districts. The community agents ranged from village heads, the headmasters and headmistresses, the healthcare workers, the police, agricultural extension workers, etc., and consists of even distribution for both genders to avoid bias. Years of residency collected for community agents also show that they have lived in the community for a reasonable time (more than five years), and thus are most likely to be knowledgeable about the human recognition provision for women farmers, the challenges, and their favorability towards adopting certain land use behavior and community participation. Table 6.1 shows the descriptive statistics of measurement variables and their corresponding latent construct.

On average, one land use practice was implemented by a woman farmer in the last cropping cycle. About 37% of women farmers reported that they were the final decision-maker for input use in the household, compared to 23% that reported spouse/partner, and 38% that reported joint

TABLE 6.1
Descriptive Summary of Measurement Variables

Measurement variables	No. of variables	Equation name	Mean	Type
Land use behavior		LUB		
Number of land use practices in the last cropping cycle	5	No. of land use practices	1.407	Count (0–5)
Decision-maker: land Improvements	1	Deci: Land improv	2.984	Categorical (1–4)
Decision-maker: agricultural input use	1	Deci: Input use	2.986	Categorical (1–4)
Self/household human recognition (SR)		Recog:S/HH		
Human recognition scores		MDI:HR	7.295	Continuous
Subjective human recognition (self)	3	SR: Subj. Self HR	9.386	Likert (1–4)
Subjective human recognition (community)	3	SR: Subj.Comm. HR	9.313	Likert (1–4)
Community human recognition (AR)		Insti		
Ease of partaking in village decisions	3	AR: Ease decision	2.444	Likert (1–4)
Ease of property rights and divorce	3	AR: Ease rights	2.743	Likert (1–4)
General and female community recognition	9	AR: G&F Comm.HR	14.383	Likert (1–4)
Participation in community discourse		Par_comm		
Comfort with public speaking: misbehavior of public officials	1	PS: Misbehavior	2.94	Likert (1–5)
Comfort with public speaking: community infrastructure	1	PS: Infra	3.094	Likert (1–5)

Notes: breakdown of all indicators is available on request; AR = Agent-reported: SR = Self-reported.

* Each respondent reported the first three main crops cultivated in the last cropping cycle. 22% of the women farmers reported they cultivated one crop type, 41% two different crop types, and only 34% three different crop types.

decision-making. About 38% of our respondents reported that they make the final decision on land improvements compared to their spouse making a sole decision at 23%, and household joint decision at 37%. Only 1% reported that someone else is the decision-maker for land improvement. Finally, 27% reported discomfort in speaking for community infrastructure, 26% reported fair comfort, while 25% reported very comfortable. Ca.78% of our respondents were between 18–48 years old, and 39% had no education at all, while 9% had education between secondary levels 1–6.

6.4 RESULTS: SELF/HOUSEHOLD AND INSTITUTIONAL HUMAN RECOGNITION ON LAND USE BEHAVIOR AND COMMUNITY PARTICIPATION

Figure 6.2 and Table 6.2 show the estimated results. We comment only on the standardized beta coefficients. We set the normalization constraints, i.e. anchor point, for self/household and institutional human recognition at the indicators for the number of land use practices implemented and agent-reported general/female community recognition respectively. The path coefficients in the model suggests a strong relationship between observed variables and the latent construct. Self/household and institutional human recognition significantly affect women farmers' land use behaviors and community participation. The beta path coefficients for self/household human recognition and institutional recognition influence on land use behaviour (LUB) are 0.319 (p = 000) and 0.607 (p = 000) respectively. That is, self/household human recognition and community human recognition increases land use behaviour by 32% and 61%, respectively. Thus, women farmers in communities with better self/household human recognition and institutional recognition are likely to take on land use behaviour to improve their farmlands. Interestingly, we see that land use behaviour has a negative influence (–27% and –25%) on the decision-maker for land improvement and input

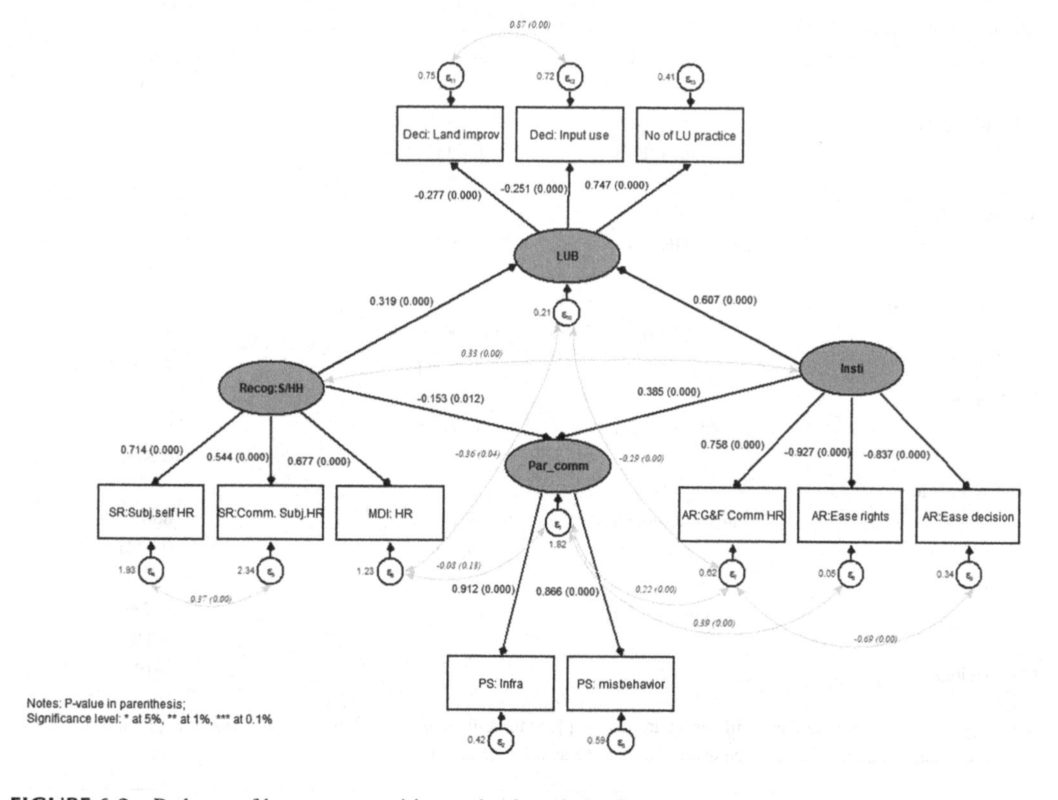

FIGURE 6.2 Pathway of human recognition on land use behavior.

TABLE 6.2
Effects of Human Recognition on Land Use Behavior and Community Participation

Structural equations		Coefficient	Standardized beta
Land use behavior (LUB) ←			
	Insti (LV)	0.166***	0.607***
		{0.033}	(0.066)
	Recog:S/HH (LV)	0.056**	0.319**
		{0.020}	(0.083)
Par_comm←			
	Insti (LV)	0.612***	0.385***
		{0.093}	(0.053)
Recog:S/HH (LV)		−0.156*	−0.153*
Measurement		{0.066}	(0.061)
Deci: Land improv←			
	LUB(LV)	−0.344***	−0.277***
		{0.066}	(0.045)
Deci: Input use←			
LUB(LV)		−0.306***	−0.251***
		{0.064}	(0.045)
No of land use practices←			
	LUB(LV)	1.000***	0.747***
		{.}	(0.069)
AR:G&F Comm. HR←			
	Insti (LV)	1.000***	0.758***
		{.}	(0.027)
AR: Ease decision←			
	Insti (LV)	−0.973***	−0.837***
		{0.027}	(0.023)
AR:Ease rights←			
	Insti (LV)	−0.635***	−0.927***
		{0.039}	(0.022)
SR: Subj.self HR←			
	Recog:S/HH (LV)	1.000	0.714***
		{.}	(0.068)
SR: Subj.comm HR←			
	Recog:S/HH (LV)	0.699***	0.544***
		{0.057}	(0.059)
MDI: HR←			
	Recog:S/HH (LV)	0.721***	0.677***
		{0.134}	(0.065)
PS: Misbehavior←			
	Par_comm (LV)	0.919***	0.866***
		{0.061}	(0.028)
PS: Infra←			
	Par_comm (LV)	1.000***	0.912***
		{.}	(0.028)
Observations		619	619

Note: point estimate standard error in curly parenthesis { }; Standardized beta coefficient standard error in parenthesis (); LV = Latent Variable; Significance level: * at 5%, ** at 1%, *** at 0.1%.

use, i.e. the land manager. That is, land use behavior to improve farm lands decreases the chances that the final decision-maker is the woman farmer alone. We also note that self/household human recognition significantly reduces community participation by 15%, while institutional recognition significantly and positively impacts community participation by 39%.

6.4.1 Drivers of Self/Household Human Recognition (Recog:S/HH)

All three variables, i.e. self-reported subjective self and community human recognition, as well as the Alkire-Foster method-measured positive human recognition, are positive and significantly influenced by self/household human recognition. Self-reported subjective self and community human recognition are 71% and 54% influenced by self/household human recognition respectively. Alkire-Foster-measured positive human recognition is 68% positively and significantly influenced by self/household human recognition.

6.4.2 Drivers of Institutional Recognition (Insti)

Institutional recognition has a positive and significant influence on agent-reported general and female community recognition by 76% in these communities. In contrast to self/household human recognition, agent-reported ease of village decision participation and ease of rights are negatively and significantly influenced by institutional recognition. In particular, institutional recognition influence on ease of rights and participation in binding village decisions are −93% and −84% respectively. It is imperative to examine all three variables side by side. Although community agents report that community members, including women, are recognized within their communities, the institutional frames in place do not support easy access to rights in property and divorce and participation in binding village decisions. This gives us a glimpse of how underlying structural challenges like norms affect women and do not allow them the full exercise of rights as community members. It also shows the discrepancy between community perception of how women are treated and valued, and the institutions in place that determine the rights to assets and decision-making.

Model fit: we assessed our model's goodness-of-fit by estimating goodness-of-fit indices. The SEM method provides a variety of indicators to verify the degree to which the hypothesized model is close to the estimated sample. Table 6.3 highlights the model fit for influencing factors for land use behavior. These fit indices are used to check model adequacy: the Chi-square test, the comparative fit index (CFI), the standardized root mean square residual (SRMR), and the root mean square

TABLE 6.3
Model Goodness-of-Fit

Fit index		Value	Criterion	Implication
Likelihood ratio	**Chi^2**	**43.722 p = 0.064**	**Non-significant ($p > 0.05$)**	**Model is a good fit**
		Population error		
Root mean squared error of approximation	RMSEA	0.026	<0.05	Acceptable
		Baseline comparison		
Tucker Lewis index	TLI	0.994	>0.90	Acceptable
Comparative fit index	CFI	0.997	>0.90	Acceptable
		Size of residuals		
Coefficient of determination	CD	0.979	>0.90	Acceptable
Standardized root mean squared residuals	SRMR	0.026	<0.05	Acceptable

error of approximation (RMSEA) (Hu and Bentler, 1998; Hu and Bentler, 1999). The CFI index shows how well the hypothesis model fits with the non-nested model. The null hypothesis of the Chi-square tests that the observed and estimated covariance matrices in the model are unbiased.

A non-significant Chi-square test ($p > 0.05$) signifies that the data provided a good fit to the model. CFI index is normally accepted at a value of at least 0.9 to indicate the model fit. The SRMR describes the differences between the observed and predicted covariance and accepts a model fit within the threshold of SRMR <0.05. The RMSEA measures the amount of variation between the observed covariance matrix and the hypothesized covariance matrix. Table 6.3 shows that the overall model fit for factors influencing land use behaviour is acceptable.

We also assess the convergent and discriminant validity of our measurement variables. Convergent or construct validity argues that the indicators chosen to measure the latent construct should interact in a way that captures the latent essence of the construct, while discriminant validity measures the extent to which measures of different latent construct are distinct (Zait and Bertea, 2011, p.217; Li et al., 2019, p.8; Bagozzi, Yi, and Phillips, 1991, p.425). The squared correlation matrix and averaged variance extracted (AVE) can be used to assess both convergent and discriminant validity. A latent construct must meet the following criteria to satisfy the appropriate degree of convergent and discriminant validity: the average extracted variance should be greater than 0.5 for convergent validity and less than the values of the squared correlations among latent constructs for discriminant validity, as shown in Table 6.4. As seen from this Table, the criteria are satisfied, indicating that the measurement model has a good discriminant and convergent validity.

6.4.3 Indirect Impact of Self/Household and Community Recognition

Table 6.5 presents the indirect effects of self/household and institutional recognition on indicators of land use behavior and community participation.

In particular, self/household and institutional recognition has an indirect and negative influence on who the decision-maker for farm land improvement is, by 6% and 17%, respectively. Better human recognition at the household and institutional level increases the chance that the final decision-maker in the household for farm land improvement and input use is not the woman farmer

TABLE 6.4
Test for Discriminant and Convergent Validity

Squared correlations (SC) among latent variables

	LUB	Insti	Recog:S/HH	Par_comm
LUB	1			
Insti	0.018	1		
Recog: S/HH	0.048	0.065	1	
Par_comm	0.003	0.088	0	1

Average variance extracted (AVE) by latent variables

Latent variable	AVE	Convergent validity	Discriminant validity
AVE_LUB	0.602	Acceptable	Acceptable
AVE_Insti	0.798	Acceptable	Acceptable
AVE_Recog:S/HH	0.520	Acceptable	Acceptable
AVE_Par_comm	0.795	Acceptable	Acceptable

Notes: when AVE values >=SC, there is no problem with discriminant validity.
when AVE values >= 0.5, there is no problem with convergent validity.

TABLE 6.5
Indirect Effects of Human Recognition on Indicators of Land Use Behavior and Community Participation

Structural equations		Indirect effects
		Coefficient
Deci: Land Improv←		
	Recog:S/HH (LV)	−0.056**
		(0.020)
	Insti (LV)	−0.166***
		(0.033)
Deci: Input use←		
	Recog:S/HH (LV)	−0.050**
		(0.019)
	Insti (LV)	−0.148***
		(0.031)
No of land use practices ←		
	Recog:S/HH (LV)	0.163***
		(0.048)
	Insti (LV)	0.482***
		(0.051)
PS: Infra←		
	Recog:S/HH (LV)	−0.157*
		(0.066)
	Insti (LV)	0.612***
		(0.093)
PS: Misbehavior←		
	Recog:S/HH (LV)	−0.144*
		(0.061)
	Insti (LV)	0.562***
		(0.086)

Note: standardized beta coefficient standard error in parenthesis (); errors (LV) = Latent Variable; Significance level: * at 5%, ** at 1%, *** at 0.1%.

alone. One may argue that it also increases the chance that such decisions are made by the household jointly.

As expected, self/household and institutional recognition has a positive and indirect influence on the number of land use practices implemented. That is, women with better self/household recognition and communities with institutional recognition indirectly increase the number of land use practices implemented on the farm by 16% and 48% respectively. This means that land use practices to improve farm lands can be indirectly supported by self/household and institutional recognition, if barriers that hinder women as recognized economic agents are removed or reconciled.

Finally, institutional recognition has an indirect and positive influence on public speaking for community infrastructure and public office abuse (61% and 52%), while self/household human recognition has an indirect and negative influence on community participation.

6.5 DISCUSSION

Using structural equation modeling (SEM), we examine the influence of human recognition on land use behaviour and community participation for women farmers in communities in five Malawian districts.

Firstly, self/household human recognition and institutional recognition has a positive influence on land use behaviour. Delving into the indicators of institutional recognition show that better agent-reported community recognition does not result in better overall ease of village decision-making and rights. This is consistent with the World Bank (2006, pp.18–30) and Lovo (2016, p.219), which argues that women's status in their community being viewed through the lens of rights is one challenge facing the adoption of improved land use, with effects on short- and long-term conservation investments.

Secondly, we note that improved land use behaviour is mostly adopted when the woman farmer does not make the final decision on input use and land improvements, i.e. does not manage the land. One path of influence could be through gendered decision-making allocation in traditional agricultural households, where decision-making for farm land management (e.g., what and how to grow) may be seen as a predominately male domain, or carried out jointly in the household regardless of the tenure system, as observed by Kishindo (2010, pp.94–95) and Djurfeldt et al. (2018, p.606). This in line with our findings.

One also cannot ignore the effect of the managerial capabilities of the woman farmer. Traditional roles in Malawi may see woman farmers as lacking the know-how to make land decisions, and thus they may leave the domain solely to their spouse or make the decision jointly. For a female headed household, financial or labor capabilities may be particularly constrained, as argued by Toma et al. (2018, p.864), Waldman et al. (2016, p.1094), and Meinzen-Dick et al. (1997, p.1312). Most land use practices are expensive and involve investments which may not yield immediate returns, as noted by Lovo (2016, p.226). Trade-offs may also exist between household short-term nutritional security and the long-term land improvement outlook as most rural household only want to have enough food until the next cropping cycle.

Finally, we observe that institutional recognition directly improves community participation for women farmers. That is, women farmers with better institutional recognition feel more comfortable speaking out on community issues. This is supported by Castleman (2016, p.142) who argues that practices that improve an individual's human recognition within the community may contribute significantly to improving community voice, accountability, and participation.

Our results are consistent with the expectation that positive human recognition influences land use behavior and community participation positively. We note that overall, women farmers with better recognition are more likely to participate in the community and to implement land use behavior. We also note that they are less likely to be the ones to take decisions on the final farm land improvement ventures. The latter could also signal the discrepancy between who makes the decision to implement, and the actual implementation.

6.6 CONCLUSION

We examined the influence of human recognition on land use behaviour and community participation for women farmers in five Malawian districts. We found that self/household human and institutional recognition has a positive influence on land use behavior and community participation.

Our findings have interesting policy implications. Firstly, human recognition has significant influence on overall land management and should not be ruled out when discussing land use management issues in community discourse. Improved land use behavior significantly influences who makes decisions on land improvements and input use. Thus, if certain land use behavior is to be adopted in these communities, future reforms should reconcile the gap between farmland owners and farm managers. As argued by Lovo (2016, p.228), this is important, as ownership and management rights still remain separate rights in Malawi. Secondly, the effect of institutional recognition cannot be overstated. The dependence of land use behaviour on how community members are viewed as productive agents and value-adders in their society is a very important intersection which has not been analyzed before. Thus, structural policies should tackle obstacles facing women in the community and improve the perception of women farmers as drivers of responsible land use management. Finally, further research to investigate the exact mechanism and pathway of human recognition on overall land management is urgently needed.

REFERENCES

Arslan, A., N. McCarthy, L. Lipper, S. Asfaw, and A. Cattaneo. 2014. "Adoption and intensity of adoption of conservation farming practices in Zambia." *Agriculture, Ecosystems & Environment* 187: 72–86.

Asfaw, S., N. McCarthy, L. Lipper, A. Arslan, and A. Cattaneo. 2016. "What determines farmers' adaptive capacity? Empirical evidence from Malawi." *Food Security* 8(3): 643–64.

Bagozzi, R. P., Y. Yi, and L. W. Phillips. 1991. "Assessing construct validity in organizational research." *Administrative Science Quarterly* 36(3): 421–58.

Carter, M. R., and C. B. Barrett. 2006. "The economics of poverty traps and persistent poverty: An asset-based approach." *Journal of Development Studies* 42(2): 178–99.

Castleman, T. 2016. "The role of human recognition in development." *Oxford Development Studies* 44(2): 135–51.

Denny, R. C., S. T. Marquart-Pyatt, A. Ligmann-Zielinska, L. S. Olabisi, L. Rivers, J. Du, and L. S. Liverpool-Tasie. 2018. "Food security in Africa: A cross-scale, empirical investigation using structural equation modeling." *Environment Systems & Decisions* 38(1): 6–22.

de Vries, W. T., and U. E. Chigbu. 2017. "Responsible land management-Concept and application in a territorial rural context." *fub. Flächenmanagement & Bodenordnung* 79(2): 65–73.

deWrachien, D. 2010. "Land use planning: A key to sustainable agriculture." In: *Conservation Agriculture*, García-Torres, L. et al. (Eds), 471–83. Dordrecht: Springer Netherlands.

Djurfeldt, A. A., E. Hillbom, W. O. Mulwafu, P. Mvula, and G. Djurfeldt. 2018. ""The family farms together, the decisions, however are made by the man" —Matrilineal land tenure systems, welfare and decision making in rural Malawi." *Land Use Policy* 70: 601–10.

FAO. 2012. "Voluntary guidelines on the responsible governance of tenure of land, fisheries and forests in the context of national food security." Rome, Italy: Food and Agricultural Organization of the United Nations (FAO). http://www.fao.org/3/a-i2801e.pdf.

Henssen, J. 1995. "Basic principles of the main cadastral systems in the world." In: *Proceedings of the One Day Seminar Held during the Annual Meeting of FIG Commission, Cadastre and Rural Land Development.* Vol. 7, 5–12. Delft: The Netherlands.

Holden, S. T., and H. Ghebru. 2016. "Land tenure reforms, tenure security and food security in poor agrarian economies: Causal linkages and research gaps." *Global Food Security* 10: 21–8.

Holden, S. T., and K. Otsuka. 2014. "The roles of land tenure reforms and land markets in the context of population growth and land use intensification in Africa." *Food Policy* 48: 88–97.

Hu, L.-t., and P. M. Bentler. 1998. "Fit indices in covariance structure modeling: Sensitivity to underparameterized model misspecification." *Psychological Methods* 3(4): 424–53.

Hu, L.-t., and P. M. Bentler. 1999. "Cutoff criteria for fit indexes in covariance structure analysis: Conventional criteria versus new alternatives." *Structural Equation Modeling: A Multidisciplinary Journal* 6(1): 1–55.

Jaijit, S., N. Paoprasert, and J. Pichitlamken. 2018. "Economic and social impact assessment of rice research funding in Thailand using the structural equation modeling technique." *Kasetsart Journal of Social Sciences*: 1–7.

Kassie, M., J. Pender, M. Yesuf, G. Kohlin, R. Bluffstone, and E. Mulugeta. 2008. "Estimating returns to soil conservation adoption in the northern Ethiopian highlands." *Agricultural Economics* 38(2): 213–32.

Kishindo, P. 2010. "The marital immigrant. Land, and Agriculture: A Malawian case study." *African Sociological Review / Revue Africaine de Sociologie/Revue Africaine de Sociologie* 14(2): 89–97.

Knowler, D., and B. Bradshaw. 2007. "Farmers' adoption of conservation agriculture: A review and synthesis of recent research." *Food Policy* 32(1): 25–48.

Kodoth, P. 2001. "Gender, family and property rights: Questions from Kerala's land reforms." *Bulletin (Centre for Women's Development Studies)* 8(2): 291–306.

Lemmen, C., P. van Oosterom, and R. Bennett. 2015. "The land administration domain model." *Land Use Policy* 49: 535–45.

Li, W., X. Wei, R. Zhu, and K. Guo. 2019. "Study on factors affecting the agricultural mechanization level in China based on structural equation modeling." *Sustainability* 11(51): 1–16.

Liu, H., and X. Luo. 2018. "Understanding farmers' perceptions and behaviors towards farmland quality change in Northeast China: A structural equation modeling approach." *Sustainability* 10(3345): 1–17.

Lovo, S. 2016. “Tenure insecurity and investment in soil conservation: Evidence from Malawi.” *World Development* 78: 219–29.

Malley, Z. J., M. Taeb, and T. Matsumoto. 2009. “Agricultural productivity and environmental insecurity in the Usangu plain, Tanzania: Policy implications for sustainability of agriculture.” *Environ Dev Sustain* 11(1): 175–95.

Meinzen-Dick, R., L. R. Brown, H. S. Feldstein, and A. R. Quisumbing. 1997. “Gender, property rights, and natural resources.” *World Development* 25(8): 1303–15.

Meinzen-Dick, R., A. Quisumbing, C. Doss, and S. Theis. 2019. “Women’s land rights as a pathway to poverty reduction: Framework and review of available evidence.” *Agricultural Systems* 172(1): 72–82.

Munthali, K., and Y. Murayama. 2013. “Interdependences between smallholder farming and environmental management in Rural Malawi: A case of agriculture-induced environmental degradation in Malingunde extension planning area (EPA).” *Land* 2(2): 158–75.

Najafabadi, M. O. 2014. “A gender sensitive analysis Towards organic agriculture: A structural equation modeling approach.” *Journal of Agricultural & Environmental Ethics* 27(2): 225–40.

Peters, P. E. 2010. ““Our daughters inherit our land, but our sons use their wives’ fields”: Matrilineal-matrilocal land tenure and the New Land Policy in Malawi.” *Journal of Eastern African Studies* 4(1): 179–99.

Snapp, S. S., D. D. Rohrbach, F. Simtowe, and H. A. Freeman. 2002. “Sustainable soil management options for Malawi: Can smallholder farmers grow more legumes?.” *Agriculture, Ecosystems & Environment* 91(1–3): 159–74.

Steiger, J. H. 1990. “Structural model evaluation and modification: An interval estimation approach.” *Multivariate Behavioral Research* 25(2): 173–80.

Sylwester, K. 2004. “Simple model of resource degradation and agricultural productivity in a subsistence economy.” *Review of Development Economics* 8(1): 128–40.

Teklewold, H., M. Kassie, and B. Shiferaw. 2013. “Adoption of multiple sustainable agricultural practices in Rural Ethiopia.” *Journal of Agricultural Economics* 64(3): 597–623.

Toma, L., A. P. Barnes, L.-A. Sutherland, S. Thomson, F. Burnett, and K. Mathews. 2018. “Impact of information transfer on farmers’ uptake of innovative crop technologies: A structural equation model applied to survey data.” *The Journal of Technology Transfer* 43(4): 864–81.

Waldman, K. B., D. L. Ortega, R. B. Richardson, D. C. Clay, and S. Snapp. 2016. “Preferences for legume attributes in maize-legume cropping systems in Malawi.” *Food Security* 8(6): 1087–99.

Wang, Z., B. Li, and J. Yang. 2015. “Impacts of land use change on the regional climate: A structural equation modeling study in Southern China.” *Advances in Meteorology* 2: 1–10.

Williamson, I., S. Enemark, J. Wallace, and A. Rajabifard. 2010. *Land Administration for Sustainable Development.* Redlands, CA: ESRI Press Academic.

World Bank. 2006. *Sustainable Land Management: Challenges, Opportunities, and Trade-Offs.* Washington, DC: World Bank.

Zait, A., and P. E. Bertea. 2011. “Methods for testing discriminant validity.” *Management & Marketing Journal* IX(2): 217–24.

Zevenbergen, J. A., W. T. de Vries, and R. M. Bennett, eds. 2015. *Advances in Responsible Land Administration.* Boca Raton, FL: Taylor & Francis, CRC Press.

Section III

Context and External Drivers of Land Management

7 Reflections toward a Responsible Fast-Track of Namibia's Redistributive Land Reform

Uchendu E. Chigbu and Peterina Sakaria

CONTENTS

It is true that they came and stole the land 100 years ago, but a white boy who was born on that land has Namibian blood.

President Hage Geingob of Namibia

7.1 INTRODUCTION

Many studies on land reform in Namibia have focused on analyzing the country's reform programs, but not many of them have particularly investigated ways of fast-tracking its procedures

(Amoo and Harring, 2012; Lohmann et al., 2014). Some of the most recent attempts at investigating the redistributive land reform (RLR) of Namibia were done by Sakaria (2016), Chigbu et al. (2017), and Falk et al. (2017). Sakaria's (2016) and Chigbu et al.'s (2017) studies explored the RLR process and concluded that it was slow. Falk et al. (2017) assessed the degree to which land rights affected the farm income of its beneficiaries. The study presented in this chapter extends Sakaria (2016) and Chigbu et al.'s (2017) approach to fast-tracking Namibia's land reform procedures. It puts the focus on how to fast-track the reform responsibly. Furthermore, it assesses whether the beneficiaries of Namibia's RLR perceive the land reform to be responsive to their needs. It also probes ways of fast-tracking the implementation of the RLR without undermining its objectives (which is to give access to land to millions of landless Namibians). The study answers two key questions: (1) Do the beneficiaries of the RLR consider the reform to be a responsible land intervention, and if so, in what ways? (2) What problems are militating against the fast implementation of the RLR, and how can these problems be improved to fast-track it? The study approaches these questions by presenting background information about Namibia's land reform. It then presents and defines "responsible" in the context of RLR. These are followed by a description of the methodology used for the study, followed by findings and emerging issues from the findings, before a conclusion.

7.2 NAMIBIA'S LAND REFORM: NATIONAL RESETTLEMENT POLICY (NRP) AND AFFIRMATIVE ACTION LOAN SCHEME (AALS) IN FOCUS

The land reform in Namibia has three components—the RLR (implemented through the NRP), tenure reform, and affirmative action. The NRP is meant to address the unequal distribution of land through equitable distribution and access to land. This is meant to give opportunities to previously disadvantaged Namibians to become self-supporting through food production and sale of such products. It is also supposed to reduce pressure from communal areas, among other objectives. The tenure reform is mainly aimed at communal areas. It entails registration of customary land rights and administration of leaseholds in such land with the aim of providing security of tenure through issuing certificates of registration, as well as aiming to develop underutilized land in communal areas into small-scale commercial farms. The Affirmative Action Loan Scheme (AALS) assists previously disadvantaged people through provision of subsidized interest rates and loan guarantees in order for them to acquire their own commercial farms with the aim of reducing the pressure on communal areas such as the grazing pastures. Under this initiative, the farmers apply for entry into the scheme following the same procedures they would have followed when applying at any other financial institution, such as commercial banks. Upon approval of such a loan and purchase of the commercial farm, Agribank registers a mortgage bond over such land for a period of 25 years to serve as security (Government of Namibia, 2015).

This chapter focuses on the RLR (being implemented under the NRP) and AALS. Hence, the chapter refers to the RLR (which comprises the NRP and AALS) as land reform (or redistributive reform or reform). In regard to the AALS, three compulsory criteria must be met by citizens who want to participate in the scheme. According to the Government of Namibia (2015), these include: being a previously disadvantaged Namibian, being in possession of 150 Large Stock Units or 800 Small Stock Units (or cash equivalent) and paying 10% of the purchase price of the farm at the time of application for such a loan. The scheme caters for full-time and part-time farmers and terms and conditions are applicable for both parties. For both parties, 10% of the purchase prices of the farm is required upon application for such a loan. But for the full-time farmers, they are exempted from repayment of the loan for the first three years, whilst for the part-time farmers they must repay the interest portion in the first three years and the capital amount after that period. The other condition applicable for full-time farmers is that repayment commences from the fourth year starting at a 2% interest rate, but it is accumulative up to 14% after the tenth year (Sakaria, 2016; Chigbu et al., 2017).

The NRP targets the poor and landless Namibians by redistributing land on state-acquired commercial farms. This acquisition is based on the preferential right of the Namibian state to purchase agricultural land whenever any owner of such land intends to dispose of it (Government of Namibia, 1995). Under the NRP, "any Namibian citizen who has been socially, economically, or educationally disadvantaged by past discriminatory laws can apply for an allotment of land acquired for resettlement" (Falk et al., 2017, p.316). Successful applicants receive a 99-year leasehold from the government. The implementation of the NRP entails two major parts: the land acquisition (involving farm offer and purchase) and land allocation (resettlement) components. The farm offer occurs when land is made available to the state for purchase (if the land is found suitable for resettlement purposes). The state purchases such land and allocates it (resettlement) to landless Namibians.

7.3 MAKING THE LAND REFORM "RESPONSIBLE" BY TACKLING THE DISPUTES OF STAKEHOLDERS

"Limited success has been achieved in the acquisition of land for the landless, but demands are increasing for a more drastic and radical approach to reform current land holding" (de Villiers, 2003, p.29). The pressure on the government is twofold. First, the privileged white landowners and international donors are putting pressure on the Namibian government to abide by a willing buyer-willing seller (WS-WB) approach, and not to adopt a Zimbabwean-style dispossession of land (Sakaria, 2016). Second, the less-privileged Namibians are calling for an approach that is comparable to that of Zimbabwe (Sakaria, 2016). Despite these two divergent reform ideologies being pushed by various stakeholders in the country, there is a common view in Namibia that fully implementing the reform is mandatory to the country's development. "The question is rather how to progress it in a manner that would balance the seemingly competing interests of the landless and landowners" (de Villiers, 2003, p.29). Concerning what connotes the best approach to RLRs, Dorner (1999) noted the emerging consensus that land reform is more difficult to implement in a neoliberal context. This led Borras (2001, p.545) to conclude that the solution to the land reform challenge is "abandoning the classic RLR project" approach.

This study argues that, institutions aside, the outcome of land reforms is determined by structures, processes, and outcomes (de Vries and Chigbu, 2017, p.68). Structures relate to administration, organizational arrangements, and institutions. Processes entail sequences of activities and inter-organizational dependencies. Outcomes are about societal changes, impacts, and products resulting from the structures and processes. The study considers being "responsible" to mean causing land reform activities (actions) to be responsive to the needs of people and taking full responsibility for emerging outcomes. It aligns with de Vries and Chigbu's (2017) idea of "responsible," which argues that land interventions should be responsive to the needs of all people in aspects of *structures*, *processes*, and *outcomes* of land management interventions.

7.4 METHODOLOGY: APPROACH AND NATURE OF THE STUDY

The approach to this study is twofold. The first part involves a perceptional survey carried out to ascertain the perception of beneficiaries on whether the RLR has been a responsible reform so far. To gain a perception of how responsible the reform has been to beneficiaries, the survey was done using a responsible land reform assessment which was adapted from de Vries and Chigbu's (2017) simple responsible land intervention assessment sheet. The simple assessment sheet consists of a checklist of three elements of responsible (*structure*, *process*, and *outcome* or *impacts*) and eight indicators of *responsible*. The eight indicators include *responsive*, *resilient*, *robust*, *reliable*, *respected*, *reflexive*, *retraceable*, and *recognizable* (otherwise referred to as the 8Rs). These indicators form the characteristics that should be present for any intervention to be considered as responsible (de Vries and Chigbu, 2017). In this study, the matrix was administered to 150 beneficiaries of the RLR. The selection of participants was non-random; rather the study used a predetermined list of beneficiaries provided by the Ministry of Land Reform (MLR). This was done in May 2018. The survey data also enabled the researchers to identify specific challenges causing the slow pace of the RLR in order to ensure appropriate recommendations for improvement. This part of the data collection was in May 2018. This was done using the responsible land reform assessment sheet. During these periods, 12 farms were visited to ascertain their pre- and post-RLR conditions. The farms were located in the *Oshikoto, Hardap, Karas, Erongo, Otjozondjupa, Omaheke, Khomas*, and Kunene regions of Namibia.

The second part involved investigating the reform processes and procedures using key-informant interviews. The experiential empirical data were accessed by administering structured interviews to 20 (n = 20) key-informants to further understand the *structures*, *processes*, and *outcomes* of the RLR. These interviews were necessary in order to grasp the specific impediments militating against the fast-track of the RLR. The selection of key-informants was predetermined based on the ease of accessibility of information. All key-informants selected fall within four categories. They included policy-makers/implementers, farm owners, previously disadvantaged Namibians who have become beneficiaries of the resettlement farms, and those who have not yet benefited from the NRP (non-beneficiaries). All the face-to-face interviews were conducted using questions that probed the structure, processes, and outcomes of the RLR. Based on the responses received, it was possible to identify specific weakness of the reform and provide recommendations for improvement. This part was conducted in 2016 and 2018. The focus of data collection was on answering the two research questions mentioned earlier in this chapter. The following sections present the outcomes of the study.

7.5 IS THE LAND REFORM "RESPONSIBLE"? PERCEPTIONS, STRUCTURE, PROCESS, AND OUTCOMES

7.5.1 Perception: Beneficiaries Rate the "Process" of the Reform as Weak in "Responsible"

In a responsible intervention context, outcome entails appropriateness in relation to targets, and impact in relation to the needs of the people (de Vries and Chigbu, 2017). This is crucial because in all land interventions (such as the RLR), "actors seek to see visible results and changes" (de Vries and Chigbu, 2017, p.69). Process connotes the logical sequence of activities (and their appropriateness) in relation to envisaged outcomes (de Vries and Chigbu, 2017). This implies procedures for carrying out the reform. Structures denotes functionality of administrative arrangements put in place to implement the land redistributive reform (de Vries and Chigbu, 2017). The perception of beneficiaries (comprising willing sellers whose lands were acquired by the government, and landless Namibians who were resettled) on the three aspects—outcome, structure, and process—were measured within the context of the eight indicators of responsible. These eight indicators (or 8Rs)—*responsive, resilient, robust, reliable, respected, reflexive, retraceable*, and *recognizable*—form the characteristics of a responsible RLR. Concerning their meanings or definitions, any land intervention (like the RLR) is:

- *Responsive*: When it embraces the needs of the people and addresses urgency of their need (in this case, land redistribution for social equality among Namibian women, youth, and men).
- *Resilient*: When it ensures the sustenance of societal structures such that it is capable of avoiding disruptions that segregate society along the lines of gender and age (e.g. either politically or culturally).
- *Robust*: When it is based on solid mechanisms that will not make it susceptible to disruptions (e.g. political, economic, environmental, or cultural).

- *Reliable*: When it is based on decisions that are trusted or are based on trust or creating trust (in this case, disadvantaged or landless Namibians living in harmony with the White population).
- *Respected*: When decisions and actions are valued positively such that decision-makers are seen as appropriate leaders or managers, the projects they undertake are esteemed, and project decisions respect the rights of all people (youth, women, and men) affected by the project.
- *Retraceable*: When all activities are documented, so history can be reconstructed and it is possible to see which activities have been taken by whom, and what still needs to occur.
- *Recognizable*: When all people (men, women, and youth alike) can identify with the decisions (e.g. collective ownership of the project or intervention)
- *Reflexive*: When its rightfulness or appropriateness are re-assessable for improvements.

Based on these indicators, all selected beneficiaries were asked to relate each of the 8Rs to the three aspects of *structure*, *process*, and *outcome* of the land reform. Out of 150 beneficiaries considered for this analysis, 95% of them think the *structure* of the reform is *responsive*, 38% said the *process* is *responsive*, and 25% noted the *outcome* to be *responsive* (refer to Figure 7.1 for the overall results).

The most striking aspect of these results are as follows: (1) the *structure* of the reform is considered *robust* (97%), but the outcome is not (46%). Both the *structure* (99%) and *outcomes* (96%) of the research are viewed to be overwhelmingly responsive, but the process is not exactly viewed the same way (69%). The beneficiaries also have a high regard for *retraceable*, *recognizable*, and *reflexive*. The lowest perception level is visible in all aspects of the *process*. The beneficiaries' views on *structure* and *outcome* are high, while their views on the *process* is low. The *processes* leading

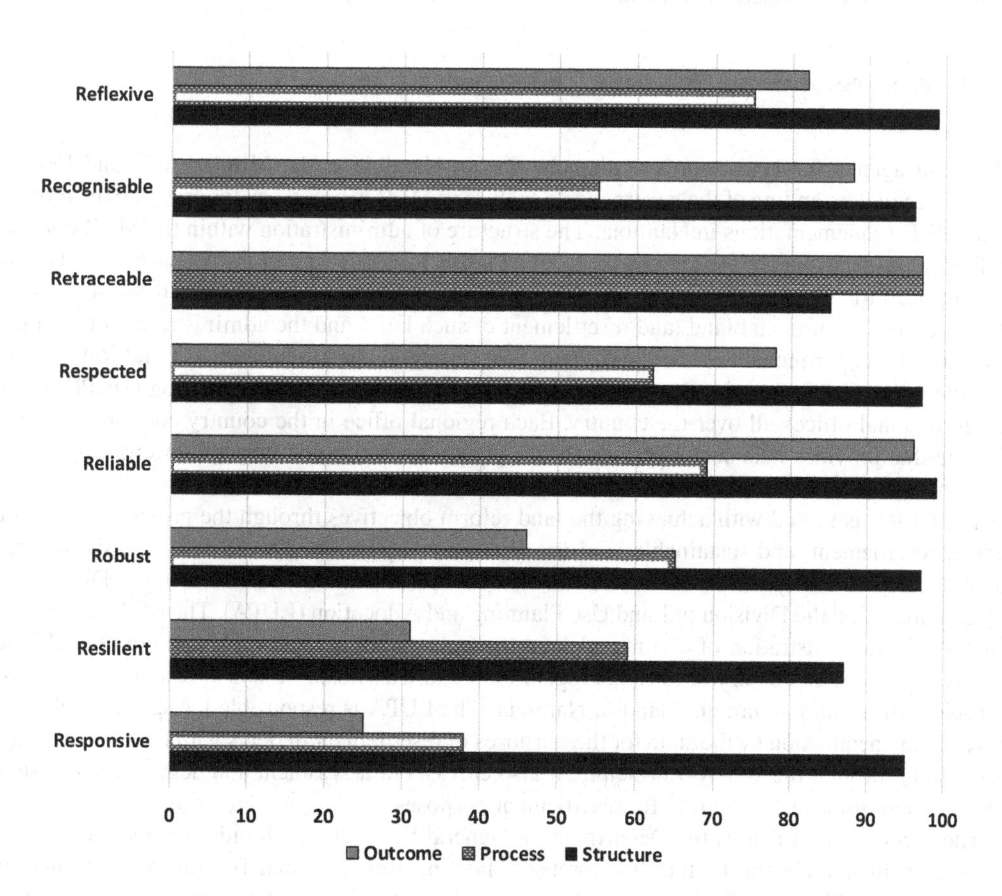

FIGURE 7.1 Respondents' perception (view) of how "responsible" the RLR is in terms of the 8Rs.

to the *outcomes* do not receive very positive responses on the *8Rs* of "responsible." Thus, whiles *structures* and *outcomes* are good, there are impediments that delay the reform *processes*, leading to low achievement of targets when viewed from the perspective of its pace. Remedial interventions are therefore required to accelerate the *process* to achieve the desired *outcome* following the dictates of *responsible* indicators.

7.5.2 Though the Outcome Is Viewed as "Responsible," There Is Dissatisfaction about the Slow Pace of the Reform

The redistributive reform targets the acquisition of 15 million ha of land by 2020. Out of this total, 10 million ha are to be acquired under AALS (through the Agricultural Bank of Namibia) and 5 million ha are to be acquired under the NRP (through the MLR) (Sakaria, 2016; Chigbu et al., 2017). Under the AALS, the Agricultural Bank of Namibia provides funding to previously disadvantaged farmers at subsidized interest rates. A total of 648 loans have been granted of about US$54 million. Of the 10 million ha to be acquired by the scheme by the year 2020, about 64% have been acquired through the scheme since 2005. This makes it mathematically impossible (considering the current pace of the program) for the remaining 36% to be achieved by 2020. Out of NRP's 5 million ha target, only 524 farms (3.2 million ha of land with 5,338 beneficiaries) has been acquired from 2005 to date. It is evident that the process has been going at a very slow pace. It is unrealistic to assume that the remaining 1.8 million ha will be achieved soon. These data justify the "widespread dissatisfaction about the pace of land redistribution" in Namibia (Werner 2015, p.321). This dissatisfaction is seen as one of the weaknesses of the land reform in Namibia (Sakaria, 2016).

7.5.3 A Snapshot of the "Structure" Responsible for the Implementation of the Land Reform

The main agency for land reform implementation in Namibia is the Ministry of Land Reform (MLR). An understanding of the governance structure of MLR is necessary for grasping the process of the RLR implementations in Namibia. The structure of administration within the MLR consists of four main departments for land governance (Figure 7.2). The first is the Department of Land Reform, Resettlement and Regional Programme Implementation which deals with the acquisition of agricultural commercial land (and resettlement of such land) and the administration of communal land. The department consists of two main Directorates—the Directorate of Land Reform and Resettlement (DLRR) and the Directorate of Regional Programme Implementation (DRPI) which has 14 regional offices all over the country. Each regional office in the country consists of such a division that provides land sector services to the people in that region through decentralization of the ministry's land administration activities.

The DLRR is tasked with achieving the land reform objectives through the provision of secure tenure, resettlement, and sustainable land use and land acquisition. The DLRR consists of three main divisions, namely the Division of Land Boards Tenure and Advice (LBTA), the Division of Resettlement, and the Division of Land Use Planning and Allocation (LUPA). The LBTA is responsible for the administration of communal land through traditional authority and communal land boards. It ensures customary land rights registration and leaseholds. It is also responsible for sustainable utilization of communal land in Namibia. The LUPA is responsible for agricultural commercial land acquisition by the state for the purposes of resettlement. It deals with farm offers made to the state through the WS-WB principle. It also carries out assessment and demarcation of such land and determines its suitability for resettlement purposes.

The second department is the Department of General Services which only provides services to other departments. The third is the Directorate of Planning, Research, and Training and Information Services (DPRTIS). The DPRTIS has the task of planning, training, and research; and information

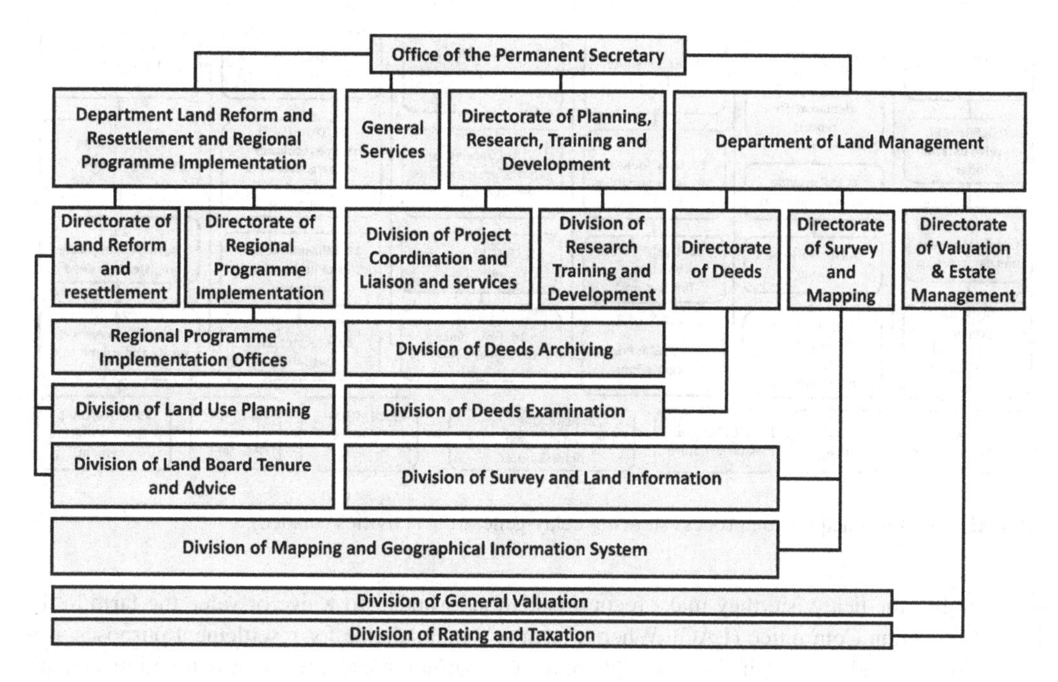

FIGURE 7.2 Illustration of the organizational structure of MLR.

services related to the work of MLR. The fourth is the Department of Land Management (DLM) which is tasked with administration of urban land through ensuring registration, valuation, and taxation purposes, recording and keeping the country-wide cadastre of such properties. It is the department responsible for a variety of land management aspects, including property registration, cadastre, and taxation just to mention a few. DLM consists of three directorates (each with two divisions). Its Directorate of Deeds (DDR) is responsible for registration of fixed properties and real rights in the country. It carries out its activities through two main divisions, namely, division of deeds examination and division of deeds archiving. One of the directorates within the DLM is the Directorate of Valuation and Estate Management (DVEM) which is responsible for the provision of valuation and land tax administration services. It carries out its mandate through two divisions—the Division of General Valuation and the Division of Rating and Taxation. With an understanding of the governance structure of MLR, it was possible to conduct critical analyses of the processes of land acquisition and allocation to identify problem areas.

7.5.4 Identifying the Delay-Generating Steps in the Land Acquisition Process in the Reform

The WS-WB is the current mode of land acquisition. It is dependent on the land made available to the market by the sellers, but the state has a preferential right to purchase. The Agricultural (Commercial) Land Reform Act 1995 (No. 6 of 1995) makes provision for the state's preferential right. This means that before any agricultural land enters the open market, the state, through the MLR, needs to express its interest. In simple terms, only when the state has waived its interest by issuing a certificate of waiver can the seller offer the farm to private buyers. The process of land acquisition starts when a willing seller makes an offer to sell agricultural land to the state (Figure 7.3).

After the offer is made, the DLRR/DRPL receive and scrutinize the offer, then forward it to the land-use planners to carry out assessment and demarcation. The planners carry out field visits to determine the farm's suitability and demarcate the farm into farming units ranging from preferably 2,500 ha and more, or less depending on the carrying capacity and agro-ecological zoning of that

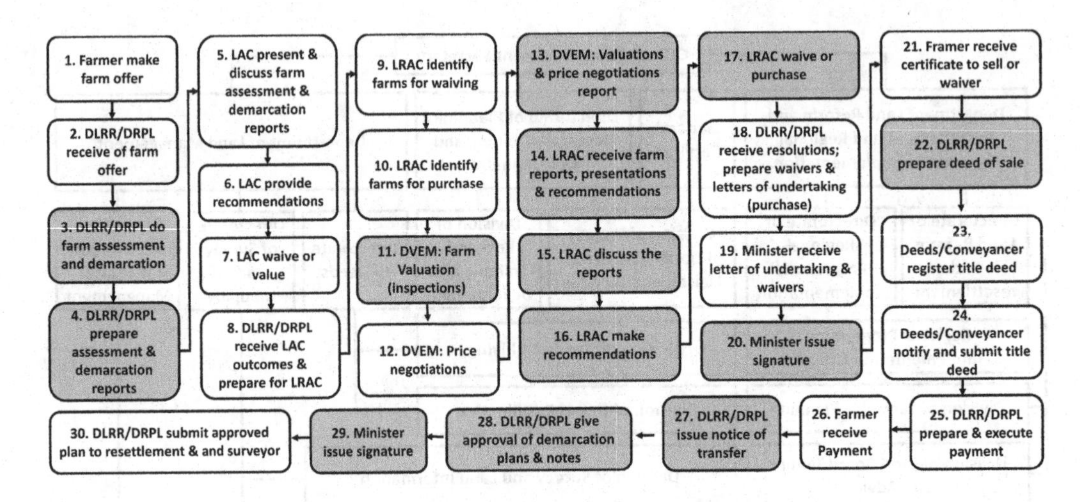

FIGURE 7.3 Land acquisition process showing delay-generating activities (shaded).

region. After the field visit, they make recommendations whether to waive or value the farm to the Land Acquisition Committee (LAC). When a farm is found suitable for resettlement purposes, it is then recommended to be valued for possible purchase, but in the event where it is found unsuitable, the LAC recommends to the Land Reform Advisory Commission (LRAC) to advise the Minister to waive such a farm. When a farm is found suitable, it is forwarded to the Directorate of Valuation, whereby a second team carries out field work to inspect the farm to determine the value. On the basis of this valuation a counter-offer is recommended to the owner and price negotiations take place. In the event where the farm owner is not in agreement with the valuation price offered by the state, they are entitled to appeal to the Lands Tribunal, as there is no room for withdrawal of such offer after a counter-offer is issued. The valuation reports, assessment reports by the land use planners, and price negotiations outcomes are then submitted and presented to the LRAC for discussions and consideration. The LRAC has the right to oppose the recommendations of both the planners and valuers. The LRAC then advises the Minister to purchase such land if they are in agreement or either to waive it. The legal compliance is that the Act stipulates different time periods for the different stages within the land acquisition process. The farm offer must be accepted within 90 days. The first stage of the process takes up to 60 days, the State must decide to either buy or not, and carry out activities such as assessment, presentation to LAC, valuation, and recommendation to LRAC. The LRAC is required within 30 days to make recommendations to the Minister. The Minister then has 14 days to accept or reject the offer. In the event where the Minister rejects the offer, a certificate of waiver (valid for 12 months) is issued as there is no present interest from the State. It is a requirement that the LRAC/Minister makes a recommendation within the stipulated timeframe. In the event where any of the parties fails to do so, the offer is deemed declined by the farm owner, and the Minister, upon the request of the farm owner, must issue a certificate of waiver. In the event where the state is interested in acquiring the farm and a counter-offer is issued, the farm owner must inform the Minister in writing whether the offer is accepted or not within 14 days from the date of notice. The Minister convenes a price negotiation committee to negotiate the purchase price with the farm owner and advise the Minister thereof. If no agreement is reached during negotiations, the owner, not later than 60 days afterward, may make an application to the Lands Tribunal for the determination of the purchase price.

7.5.5 Exposing the Delay-Generating Steps in Land Allocation (Resettlement) Process in the Reform

The other aspect of the reform implementation is the process of land allocation, which is the physical allocation of such land to the beneficiaries (previously disadvantaged Namibians). The process

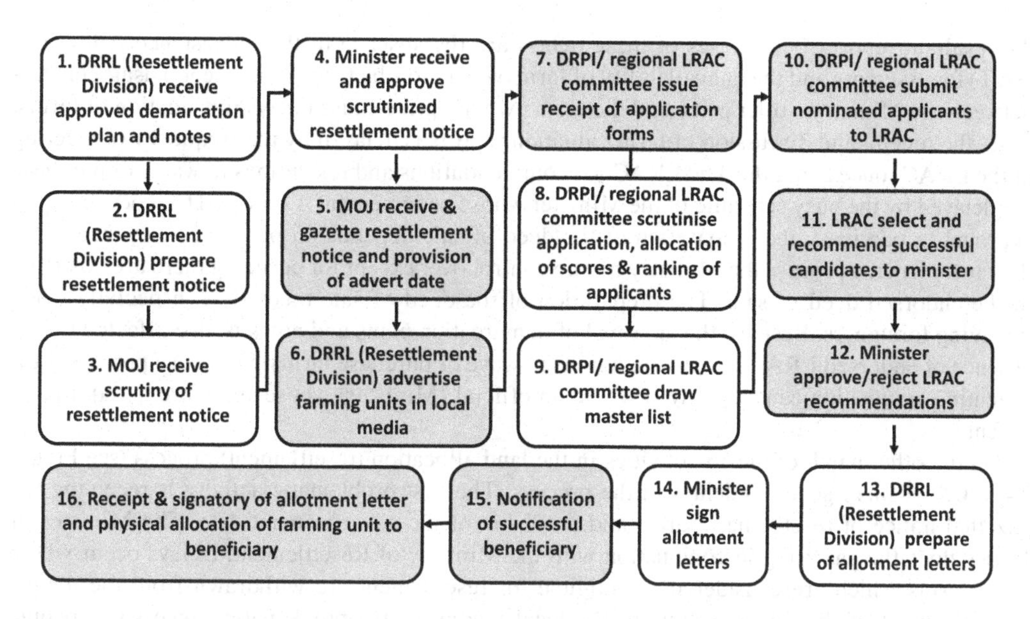

FIGURE 7.4 Land allocation (resettlement) process showing delay-generating activities (shaded).

starts off when the Division of Resettlement receives the approved allotment (demarcation) plan and notes from the LUPA division (refer to Figure 7.4). After the first activity, resettlement (legal) notices for the approved allotment plans are prepared. It is compulsory that the notices must be forwarded to Ministry of Justice (MOJ) for scrutiny by the legal drafters to ensure correctness. The scrutinized notices are then sent back to the MLR for the Minister's signature and approval. After approval, the allotments are gazette, and thereafter advertisement follows. The farming units available for resettlement are advertised in the local media: in the newspapers, advertising boards at the head office, as well as regional offices, radio stations, and so forth. The advert runs for a period of one month. The prospective beneficiaries then apply for the available farming units, using a prescribed resettlement form available from the MLR.

After the advertisement, the Resettlement (LRAC) Regional Committee (RRC) convenes not later than two weeks from the date of closure of the advert to screen applications and recommend applicants to the LRAC. The LRAC sits once a month and selects, makes final recommendations, and submits the names of the successful applicants to the Minister. The Minister has the mandate to either reject or approve the LRAC recommendations. In a case where the Minister approves the LRAC recommendations, the LRAC minutes are signed by the chairperson or Permanent Secretary and preparation of the allotment letters commence. The allotment letters are prepared and forwarded to the Minister for signature. The signed allotment letters are then delivered to relevant regional offices for notification and issued to the approved/successful applicants. The lease agreement is the legally binding contract (between the state and the beneficiary) because it stipulates the conditions of the lease to be agreed by both parties (Permanent Technical Team on Land Reform of Namibia, 2005). The agreement is prepared and signed by lessee (beneficiary) and the Minister on behalf of the state (lessor). After signing of the agreement, the beneficiaries are inducted and physically resettled on their farming units by resettlement officers of the MLR.

7.6 THE IMPEDIMENTS THAT CAUSE DELAYS IN THE PACE OF THE LAND REFORM PROCESSES

7.6.1 Land Administrative and Land Management Impediments

Out of the 30 activities involved in the land acquisition (see Figure 7.3), 43% of them cause delays in the process. There are delays in the assessment and demarcation of farms and preparation of the resultant reports. The causes of these delays are the absence of the Subsistence Allowance (S&T) for assessors and the unavailability of farm owners for the farm assessment. Disagreements between the Ministry's offer price and the farm owner's price cause delays in valuation of farms. Also, the receipt and discussion of farm valuation reports is delayed by the frequency of meeting of the LRAC (once a month). The LRAC's recommendations and resolutions to waive or purchase are delayed by the busy schedule of the Minister whose endorsement is required to proceed. Upon approval to purchase, the preparation of the deed of sale depends on private conveyancers and the farm owner, a process the Ministry has no control over except for ensuring there are no errors in the endorsed deed of sale. The preparation of the deed of sale takes time, especially when involving foreigners. Besides, the approval of demarcation plans and notes is also affected by the absence of endorsed LRAC recommendations to wave or purchase farms. Signatory processes for executing acquisition can be delayed where an official (Minister) whose signature is required is absent.

On the other hand, of the 16 activities in the land allocation (resettlement) process (see Figure 7.4), 50% of them generate delays in the process. The first problematic activity is receiving the gazetted notice of resettlement farms and provision of the advertisement date. The Ministry of Justice does the gazetting in conjunction with the Ministry of Resettlement. Delays occur where some farms which are considered not eligible for resettlement are withdrawn from the notice. Advertisement of the farming units in the local media is also problematic, since some people (especially in rural communities) might miss out on the 30 days advert. During the issuance and receipt of application forms, large volumes of applications (approximately 7,000 per farming unit) received may compromise scrutiny and scoring of applications, which in turn may delay the drawing up of the master list. Again, the frequency of meeting of the LRAC delays the RRC's submission of nominated applicants to the LRAC, and the LRAC's selection and recommendation of the successful candidates to the Minister. Where the Minister's power to reject is exercised, it delays the approval of the LRAC recommendations and might delay the process of allocation. This aside, the notification of successful beneficiaries takes time as the regional secretariat must inform beneficiaries individually. Finally, the receipt and signing of allotment letter by the Minister and Beneficiaries is also affected by the absence of the official signatory at any point in time, as seen in the land acquisition process. Also, the physical allocation of farming units to the beneficiaries is not time-bound.

7.6.2 Legislative Impediments

The law regulates the timeframe allowed for the different stakeholders to make decisions with the NRP implementation process. Within the process of land acquisition, the law makes provision that an offer should be accepted within a period of 90 days. A farm offer is made to the Minister through the office of the Permanent Secretary starting from activity one (refer to Figure 7.4) which has 60 days to look at the offer until activity four when the permanent secretary makes recommendations to the LRAC. Then 30 days is given in which the LRAC must give recommendations to the Minister. The Minister has 14 days to either accept or reject the offer. The farm owner is given 60 days to respond to the counter-offer; in cases where the farm owner declines the counter-offer, they have liberty to appeal at the Lands Tribunal. However, the Lands Tribunal is not regulated in terms of timeframe; it is almost indefinite. This process of land acquisition as stipulated in the legislation could take up to 194 days (nearly six and half months) to finalize one farm offer. The reason is because there is no specific timeframe or limit allotted to the LRAC, the Minister, the DRPI, and the RRCs in the Land Allocation (Resettlement) process. According to the Permanent Technical Team on Land Reform of Namibia (2005), the average time between offer and transfer, which is the land acquisition process, takes approximately 301 days (nearly ten months), but could vary as from little as 130 days to as much as 789 days. The Act does not in any way compel the different stakeholders to execute their duties within the stipulated timeframe, apart from regulating the timeframe.

When it comes to the process of land allocation (resettlement), the Act only regulates the time for offers, the time for the decision by the Minister, and the time for appeal by the farm owner, but after that, there is no timeframe stipulated for the LRAC to resettle beneficiaries.

7.6.3 Market and Leadership Impediments

The farms which are waived by the government are sold on the open market especially at very high valuation prices (usually too high for the government). There is no restriction on the open market prices, hence land might be available, but not affordable for acquisition by the state, leading to delays in acquisition. Additionally, proactive leadership is essential for the success of any program. A lack of leadership to champion the land reform program contributes to its slow pace. Some stakeholders, especially the ministerial staff, lack the culture of urgency and concern. This leads to cases where duties are not executed on time.

7.6.4 Human Resource and Awareness Impediments

There is a lack of human capacity, especially assessors and valuers, to enable faster land suitability assessment and price determination. There are cases where farm offers expire, and the government is forced to waive them because of lack of timely execution due to lack of human capacity. Also, there is need to create awareness so that the land owners understand what the main objectives and importance of the program are, and this will entice them to freely participate through the WS-WB instrument with no fear and speculation. Currently, potential beneficiaries remain poorly informed about the program which leads to delays in processing their applications.

7.6.5 Impediments Due to Excessive Waiving of Forthcoming Offers

According to the annual ministerial reports, one of the contributing factors to the slow pace of the program is that farm offers are not forthcoming. The ministry has indicated that the WS-WB mode of land acquisition is based on the offers made by the willing sellers and if there are no offers, there will be nothing there to buy willingly. From 2011 to 2016, a total of 5.6 million ha of land has been made available to the state. This is a little over the 5 million ha target set under the NRP. However, only 10% (582,390 ha) has been acquired for resettlement. 65% (3,675,391 ha) was either waived because the land was unsuitable or it was waived in favor of affirmative action candidates. The remaining were either withdrawn, sent back, or pending purchase. The number of hectares waived yearly is much higher than the number of hectares purchased. So, offers are forthcoming, but are mostly waived. This leads to delays.

7.7 Recommendations Towards Fast-Tracking the Land Reform in a "Responsible" Manner

With the current pace of the process, it is evident that the target set by the government cannot be reached, and this will result in the failure of the reform in terms of *outcome*. Therefore, it is important to mitigate the causes hampering effectiveness and efficiency of the program, particularly as they relate to the *process* (which has been slowing the pace of the reform). To specifically fast-track the reform in line with the concept and praxis of *responsible*, this chapter suggests the following five steps.

Firstly, there is need to pair up activities which are carried out by the ministerial staff and invest in more capacity-building. The activities can be done simultaneously, provided that proper desktop study analysis and checks are done to determine the probability of the farm being suitable beforehand. For example, activities by assessor and valuers can be carried out on one field trip; this way it reduces time as well as the cost incurred by the ministry in sending out the officials on two different field trips. There is also a need to capacitate the government officials involved in the NRP implementation. The lack of the necessary equipment, instruments, knowledge, and know-how leads to slow pace of the implementation. The assessors and valuers need to be supplied with the entire necessary tools to carry out proper assessment and valuations of the farms.

Secondly, it is necessary to embrace options to purchase resettlement farming units and demand-driven approaches. The 99 years' leasehold provides the necessary security of tenure which is the state objective, but it is restrictive in a sense that it does not grant beneficiaries participation in the open market, and therefore beneficiaries are not keen to invest in resettlement farms. The option to purchase the farming units would boost beneficiaries to invest more and be more productive, and the preferential right should then apply in this case when the beneficiary wishes to alienate their farming unit so that the land reverts to the state. The government, based on more than 24 years' experience, should undertake a different approach to the land reform program. A demand-driven approach would enable the government to know the market targets, determine the demand, preferences, and need of the nation, then only acquire land after ascertaining the above-mentioned. That way the government would acquire land in regions for different agricultural activities based on the need and demands of the market as determined by the farm offerors and offerees.

Thirdly, there is need to improve on information services and public awareness, and introduce regular LRAC meetings. Proper communication mechanisms should be put in place to inform the public at large in terms of the objectives and aims of the land reform program. Platforms for discussions should be open to get opinions and views of the public. The ministry should also undertake open-door policies, share newsletters, and provide information to the public, such as educating them on the provisions of the law, the role of the stakeholders, expectations of government from the different stakeholders, and so forth. This way the process will be more transparent and will lead to the stability of the reform—a situation whereby stakeholders and beneficiaries who are involved are free to give opinions. It creates a substantial ground to build opinions which can then be used by the ministry as a tool or instrument to improve and accelerate the process. Regular LRAC meetings are needed to reduce waiting time for farm offers. It is in these meetings that farm offers are presented. The once-a-month meetings are not sufficient to handle incoming offers, because farm offers which come in after the LRAC meeting will have to wait for the next LRAC meeting to have them discussed. This process takes too much time, with farm offers ready to be presented but the waiting periods for meetings leading to the process being prolonged.

Fourthly, introducing mechanisms to immediately and quickly allocate acquired land is crucial to fast-tracking the reform. The government needs to ensure that acquired farms are allocated immediately after acquisition. Most of the acquired farms are allocated after a very long period (sometimes years), resulting in vandalism of the property. Although the government has a strategy for maintaining the farms prior to (re)allocation, those who are charged with the task of maintaining these farms are usually not experienced in farm infrastructure maintenance. The poor maintenance of acquired farms leads to dilapidation of farm infrastructure even before allocation of such land. Advertisement of farms (refer to Figure 7.4), for example, should be done when the state is busy with the process of acquiring the farm; that way the transfer beneficiaries can be allocated immediately. The government can also outsource and have a database consisting of records of people in need of land. The database can consist of all relevant applicants' information such as areas of preference, resettlement model, requirements, and experience. When a farm is bought, it can immediately be allocated to an appropriate beneficiary based on the requirements within a click of a button. This would enable the government to cut down on the period it takes for land allocation and the use of caretakers.

Fifthly (and lastly), reducing signature bureaucracy and providing or enabling pre- and post-settlement support is crucial. Matters involving no financial or social implications, such as acknowledgment letters, can be signed off by junior government officials and do not have to wait for the senior officials to consent to them. This practice can help minimize the number of days spent on waiting for signatures from top government officials, and in a way, this would help accelerate the pace of implementation. The government should prioritize the necessary support for the beneficiaries. Training and instruments are necessary. Funding for beneficiaries should be on the government's priority list to ensure that resettlement beneficiaries are not left to feed themselves, but supported to attain maximum productivity from the resettlement farms.

7.8 CONCLUSION

> I believe that we should have difficult conversations, as Namibians, with the aim of finding peaceful and sustainable solutions to the challenges of inequality, landlessness and outstanding pains of genocide. If we don't correct the wrongs of the past through appropriate policies and actions, our peace will not be sustainable … It is true that they came and stole the land 100 years ago, but a white boy who was born on that land has Namibian blood.

The above statement (the words of Namibia's current President Hage Geingob of Namibia) (AfricaNews, 2018), reflects the emotive nature of land reform debate in Namibia, where land reform has been a policy challenge since its political independence in 1990. It also reflects the need to conduct the *process* in a *responsible* manner. In the context of the study presented in this chapter, this implies ensuring that the reform process aligns with the 8Rs (de Vries and Chigbu, 2017). It also means that it has to be done faster to ensure that the already-disadvantaged Namibians do not fall further into the poverty.

This study not only responds to that call to make the reform responsible, it presents a responsible path to improving the reform in Namibia. It revealed the time-consuming activities in the reform process. It then recommended measures that can be taken to accelerate action for the future. Bureaucracy and poor communication also pose problems to its timely implementation. Lastly, it has shown that even the presence of a good land intervention *structure* may not necessarily lead to desired *outcome*, if the *process* lacks the proactive steps.

REFERENCES

AfricaNews Agency. 2018. After South Africa, Namibia wants to embark on land reform. URL: http://www.africanews.com/2018/08/27/after-south-africa-namibia-wants-to-embark-on-land-reform/. accessed on 22 October 2019.

Amoo, S.K. and Harring, S.L. 2012. Property rights and land reform in Namibia. In: Chigara, B. (Ed.), *Southern African Development Community Land Issues: Towards a New Sustainable Land Relations Policy*, p. 222. New York: Routledge.

Borras, S.M. 2001. State-society relations in land reform implementation in the Philippines. *Development and Change*, 32(3): 545–575.

Chigbu, U.E., Sakaria, P., de Vries, W. and Masum, F. 2017. Land governance and redistributive reform – For the sake of accelerating the national resettlement programme of Namibia. In: *Proceedings of 18th Annual World Bank Conference on Land and Poverty*, Washington, DC: The World Bank.

de Villiers, B. 2003. *Land Reform: Issues and Challenges. A Comparative Overview of Experiences in Zimbabwe, Namibia, South Africa and Australia*. Johannesburg: Konrad-Adenauer-Stiftung.

de Vries, W.T. and Chigbu, U.E. 2017. Responsible land management-Concept and application in a territorial rural context. *Fub - Flächenmanagement und Bodenordnung*, 79: 65–73.

Dorner, P. 1999. *Technology and Globalization: Modern-Era Constraints on Initiatives for Land Reform*. Switzerland: United Nations Research Institute of Social Development.

Falk, T., Kirk, M., Lohmann, D., Kruger, B., Hüttich, C. and Kamukuenjandje, R. 2017. The profits of excludability and transferability in redistributive land reform in central Namibia. *Development Southern Africa*, 34(3): 314–329.

Government of Namibia. 1995. *Agricultural Commercial Land Reform Act No. 6 of 1995 - Government Gazette No. 1040)*. Windhoek: Government of Namibia.

Government of Namibia. 2015. *Ministerial Structure*. Namibia: Ministry of Land Reform.

Lohmann, D., Falk, T., Geissler, K., Blaum, N. and Jeltsch, F. 2014. Determinants of semi-arid rangeland management in a land reform setting in Namibia. *Journal of Arid Environments*, 100: 23–30.
Permanent Technical Team on Land Reform of Namibia. 2005. *Report Review of Namibia's Land Reform Programme*. Namibia: Ministry of Land Reform.
Sakaria, P. 2016. Redistributive land reform: towards accelerating the national resettlement programme of Namibia. Master's thesis, Technical University of Munich.
Werner, W. 2015. Land, livelihoods and housing Programme 2015–18: 25 years of land reform. *Working Paper No. 1*. Namibia: Integrated Land Management Institute, Namibia University of Science and Technology.

8 The Effect of Urban Sprawl on Land Rights
The Case of Nairobi—Mavoko Corridor

Luke Obala

CONTENTS

8.1 INTRODUCTION

East African countries are rapidly urbanizing. This process is manifested in a rapid increase in population in urban areas, especially in the main cities such as Nairobi, Kampala, Kigali, Addis Ababa, and Dar es Salaam. The increase in population has largely been driven by social, political, and economic dynamics (Saghir and Santoro, 2018). A key dynamic/cause is that cities are more attractive than rural areas due to the quality of infrastructural facilities and their perceived potential and economic opportunities for the population. The rural areas on the other hand have long remained neglected, manifested in limited improvement in basic services such as health, education, and transport. Furthermore, agriculture, for long the main economic activity, is neither providing adequate employment nor better pay for the population. The youth who are keen on improved quality are thus left with no option than to move to urban areas. In addition, it is important to acknowledge that urban areas are today home to more than half the population in many parts of the world including East Africa. At the same time, the projections are that urban populations of many countries, including those in the East African countries, are expected to double by the year 2050.

Interestingly, urban areas have been the least prepared for the rapid population growth rates. The fast growth has seen an increase in congestion in housing and proliferation of informal settlements in areas hitherto considered unsuitable for human settlements, such riverbanks, wayleaves for oil pipelines, and road reserves among others. In addition, the cities are providing limited formal employment opportunities. Thus, the majority of the arrivals end up in informal employment with its uncertainties—poor and irregular pay and consequently, the inability to pay for decent accommodation. Despite the uncertainties, the situation in rural areas leads to further movement of people to urban areas, worsening the situation for housing provision.

Amidst the uncertainty for the new arrivals and the poor in the city, a large proportion of the emerging largely middle-class population are forced for social and economic reasons, and by changing tastes to move to the city fringes. At the city fringes they are able to enjoy the two worlds—a relatively quiet environment for a home and exploitation of economic opportunities provided by the cities. However, the arrival of these groups on the city fringes also presents new challenges. In particular, it brings into question land ownership rights by the different members of households that were original part of the rural area neighboring the city, but it also upsets their production of food items for both the rural households and the city residents.

There is an emerging consensus that the spread of urban areas into the periphery does not only lead to reduction in food supply and poor nutrition to the low-income groups, but it also contributes to intensified conflicts over land rights as well (Larbi, 1997; Owusu, 2007; Obala, 2011). These situations have been observed in the cases of Accra and Nairobi. The urban fringe further faces problems of poor planning and service delivery, and requires residents to be in a permanent state of patience, for there may be improvement in services in one's lifetime.

In this chapter, I present the results of studies in an area of immense importance. This is because the future is more urban in most of Africa, including Kenya, with more people projected to live in urban areas by 2030. In addition, urban areas in East Africa are rapidly expanding into hitherto rural areas, thereby disrupting existing social and economic relationships. This is more evident in relation to land ownership, access, and use in the urban periphery. Yet land remains important in developing countries as not only source of wealth and political power, but also as a symbol of continuity in many communities (Obala, 2011; Kenyatta, 1967).

It needs to be appreciated that land has diverse meanings and perceptions among different communities. The meanings and perceptions are central in shaping man and land relations in these societies. A disruption of these relations thus triggers new realities, such as emerging new tenure relations. This chapter brings into focus an area that is often ignored in responsible urban land management discourse—the peri-urban. Yet it is an area of great importance to the sustainable urban development, as it provides both food supply and a key area for outflow of excess urban population. Thus, it calls for rethinking of responsible land management to enable it to develop tools that would be useful in ensuring sustainable development and resilience in the urban fringe. In which case, this chapter will further deepen our understanding of the interactions between land management and urban sprawl in a broader sense. It will thus highlight the potential contribution of responsible land management to orderly and sustainable urban development.

8.2 URBANIZATION AND SPRAWL

Urban sprawl is a multifaceted concept, generally viewed as elusive, that is manifested in fragmented and leapfrog development (OECD, 2018; Makinde, 2012). The perception of sprawl as elusive is largely attributed to the difficulty of pin-pointing a single definitive driver. It is clear that several factors drive the phenomenon. The factors as outlined by the OECD (2018) include social, economic, and demographic as well as technological factors contributing to serious social costs. A critical factor related to land that promotes urban sprawl relates to land use regulations and policies. In particular, building regulations, land taxation, and urban containment policies although generally viewed as critical to promoting harmonious and coordinated urban development, the OECD (2018) report argues that such policies facilitate urban sprawl as they are barriers to densification. The OECD (2018) report thus provides a clear link between responsible land management and sustainable urban development on the one hand, and urban sprawl on the other.

Urban sprawl is further promoted by the rapid urban population increase that is evident in many parts of the developing world. Makinde (2012) uses the cases of London and Lagos to explain the influence of rapid population growth on urban sprawl. In the case of London, the city population grew from 1 to 8 million in about 150 years. The population of Lagos, on the other hand, grew from 300,000 people to 15 million in about 50 years. Given the limited resources in developing countries,

Lagos could ill-afford to deal with emerging problems resulting from the rapid population growth, and problems increased and the city in the end became unattractive to many. Thus, many were forced move out to live in hitherto rural lands in search of quality environment, but without any improvement in the infrastructure to accommodate the increasing population. Nairobi finds itself facing a similar situation—rapid population increase leading to housing shortage, congestion, and crime among other problems. This partly explains the sprawl being witnessed today.

It is therefore not surprising that the United Nations, projected Africa's urban population to grow to more than triple in over 40 years. The population is expected to grow from 395 million in 2010 to 1.339 billion by 2050. At the same time, the United Nations report in 2008 had projected that East African countries would experience urban growth rates significantly higher than the African average. This position was reinforced by the findings of the RUAF Foundation (2010) that indicated several East African cities, including Dar es Salaam, had a growth rate of about 4.39% per annum, Kampala at 4.03%, Nairobi 3.87%, and Ethiopia 3.4% per annum (RUAF Foundation, 2010).

The migration to urban areas is partly because agriculture, long considered the main economic activity, is neither providing adequate employment opportunities nor better pay for the population. As such, rural life is either static or deteriorating. The youth keen on improved quality of life are therefore left with no option than to move to urban areas in search of economic and social opportunities. Yet, as Smith (1996) observed in the 1990s, developing countries were characterized by exploding cities amidst stagnating economies. Similarly, Olima and Kreibich (2002) observed that urban areas in Eastern Africa were growing under poverty, appropriately summarized by Smith (1996) as urbanization of poverty.

Indeed, Devas and Rakodi (1993) in their work also concluded that between 1990 and 2020, the urban population is anticipated to double to 4.6 billion, with 93% of the growth being in developing countries. These observations further confirm the reality of population pressure on cities in developing countries. The urban areas are ill-prepared to provide food, housing, and other necessities for the increasing population. And as Dabie (2015) established in his study on the impact of urban sprawl on agricultural land use and food security in the Shai Osudoki District in Ghana, urban sprawl in the Accra metropolitan area led to reduced farm land sizes and diversion from farming to non-farming activities, thereby reducing food supply. Yet as the RUAF Foundation (2010) posits, food insecurity in Africa remains an issue, and it projects that the number of food insecure will exceed 500 million by 2020. In addition, housing remains an issue as the slum population continues to grow—thus 69% of households in Addis Ababa, 65% in Dar es Salaam, and 50% in Kampala and Nairobi are slum households (UN-HABITAT, 2008 and RUAF Foundation, 2010).

8.3 LAND RIGHTS AND RESPONSIBLE LAND MANAGEMENT

Central to the provision of adequate shelter and infrastructure is land and its management. And as Mattingly (1996) aptly put it: "land management should be understood to constitute a process aiming at achieving effective functioning of urban settlements which include among others – sustaining and boosting the social, economic, physical and cultural wellbeing of the people." He adds that another way that "land management further contributes maintenance for land is sustainability, search for land development resources as well as facilitating operation of activities and use." It also needs to be understood that the achievement of *responsible* land management objectives rests on respect for existing land rights. This calls for interactions between land management processes aiming at land interventions, and preservation and recognition of rights. It is understood that land rights refer to the relations between man and land, and as Bhalla (1996) posits, this relationship remains central to human existence. Bhalla (1996) further emphasizes that to deprive one of land or property is to deprive him or her of life. Hence, *responsible* land management also needs to rely on legitimate institutions which take into account everyone's interests in land.

The literature in mainstream land management institutions and scholars, however, points towards an urban bias. This is more evident in East African countries, as works by a number of scholars such as Olima and Obala (1997) as well as Kombe (2002), among others, focus on urban land management with no attention to food security. This is reinforced by Dekolo et al.'s (2015) argument that policies on urban sprawl in Africa are not well-documented, while their study in Nigeria revealed that urban sprawl led to loss of up to about 25% of agricultural land in the periphery, thereby putting food security at risk for the residents. Thus Dekolo et al. (2015) proposed that countries in Africa should incorporate real-time monitoring of urban expansion to limit the risk of losing more agricultural land. Their concerns were validated by Thebo et al. (2014) in an earlier study, when they undertook a global assessment of urban and peri-urban agriculture focusing on both irrigated and rain-fed croplands.

The study by Martellozzo et al. (2014) that focused on analysis of space constraints to meet urban demand for vegetables further points to the critical role urban sprawl has on food security, thereby explaining the relationship between urban and rural areas. These works by Thebo et al. (2014) and Martellozzo et al. (2014) both reinforce the argument that urban and rural areas play complementary roles in ensuring human existence. Rural areas in many parts of the world have been the main source of food. On the other hand, urban areas spur growth through economies of scale, innovation, and positive externalities.

Indeed, many authors including Page (2012) and Miller (2014) see urbanization as critical to improvement in productivity. This is premised on the thesis that urbanization contributes to agglomeration economies, concentration of capital and skills, thereby providing the ground for innovation and efficiency. Indeed, they further argue that it leads to improved infrastructure necessary for the generation of economies of scale, opportunity information, and knowledge spillover. Nairobi City and Mavoko Municipality attests to these. However, as the results from the cases reveal, they have faced equal challenges emanating from the effects of rapid urbanization. Interestingly, Page (2012) and Miller (2014) further argue that rapid urbanization observable in most countries has several negative effects; that is, it may lead to diminished quality of life. This is a situation obtainable in East African cities and the causes include: overcrowding, congestion, and destabilization of social life through quick transformation of the urban periphery (Revi and Rosenzweig, 2013).

8.4 URBAN SPRAWL AND LAND RIGHTS

The concept of urban sprawl is contentious as definitions differ. For instance, some authors like Adebayo (2012) and Sag and Karaman (2011) both see sprawl as a product of hyperphysical urban growth. They further see it as presenting both challenges and opportunities. This is a position that is presented by Dekolo et al. (2016) that associates sprawl with the loss of agricultural land. Similarly, Thebo et al. (2014) in their global assessment saw urban sprawl as having an important impact on food security. As has been observed, the problems associated with rapid growth of urbanization in Nairobi and the neighboring town of Mavoko include air pollution, water pollution, noise pollution, and the development of slums and informal settlements (UN-HABITAT, 2006, 2007; Chege, 2010; Mitulla, 2003; Mutisya and Yasmin, 2011).

Many scholars thus attribute rapid sprawling to the inability of urban management institutions to efficiently manage land in their control (Obala, 2011; Kombe, 2000). It is for this reason that Locke and Giles (2016) argue that *responsible* land management is key to the realization and/or maximization of the potential of urbanization whilst minimizing the negative effects on poor and vulnerable groups. They further posit that poor management is likely to contribute tensions, violence, and destabilization of the relations in the urban periphery. Past studies in various African countries also attest to this argument (Obala and Mattingly, 2014).

This chapter aims at deepening our understanding of the influence of urban sprawl on property rights in the urban fringes. In the process, it highlights the interactions between the various forces that influence land ownership, access to land, and land use in the urban periphery. It brings to focus the relationship between rapid urbanization and emerging land rights. As such it highlights the critical role played by land in urban growth and development. It further highlights the disruptions and their effects on land tenure in the urban periphery. In the end, it will help in recommending appropriate policy interventions, particularly in developing countries.

This calls for a more inclusive land management approach, as expounded de Vries and Chigbu (2017). De Vries and Chigbu (2017) see land management as the science and practice related to conceptualization, design, implementation, and evaluation of socio-spatial interventions with the purpose of improving the quality of life and resilience of livelihoods in a responsible, effective, efficient, consensual, and smart manner. This captures the whole spectrum of land management, and if appropriately adopted, promises to deliver on various aspects of responsible land management objectives.

The third African Ministerial Conference on Housing and Urban Development held in Bamako, Mali concluded broadly that provision of easy access to land and improvement of security of tenure is central to improving the social and economic situation of a significant proportion of the urban population. Thus, it called for commitment and action by the various governments. Suffice to add, the call for commitment saw the birth of land policy initiative. And as Obala (2011) noted rapid urbanization coupled with poor policies as well as institutional failure have contributed to inequitable access to land, particularly for the poor.

This has contributed to a large proportion of the population moving to the urban edges, leading to urban sprawl. This is evident in the cases of urban areas around Nairobi or the Nairobi metropolitan area. The desire for more livable environment has been partly responsible for the movement outwards from the cities. Besides, a significant number of the urban population find it impossible to access affordable housing in the city and opt to live in the periphery. It is this that has contributed to urban sprawl, particularly in the cases of Nairobi and Mavoko.

8.5 METHODOLOGY

Data for the study was obtained between 2014 and 2019. The data was collected using diverse approaches, but was largely participatory. As such, at the beginning a reconnaissance survey of the study area was undertaken that provided that provided the initial information on the existing land uses, access to land, density, types of development, and challenges. During the reconnaissance, the main data collection approach was observation, and limited discussions were undertaken with residents, although on an ad hoc basis. The results of the observation and limited interactions were recorded. The results were useful for development of data collection and the selection of data collection techniques.

The main sources of information were interviews with property owners, squatters, professional elites, and officials from government departments, Nairobi City Council, and civil society organizations, plus focus group discussions with diverse groups including the youth. Earlier reports and surveys provided some quantitative data. The heavy dependence on qualitative data required corroboration and triangulation to establish their validity. Verification of some data analysis was achieved in a workshop involving policy-makers, public officials, community leaders, and civil society representatives. In summary, data for the study was collected using a mixed approach, in particular the explanatory sequential approach. The approach involved interviews with key informants that largely comprised public officials in the Department of Lands and Urban Development and local community leaders. This helped in drawing appropriate conclusions for the study.

8.6 THE CASE STUDY AREA

Mavoko, largely known as Athi River, is situated in Machakos County some 25 km southeast of Nairobi. It is one of the fastest-growing towns in Kenya. The is partly due to Nairobi's industrial area expansion along the Nairobi–Mombasa highway that traverses the town. Until the mid-1990s, the area around Mavoko saw comparatively little development. Recent years have, however, seen rapid growth of the town resulting mainly from industrial expansion, residential development, and

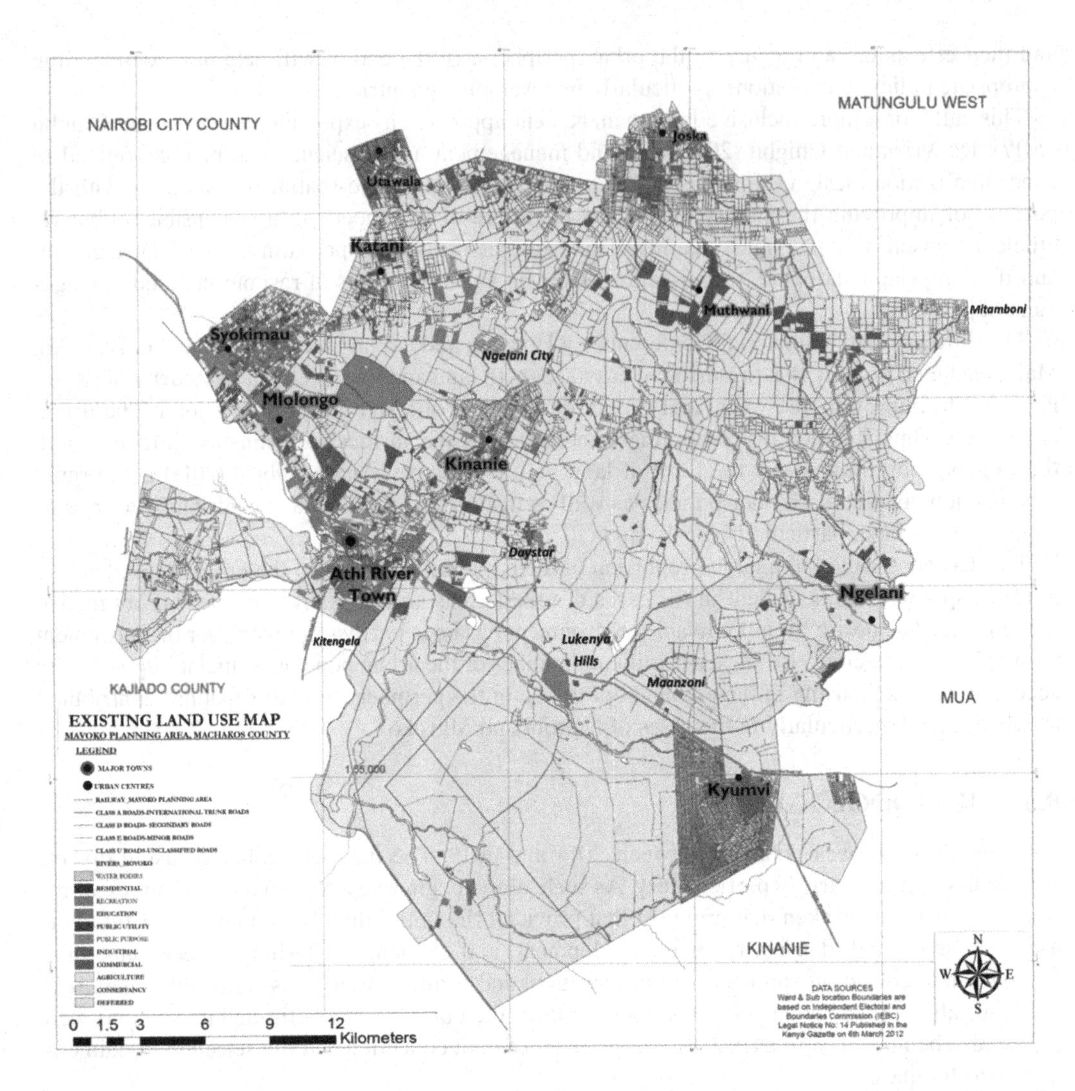

FIGURE 8.1 Land use map for Mavoko Sub-County. Source: Draft Mavoko Sub-County Integrated Strategic Urban Development Plan, 2019.

the concomitant services. Interestingly, most residential properties are occupied by people working in Nairobi and Machakos town. This is because they find the town more affordable, in spite of the commuting distance.

Mavoko had a population of 139,380 in 2009 (GoK, 2010) and it was the fifth-largest urban center amongst the 24 centers in the Nairobi Metropolitan Region (NMR). The growth rate of the area is about 17.1% (GoK, 2013). It is the fourth-fastest-growing urban center in the NMR. The other three fastest-growing are Juja (21.09%), neighboring Kitengela (20.09%), and Ngong (17.87%). In addition, the cumulative annual growth rate (CAGR) was more than five times faster than the entire NMR, with CAGR = 3.32% during the same period (CGM, 2019) (Figure 8.1).

8.7 STUDY FINDINGS

Given the nature of the study phenomenon, the results were diverse and could be grouped as legal, institutional, social, economic, and spatial. In particular, the results include: changing land uses, conflicts over explanatory sequential approach to land access and land uses, inequitable access, ownership disputes, and institutional failure among others. The specific results were as follows:

i) **Changing Land Use Patterns**

It emerged from the study that, given the increasing population along the corridor and the Mavoko area, land use patterns have witnessed steady change. The area witnessed a change from cattle ranching in the 1950s to intensive poultry and livestock farming. This finding concurs with findings by Larbi (1995) and Hitte (1998) respectively. It further emerged that these uses keep shifting further away from the city fringes due to increasing competition and/or demand for the land in the areas closer to the urban fringes by more paying for land uses such residential, commercial, and industrial land uses.

In addition, the change from farming to residential, commercial, and industrial land uses has also contributed to increasing the population in the urban fringe, consequently increasing demand for more residential housing and spaces for commercial as well as industrial activities. Indeed, the last two decades have seen major industrial complexes set up in the town. This explains the high cumulative growth rate of 3.32%. This essentially points to a high in-migration absorption rate for the study area (CGM, 2019).

Indeed, the setting up of an Export Processing Zone (EPZ) that resulted in the establishment of a number of other industrial entities, such as cement producers, distillers, and quarrying, among others. These industries are the main employers in the area and attract large numbers of semi-skilled and unskilled workers from the area and around the country. The findings confirm the argument that demand for land remains a derived demand (Warren, 2000).

ii) **Competition for Land**

The study established that a significant proportion of the population (largely new arrivals in Nairobi and poor residents) are forced to the periphery in search of reasonably priced housing. Interestingly, even in the periphery, this type of accommodation is found largely in informal settlements, on the one hand. On the other hand, it is a consequence of the inability of the City of Nairobi to provide adequate accommodation for the population at affordable prices.

Similarly, the study established from a literature review and spatial data that a large proportion of the upcoming middle-income population are forced to the fringes by their inability to access land within the city (Obala, 2011; Draft Mavoko Integrated Urban Strategic Development Plan, 2019). This partly explains the rapid expansion into the hitherto agricultural and/or UNCHS (1982), ranch land. This is because at the city fringes, they are able to enjoy the two worlds—a relatively quiet environment for a home and economic opportunities provided by the city. However, the arrival of these groups on the city fringes also presents new challenges. While most practitioners are quick to see a strain on infrastructure and management of the areas, other important challenges, such as potential and real conflicts, remain unexplored.

iii) **Land Subdivision and Change of Use**

The study results further revealed that alongside population increase in the study area, social and economic processes resulted in land use changes and subdivisions. The social processes included weakening governance structures, natural household formations, changing family structures, and community dynamics, including changing aspirations, among others. On the one hand, economic processes included changing income levels, economic changes, and changes in tastes and preferences. These factors are similar to those Larbi (1995) observed in the case of Accra and Kombe and Kreibich (2000) observed in the case of Dar es Salaam. In the study area, it was revealed that these processes peaked in the 1990s and saw large parcels that were largely owned by local cooperative societies and private limited liability companies in the area being subdivided into smaller portions.

Examples are cited, such as Kinanie, Kyumvi, Mlolongo, and the hills near Daystar University that were subdivided into smaller portions and the uses were changed.

The most notable was the conversion of land in the area by Farmer's Choice, a chicken and pig-rearing company that had for decades been located in Mlolongo area of Mavoko. It relocated its farming activities about 40 km from Mlolongo. It converted the land use in Mlolongo from agriculture to residential, commercial, and industrial uses. The movement by Farmer's Choice was informed by largely economic considerations, as the land use in the areas was quickly changing from agricultural to other uses (residential, commercial, and industrial). Cumulatively such happenings lead to densification of the areas as well as facilitating the supply of land for other uses, thereby fueling the land market.

Similarly, public institutions in Mavoko were largely allocated land by the state. Like private individuals, the public bodies too are occupying land parcels without titles or registration of ownership. The public institutions include schools, health centers, and administrative offices, amongst others. The lack of ownership documents poses a major challenge for proof of ownership of land and is likely to lead to increased land conflicts. In addition, this state of affairs unfortunately provides a fertile ground for fraudulent access to land owned by either individuals or public institutions.

iv) **Access to Land and Ownership**

There are two main categories of land in the study area, that is, public and private. Access to land can be mainly achieved through two main ways—by public allocation and as a gradual market process (Government of Kenya, 1991). Results from the study interviews revealed that others accessed land through inheritance and/or succession. The category that accessed through inheritance and/or succession received from their relatives (parents, wives, or husbands, etc.). It is understood that access and ownership of land entails having secure rights. This presumes acquisition of land through a legally recognized process that includes: acquisition through the market, public allocation, and/or inheritance or succession. Acquisition through the market implies a willing buyer-willing seller transaction process leading to the buyer being given full rights. Ordinarily, this would mean secure tenure rights for the holder. The same applies to public land acquisition processes—it assumes that the right and legal procedures were followed in the allocation process.

Official land allocation processes, the study revealed, have been skewed towards a clique with political networks. They are thus neither able nor capable of allocating land to those either in need or capable of developing the same. On the other hand, access to land through market delivery mechanisms has facilitated access by the rich who are keen on hoarding land for prices to rise before selling. This partly explains the housing deficit, particularly for the low-income groups.

v) **Illegal Land Delivery Mechanisms**

The failure of the market and public land delivery processes to satisfy a significant proportion of the population has partly led to the evolution of illegal land delivery processes. This is an approach based on fraudulent processes. The fraudsters mimic official land delivery approaches—they register a self-group that is a community-based organization, then in collusion with public officials invade the land, subdivide the land, and allocate it to a few members while, selling the remainder to unsuspecting members of the public. From the interview results, it emerged that the unsuspecting individuals are then issued with share certificates as evidence of ownership. Often land owners are unable to evict the invaders even with court orders, given the web of interests. In most cases, the leaders of the groups are often powerful and well-connected (See Obala, 2011; Obala and Mattingly, 2016).

It is evident that public land parcels are more susceptible to land grabbing because they are often not clearly demarcated and/or fenced, or because the heads of these institutions do not intervene in time to stop the land grabbing. Land grabbing has thus continued to mainly affect public institutions. Illegal land occupations lead to lengthy legal actions. In the process, legitimate landowners incur huge unwarranted legal costs. Yet in reality, this points to the failure of local authorities to effectively administer and manage both public and private land parcels.

vi) **Land Conflicts and Insecure Land Tenure Rights**

Ownership of land entails having secure rights that ensure the property owner has the right to use, sell, and undertake any permissible development on the property. This presumes acquisition of land through legally recognized processes that include: acquisition through the market, public allocation, and/or inheritance or succession. Acquisition through the market implies a willing buyer-willing seller transaction process, leading to the buyer being given full rights. Ordinarily, this would mean secure tenure rights to the holder. However, this is not often the case in the study area, where landowners, particularly those whose land parcels are undeveloped, are faced with litigation. These undeveloped land parcels in the area are subject to uncertainties over ownership, because of the risks of either forceful occupation and/or uncertainty over authenticity of ownership documents. This is particularly the case because of fraud in the land management processes—where at times one parcel of land has several sets of ownership documents.

The responses from interviews/documentation of transactions indicate that land cases take a long time to be concluded, given the huge backlog of cases, and in the meantime the legitimate property owners are unable to undertake any transactions on their land parcels. Yet in many cases, the litigant may be using fake papers. Such cases are not uncommon in Kenya. For instance, Ouma Wanzala reported in the Daily Nation Newspaper of February 18, 2019 that the Ethics and Anti-Corruption Commission (EACC) was due to hand over the title of a property that had been subject of a court dispute between the University of Nairobi and a private company for decades. In this case, the involvement of the Directorate of Public Prosecutions, EACC, and other government agencies may have forced the private company to drop the case and seek consent.

Similar cases are common in the study area; for instance, on December 6, 2018, George Owiti reported in the Star newspaper that the Cabinet Secretary in Charge of Interior Affairs in the Country had ordered groups that had forcefully invaded land parcels owned by private and public companies to move out of those land parcels. It was further reported that the local area administrator pointed out that the groups had changed their approach—they first invaded the land, then posed as land agents and/or brokers. They then sold the land to unsuspecting individuals.

Although the study did not obtain quantitative data on fraudulent land transactions, qualitative data points to this. The attempt to rectify the situation by the establishment of several tasks forces to address the question of fraudulent transactions confirms that when the problem is coupled with forceful invasions and official complicity, this contributes uncertainty on the security of tenure for landowners. It is this collusion with public officials in the lands department that has seen both public and private institutions and individuals lose their land parcels to unscrupulous gangs and individuals. Thus, public institutions like the Meat Training Institute, the Numerical Machining Complex, East African Portland Cement, numerous cooperative societies (including those in Ngelani, Lukenya, and Drumvile), and private individuals lose large swathes of land to fraudsters and land grabbers. These have interfered with their ability to meet their mandates.

vii) **Proliferation of Squatter and Illegal Settlements**

Slums and settlements are quickly sprouting on land illegally acquired by the land fraudsters in the study area. In a span of about 20 years, more than ten parcels have had settlements illegally established on them by organized groups, who in addition obtain illegal subdivision approvals and documents that purport to show their ownership of the land parcels. This is attributed to the inability of local administration to rein in unapproved housing development. On almost all undeveloped public land parcels in the area are "squatters." Squatting has a new way of accessing both public and private land. Land transactions in these settlements are largely through informal processes based on local networks and trust, similar to what Obala and Mattingly (2014) observed in case of informal settlements in Nairobi.

In addition, land management in the informal settlements mimics formal practices, with sale agreements drawn by lawyers and creation of ownership registers specific to each settlement. This is fairly adaptive and responsive to the existing situation.

The informal settlements play a critical role in the provision of accommodation to the large workforces, particularly laborers working in the industries and for logistics companies. This is because they would find it difficult to pay rent. It is evident from spatial data that informal settlements are either located close to high-end residential neighborhoods and/or near manufacturing concerns. Thus, the illegal and squatter settlements are largely located close to sources of labor. This in itself confirms the centrality and symbiotic relationships between illegal and formal settlements on the one hand, and on the other, it reveals a positive contribution at the macro-economic level facilitating urban economic development.

Besides, there are inadequate housing units available in the area to accommodate the increasing population. This role is significant, and must be looked at against the argument of UNCHS (1982) that informal settlements are part of the important links between rural and urban development.

8.7 SUMMARY AND CONCLUSIONS

It is evident from all accounts that there is rapid expansion of urban areas into hitherto rural areas. This is evident in the case of the study area that is today home to many industries, commercial ventures, and a mixed population of both locals and immigrants from other parts of the world. The impact of the spread is varied and of various dimensions. On the one hand, it leads to economies of scale as the concentration of large population and firms locate in the area. This invariably contributes to efficiency in production and consequently high revenue streams for both the state and individuals. But as Smith (1996), Kreibich and Olima (2000), and Kakumu (2016) posit, urbanization in developing countries is poverty—thus contradicting the logic. Evidence from observation indicate a number of positive signs: for instance, improvement in the quality of housing. In addition, the area has several new industries producing for both local and export markets.

The rapid expansion of urban areas and activities into hitherto rural areas further brings into focus the capacity of existing local institutions to manage both private and public land in their areas of jurisdiction. This is particularly with respect to upholding of tenure relations, which is threatened in the case of the study area to a great extent. It would require that capacity of the institutions improved with focus on both human and other related resources.

Rapid expansion urban areas do not only lead to reduction in food supply and poor nutrition to the low-income groups (RUAF Foundation, 2010), but they also contribute to intensified conflicts over land rights (Larbi, 1997; Owusu, 2007; Obala 2011). These situations have been observed in the cases of both Accra and Nairobi. The urban fringe further faces problems of poor planning and service delivery, and requires residents to be in a permanent state of patience—for there may be improvement in services in one's lifetime.

Consequently, it is clear that there is an urgent need to rethink how to improve the effectiveness of land management. The key principles of responsible land management may need further clarification. In this respect, de Vries and Chigbu's (2017) definition and expounded principles for responsible land management are appropriate. An appreciation of land management as the science and practice related to conceptualization, design, implementation, and evaluation of socio-spatial interventions with the purpose of improving the quality of life and resilience of livelihoods in a responsible, effective, efficient, consensual, and smart manner as espoused by de Vries and Chigbu (2017) further makes sense.

In rethinking land management, there is a clearly need for adoption of the 8Rs, namely: Resilience, Robust, Reliable, Retraceable, Respected, Reflexive, Recognizable, and Responsible, which if adopted would contribute towards the achievement of effective land management and minimize negative effects of urban sprawl, while allowing enjoyment of the benefits of urbanization. It needs to be recognized that to successfully adopt responsible land management, there is a need first to recognize that its key features require a combination of skills that facilitate participation and technical knowledge. But more importantly, it is apparent that land management in the urban periphery will require having clearly defined structures, as well as decision-makers that are recognizable and respected.

REFERENCES

Adebayo, A.A. (2012): A tale of two African cities: Hyper growth, sprawl and compact city development. In: *48th ISOCARP Congress: Towards the Development of a Sustainable Future City* (pp. 1–13).

Bhalla, R.S. (1996): Property rights, public interest and environment. In: Juma, C. and J.B. Ojwang, (eds.), *Land We Trust: Environment, Private Property and Constitutional Change*, Initiatives Publishers, Nairobi, Kenya.

Chege, P. (2010): Action planning: Urban regeneration and slum/ Peri – Urban upgrading, unpublished, Nairobi.

County Government of Machakos. (2019): *Draft Integrated Urban Strategic Development Plan for Mavoko, Machakos.*

Dabie, P.K. (2015): Assessing the impact of urban sprawl on agricultural land use and food security in Shai Osudoku District. Doctoral dissertation, University of Ghana.

de Vries, W.T. and U.E. Chigbu (2017): Responsible land management - Concept and application in a territorial rural context. *fub. Flächenmanagement und Bodenordnung*, 79(2 - April), 65–73.

Dekolo, S. et al. (2015): Urban sprawl and loss of agricultural land in peri-urban areas of Lagos. *Regional Statistics*, 5(2): 20–33.

Dekolo, S., L. Oduwaye and I. Nwokoro (2016): Urban sprawl and loss of agricultural land in peri-urban areas of Lagos. *Munich Personal RePEc Archive*, 5(2), 20–33.

Devas, N. and C. Rakodi (eds.) (1993): *Managing Fast Growing Cities. New Approaches to Urban Planning and Management in Developing World*, Longman, New York.

Drakakis-Smith, David (1996): Third world cities: Sustainable urban development II—population, labour and poverty. *Urban Studies*, 33(4–5), 673–701.

Government of Kenya. (1991): *A Handbook on Land Administration*, Government Printer, Nairobi.

Government of Kenya. (2010): *The 2009 Kenya Population and Housing Census*, Government Printer, Nairobi.

Government of Kenya (GoK) (2013): *Kenya Population Situation Analysis.* https://www.unfpa.org/sites/default/files/admin-resource/FINALPSAREPORT_0.pdf.

Habitat, U.N. (2008): *State of the World's Cities 2008/2009: Harmonious Cities*, Earthscan, London. 264pp.

Kakumu, E. (2016): Challenges contributing to unaffordable housing in Kenya. Doctoral dissertation, United States International University-Africa.

Kenyatta, J. (1967): *Facing Mount Kenya*, East African Publishers, Nairobi.

Kombe, W.J. (2000): Regularizing housing land development during the transition to market-led supply in Tanzania. *Habitat International*, 24(2), 167–184.

Kombe, W.J. (2002): *Land Use Planning Challenges in Peri-Urban Areas in Tanzania.* Spring Research Series, 29.

Kombe, W.J. and V. Kreibich (2000): *Informal Land Management in Tanzania*, Zeitdruck, D. Kolander and K. Poggel GbR, Dortmund.

Larbi, W.O. (1995): *The Urban Land Development Process and Urban Policies in Ghana*, Royal Institute of Chartered Surveyors, London.

Larbi, W.O. (1997): Changing livelihoods in Peri-Urban Accra: breakdown of customary land ownership. In: *DPU International Conference on Rural-Urban Encounters: Managing the Environment of the Peri-Urban Interface.*

Locke, A. and Giles Henley (2016): Urbanisation, land and property rights: The need to refocus attention, ODI Report, London.

Makinde, Olusola Oladapo (2012): Urbanization, housing and environment: Megacities of Africa. *International Journal of Development and Sustainability*, 1(3), 976–993.

Martellozzo, F. et al. (2014): Urban Agriculture: A global analysis of space constraint to meet urban vegetables demand. *Environmental Research Letter*, 9. http://www.ncbi.nlm.nih.gov/pubmed/064025.

Mattingly, M. (1996): Private development and public management of urban land: A case study of Nepal. *Land Use Policy*, 13(2), 115–127.

Miller, H. (2014): What are the features of urbanisation and cities that promote productivity, employment and salaries. *Helpdesk Request. London UK: EPS-PEAKS.*

Mitullah, W. (2003): Understanding slums: Case studies for the global report on human settlements 2003: The case of Nairobi, Kenya. *UNHABITAT, Nairobi.*

Mutisya, E. and Masaru Yarime (2011): Understanding the grassroots dynamics of slums in Nairobi: The dilemma of Kibera informal settlements. *International Transaction Journal of Engineering, Management and Applied Sciences and Technologies,* 197–213.

Obala, L.M. (1997): The politics of accessibility of land in Kenya. A case of Nairobi.

Obala, L.M. (2011): The relationship between urban land conflicts and inequity: The case of Nairobi. Unpublished PhD thesis, University of the Witwatersrand, Johannesburg.

Obala, L.M. and Mattingly, M. (2014): Ethnicity, corruption and violence in urban land in Kenya. *Urban Studies*, 51 (13), 2735–2751.

OECD (2018): *Rethinking Urban Sprawl: Moving Towards Sustainable Cities*, OECD Publishing, Paris, https ://doi.org/10.1787/9789264189881-en.

Olima, W.H.A. (1997): The conflicts, shortcomings, and implications of the urban land management system in Kenya. *Habitat International*, 21(3), 319–331.

Olima, W.H. and V. Kreibich (2002): Land management for rapid urbanization under poverty: An introduction. *Urban Land Management in Africa, Faculty of Spatial Planning, Spring Research Series*, (40), 3–10.

Ouma, Wanzala (2019): University of Nairobi Reclaims Kshs 2 Billion Land from a Private Developer, Daily Nation of 28th February.

Owiti, George (2018): Cabinet secretary for internal security warns land cartels in Mavoko. The Star 6th December.

Owusu, F. (2007): Conceptualizing livelihood strategies in African cities: Planning and development implications of multiple livelihood strategies. *Journal of Planning Education and Research*, 26(4), 450–465.

Revi, A. and Cynthia Rosenzweig (2013): The urban opportunity: Enabling transformative and sustainable development, a background research paper submitted to the high-level research committee on post –2015 development agenda.

RUAF Foundation. (2010): The growth of cities in East Africa: Consequences for urban food supply. unpublished, Research Paper.

Sag, N. and A. Karaman (2011): A solution to urban sprawl: Management of urban regeneration by smart growth. Unpublished paper, Institute of Science and Technology, Secluk University.

Saghir, Jamal and Jena Santoro (2018): Urbanization in sub-Saharan Africa. *Center for Strategic & International Studies Report, Washington, DC, USA.* www.csis.org.

Thebo, A.L. et al. (2014): Global assessment of urban and peri urban agriculture: irrigated and rainfed croplands. *Environmental Research Letters*, 9(11): 1–9.

UN-HABITAT. (2006): Annual report, Nairobi, Kenya.

UN-Habitat (2007): Global report on human settlements 2007: Enhancing urban safety and security, Earthscan, London.

UNHCS (1982): *Directory 6: Finance Institutions in the field of Human Settlements in Developing Countries*, Nairobi, Kenya, UN Publication.

9 Integrating Customary Land Tenure through the Statutory Land Tenure System in Tanzania

The Case of the Iringa and Babati Land Formalization Process

Bupe Kabigi, Walter Timo de Vries, and Haule Kelvin

CONTENTS

9.1 INTRODUCTION

The integration of the customary land tenure system into the statutory system is embedded within the debates surrounding customary titling. The customary land tenure system refers to channels of accessing land through traditional and custom procedures. It is characterized by cultural norms, customs, and environment which influence the means through which people access and use land (Kironde, 1995; Manji, 2006). In the customary system, people access land through complying with the traditional conditions for accessing land. The customary land tenure system has been the dominant channel through which the majority of people access land in Africa (Dale and Baldwin, 2000).

With the increasing population pressure and commercialization of production, there is a growing tendency for these customary tenure systems to be disintegrated (Peter, 2004). As such, customary rights are seen to be weak and often in conflict with statutory laws (Wily, 2011). Customary systems are also considered inadequate in providing the required security to ensure land-based investments and productive use of land, and are also inconsistent with the requirements of modern market economies (Binswanger and Deininger, 2007). They are characterized by insecurity which lies in the absence of clearly defined and enforceable property rights (Byamugisha, 2013). As a result, the World Bank and other international and bilateral initiatives have been collaborating with state governments in developing countries to restructure customary land tenure systems through land reforms that have a particular focus on land formalization (Platteau, 1996; Knight, 2010). The process and context within which land formalization has been carried out have evolved and varied over time and space.

Between the 1950s and 1980s, land formalization in Africa was concerned with the redistribution of land asset and rights from wealthier segments of the society and people with larger plots of land to those with small plots (Smith, 2003). Formalization in these years was also aimed at addressing the challenges of historical inequality as a result of the colonial legacy. The examples of this formalization epoch persisted in Kenya, Zimbabwe, South Africa, Ethiopia, and Tanzania. Strategies such as breaking up of large estates, expropriation, and distribution of land among former tenants were used (Bruce and Migot-Adholla, 1994). Another strategy was the government acquisition of land for the promotion of a socialist form of tenure (Bruce, 1993). Nevertheless, formalization did not address injustices of inequality, access to land rights, and security. The poor continued to sell their land rights to the wealthier people resulting in land concentration by the wealthier segments of the society (Barrows and Roth, 1990).

The second period of land formalization was between the 1980s and 1990s. This era emphasized individualization, privatization, and promotion of land markets. The proponents of formalization, such as the World Bank (2003), emphasized the improvements in accessibility and productivity of land asset through transactions. Formalization implied that security of tenure and access to land were neither limited to the state-led legislation nor customary practices. Rather, it would be a combination of the two systems (Wily, 2011; Ampadu, 2013; Pedersen, 2015). As Cotula and Mathieu (2008) argue, many African countries have, since the mid-1990s and 2000s, attempted to put in place systems that allow legal registration of customary rights while also increasing the responsibilities and roles of decentralized administrative entities and government technical services. In Tanzania, dual land tenure systems were established through the Land Act 1999 (No. 4 of 1999) to cover land rights in general land, i.e. outside village and reserve land, and the Village Land Act 1999 (No. 5 of 1999) which deals with the land within village areas (URT, 1999). Krantz (2015) suggested that even when customary rights are recognized in national legislation, it may still be difficult for local people to defend their rights against outside claims simply because their land holdings are not demarcated and registered and therefore not identified on official cadastres. It is thus difficult to protect land rights that are not clearly established and documented in writing in accordance with legal procedures (Cotula and Mathieu, 2008). This implies that villagers will be encouraged by donors, investors, and financial institutions such as the World Bank to formalize their rights via land titling and registration, mainly with the objective of enhancing land tenure security and facilitating access to financial capital, avoiding conflicts, encouraging land investment, and promoting the land market (URT, 1999; Pedersen, 2010).

In the course of formalizing land rights, a number of scale-up projects were established since 2004, particularly in the Mbozi, Babati, and Iringa Districts. The integration of customary land tenure systems into statutory systems is realized through the processes and practices of recognizing and applying customary practices together with statutory systems. Integration in this chapter entails applying together, having similar values, and recognizing customary land ownership and use analogous to statutory practices. Integration entails customary practices being used, applied, and recognized by existing land laws, policies, and regulations. The chapter mainly assesses the extent to which the customary land tenure system has been integrated into the statutory system to enhance security of land tenure. Three analytical questions aim to respond to the general objective. Firstly, how is this integration empowering village governments to plan, administer and recognize the village land? Secondly, in what ways do integrating customary land rights into statutory systems reduce land disputes, improve information access, record-keeping, and registry? Thirdly, to what extent is customary land registration and ownership being integrated into statutory land tenure systems?

Relying on both primary data collected from land formalization practices in Babati and Iringa districts in Tanzania and documentary review of other national formalization reports, policies, laws, regulations, and program reports, these questions are answered in Section 9.3. The first question is answered through household data on establishment of village boundaries and certificates, benefits of issuance of certificates of village land, and establishment of village land use plans. The second question relies on documentary review and interviews of subject matter specialists. Specific issues of analysis are reducing land conflicts and the establishment of land registry to enhance information access. The last research question draws its analysis from both documentary reviews and primary data of adjudication, registration processes, and types of CCROs that culminate into statutory ownership. The chapter argues that if customary land tenure is integrated, recognized, and valued in tandem with the statutory systems, challenges of land access, allocation, and management will be reduced. Answering these questions contributes to improving knowledge and skills on smart land management. This chapter uncovers the untapped practices, knowledge, and techniques required to design a responsible land management system specifically in developing countries where customary land tenure systems have been dominant for ages.

The results are chronologically organized in terms of specific research questions. Section 9.3.1 is devoted to results from respondents to question one, followed by results on research question two which are placed in Section 9.3.2. The last subsection of this section (Section 9.3.3) elucidates results for research question three.

9.2 METHODOLOGY

The chapter is based on primary data collected from the Babati and Iringa districts in Tanzania. In Iringa, Ndiwili and Ilalasimba villages were involved. In Babati, Mamire and Mwada villages were used. These districts were engaged because of three reasons. Firstly, these were among the districts where the first ten land formalization scale-up projects were undertaken between 2006 and 2010. Secondly, these are among the districts characterized by a high number of the land use conflicts registered with the District Land and Housing Tribunal (DLHT). For example, by 2013 about 375 land disputes were registered with the DLHT (Babati DLHT, 2013). Thirdly, formalization projects in these districts adopted different approaches, and differed in terms of technology of adjudication. For example, Ilalasimba village in Iringa DC, was selected because it was one of the first pilot projects using the Mobile Application to Secure Tenure (MAST) in Tanzania, and the project was scaled-up into other villages in the Southern Agricultural Growth Corridor of Tanzania (SAGCOT) (USAID, 2011). The MAST project was implemented using mobile phone technology to facilitate the process of land registration and administration. It also employed village youths as Trusted Intermediaries engaged in mapping the village land.

Three methods of data collection were adopted. Firstly, household surveys were used to collect data on research question one and two. Secondly, interviews done with Land Officers and Village Executive Officers were used to inform on the registration processes, adjudication, and statutory change of customary arrangements of land use planning, ownership, and registration. Lastly, documentary review was used to review the available data on land use conflicts, land use plans, and registry. Regarding the data analysis, the paper adopted an interpretative analysis.

9.3 INTEGRATING CUSTOMARY LAND USE PLANNING AND STATUTORY RECOGNITION OF THE VILLAGE LAND

9.3.1 Village Boundary and Certification for Village Recognition

Establishment of village boundaries and certificate is among the strategies to recognize and integrate customary land arrangements into statutory rights. Under customary arrangements, villages were established and recognized through traditional features and general boundaries such as hills,

TABLE 9.1
Benefits of Issuance of Certificate of Village Land

Villages	Assurance of village existence	Reduced land use conflict	Protect village boundary	Allocation of land	Total
Ilalasimba	25 (47%)	14 (26)	8 (16).	6 (11)	53 (23.6)
Ndiwili	41 (51)	11 (14)	18 (22)	11 (13)	81(15.6)
Mwada	57 (54)	13 (12)	16 (15)	19 (19)	105 (29.9)
Mamire	62 (61)	9 (9)	10 (10)	20 (20)	102 (30.7)

Source: Field Data, 2016.

rivers, and forests (TFCG, 2015). These were used to define community and village land rights. With land formalization, village boundaries and jurisdiction are recognized through a Certificate of Village Land (CVL) and specific boundaries are demarcated by beacons. Section 7 of the Village Land Act 1999 in Tanzania, requires that before land formalization proceeds, the area to be formalized has to be well agreed upon by neighboring villages, and boundaries demarcated. In this regard, each village confirms its boundaries before being issued with a CVL. In case there are boundary conflicts, settlements must be done (URT, 1999). It is the Village Land Act 1999, Section 8 that also empowers the village council to manage and allocate the village land.

In the view of household respondents, establishment of CVL has various benefits and implications to customary land rights. The results in Table 9.1 show that 61% of people involved in the study perceived that CVL assures the existence of village, 10% felt it protects the boundaries of the village, 20% that it is useful in land allocation, and 9% thought it reduces land use conflicts. These benefits show that formalization legally recognizes the existence of a village and it empowers local leaders to register and allocate land. Local community members are integrated in the statutory system of land ownership and recognition. Despite such integration, formalization has just offered a non-pragmatic solution to village autonomy in land ownership and registration. The central government is still a supreme organ in surveying, registration, and approval of land rights.

9.3.2 Village Land Use Plan and Integration of Customary Rights into Statutory System

Through village land use planning (VLUP), land management is effected (Kushoka, 2011). Land use planning under the customary system is done through the traditional arrangement of land uses for grazing, worshiping, and cemetery. These are separated from residential uses, although farming has sometimes been integrated with residential uses (Huggins, 2016). Chiefdoms and clanships were responsible for organization of land uses into different categories. To date, these land uses have been formalized through the enactment of the Land Use Planning Act 2007 (No. 6 of 2007) and the Village Land Act 1999, and the land policy (1995) (Huggins, 2016). These regulatory and management instruments have established village level institutions. These are: the Village Land Council (VLC), Village Assembly (VA), Village Land Use Management Committee (VLUMC), Village Technician (Para-Technician) (VT), and Village Land Use Plan Team (VLUP Team). These institutions have different roles in the process, planning, and enforcement of a VLUP.

VLUPs are prepared with different objectives ranging from conflict resolution, enhancement of land tenure security, and increased agricultural productivity through identification of areas for commercial agricultural investment and ensuring land allocation for investment. Others are to facilitate conservation of sensitive ecosystems and critical habitat and mitigate risks of land grabbing (UCRT, 2010; Mango and Kalenzi 2011; Milder et al., 2013; Hart et al., 2014). So far, customary land uses have been integrated into statutory land tenure systems through established and designed land use plans and maps, as presented in Figures 9.1 and 9.2.

In an attempt to compare the benefits and challenges of land use planning, the analysis from this study shows that although there are positive implications for improving land uses and administration, the challenge of enhancing security of tenure manifests through the limited availability of land use maps on the village and hamlet levels. Land use maps, although they are available at district level, are not available at village and hamlet level. Huggin (2016) asserted that VLUPs were not available at district headquarters. Deininger, Hilhors, and Songwe (2014) revealed that in African countries, land use plans were often lacking or outdated, and if they existed, their elaboration was non-participatory. Centralized planning is one of the factors contributing to this challenge. One of the district land officers in Babati emphasized:

> Due to the bureaucratic and centralized procedures of the preparation of VLUP, the land use map for Mwada village was not brought back to the respective village and it cannot be found even at the district level

Furthermore, the land use planning process has not managed to ensure that sufficient pasture land is allocated for the pastoralists. Huggins (2016) and Hart et al. (2014) observed that in the Arusha region and the Mbarali district, Mbeya region, a small portion of land is allocated for pastures. Also, the stocks routes, water points, and cattle dip within the grazing zone are missing (Kosyando, 2006; Kimbi et al., 2014). Pastoralists are still facing a shortage of pastureland compelling them to invade the farming land. Also, the shortage of pastureland in Babati district occurs because of the seasonal migration of the Iraqw tribe (agro-pastoralists) and Barabaig tribe (semi-nomadic pastoralists) who come with a huge amount of livestock that results in scarcity and inadequacy of the pastures. Climate change was also associated with shortage of pastures. For example, *Brachiaria spp.* grass which used to be found near Lake Burunge which was very important during dry season has nowadays disappeared due to climate change. The same experiences occur in Kilosa, Kilindi, Longido, Ngorongoro,

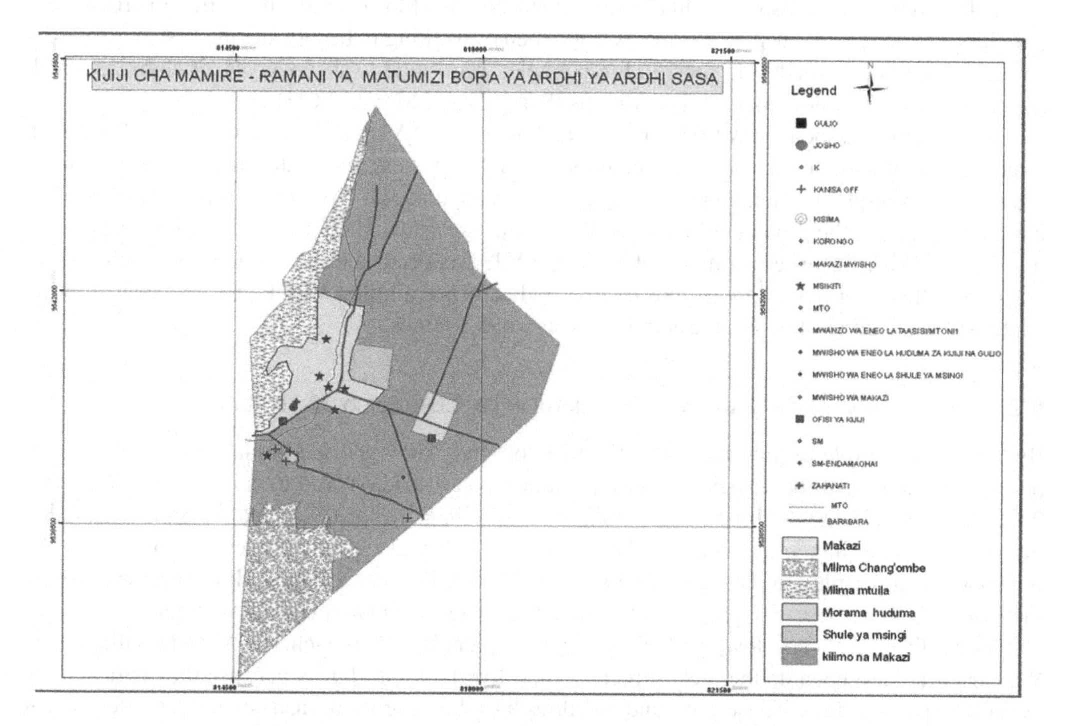

FIGURE 9.1 Mamire Village Land Use Plan showing existing land use. Source: Mamire Village Government, 2010.

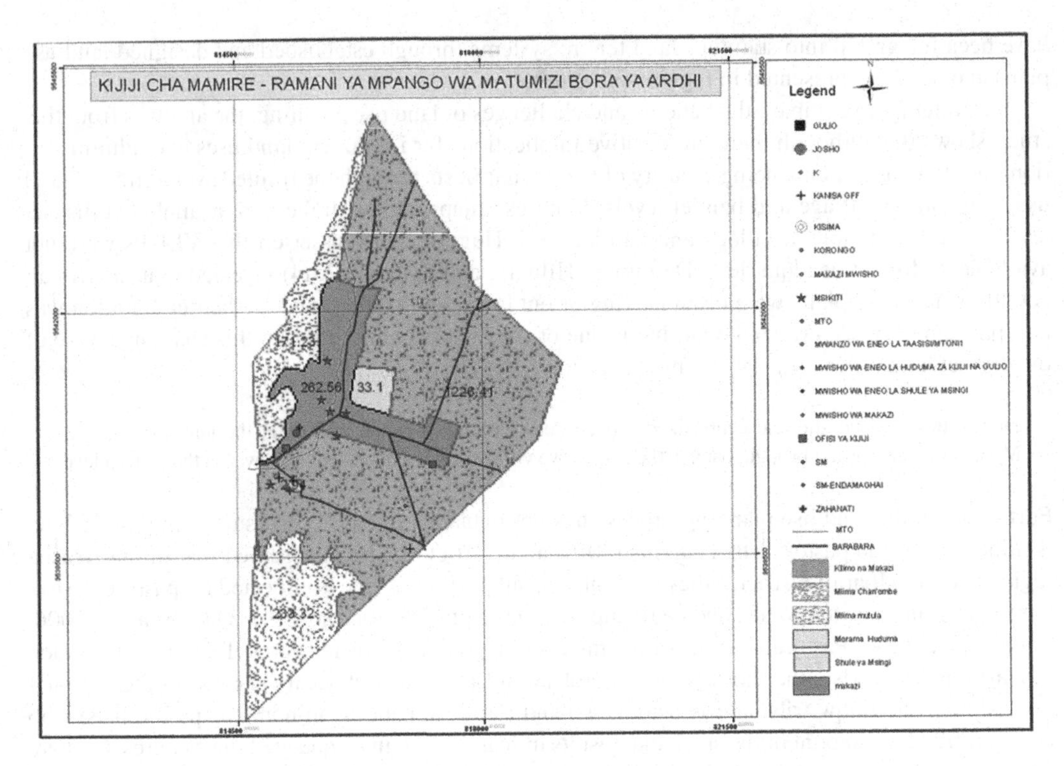

FIGURE 9.2 Mamire Village Land Use Plan showing future land use. Source: Mamire Village Government, 2010.

Monduli, and Simanjiro districts. This has also contributed to Masai youths migrating to urban areas searching for petty jobs, such as serving as watchmen or working in beauty salons. Also, pastoralists switch to alternative breeds and species, for example, from cattle to sheep, goats, or chickens which are better adapted to more marginal conditions (Sangeda and Malole, 2013).

The challenges associated with the process and practices of VLUP imply that the VLUP is not an effective instrument for ensuring land tenure security for farmers and pastoralists in rural areas, in case it is not well-planned and not involving primary stakeholders. The current practice and process have not addressed the problem of land conflicts; it has not established the mechanisms for motivating the VLUMC. Because of this problem, ILC (2013) reported that members of VLUMC opt to engage in other livelihood opportunities. Other indicators of a failed VLUP process occur through continued conducting of economic activities near water sources.

9.3.3 Enactment of By-Laws and Integration of Customary Land Rights

By-laws have the role of guiding appropriate use of land. They provide guidance for what is supposed to be done, and the associated fines, penalties, and restricted land uses in villages (NLUPC, 2011). For example, the established by-laws in Ndiwili village Iringa described the prohibited farming practices, building, and burning in the pastureland. They also prohibited cutting of trees in the forests without permission. By-laws were also established in other villages. However, the enforcement of by-laws in these villages is weak, and is associated with two major challenges.

Firstly, fines are very low, giving openings for deliberate breach. In Mwada village, the Wamang'ati* have been deliberately directing their livestock into farms because they believe they are able to pay the fines. These fines and penalties have become weak instruments for enforcement

* One of the pastoralist tribes in Tanzania whose main livelihood is livestock keeping, they move from place to another in search for pastureland.

FIGURE 9.3 Traditional irrigation at Ilalasimba Village (Vinyungu).

FIGURE 9.4 Traditional irrigation at Ilalasimba Village (Vinyungu).

of VLUP. According to Lugalla (2010), low fines and penalties pave the way for violating the existing by-laws and regulations. Weak by-laws have also been contributing to some local leaders taking bribes from pastoralists and allowing them to graze in restricted areas. In Ilalasimba village, it was reported that village leaders deliberately avoid enforcing by-laws because they are also engaged in owning and cultivating near water sources and engage in a traditional farming practice known as *Vinyungu** (see Figures 9.3 and 9.4).

Lastly, enforcement of by-laws at village level is affected by political interference. Sometimes contradictions occur between what the law stipulates and the interest of politicians. For example,

* *Vinyungu* refers to irrigation potholes locally made through traditional materials such as mud, sands, canals, and grasses.

conducting socio-economic activities within 60 m of the water sources is restricted by the Water Resource Management Act 2009 section 34 and the Environmental Management Act 2004 section 57(1) (URT, 2004; 2009) which stipulate that

> No human activities of a permanent nature or which may, by their nature, likely to compromise or adversely affect conservation and or the protection of ocean, river bank, water bank or reservoir, shall be conducted within sixty meters.

Contrary to these laws, politicians have been allowing people to conduct human activities such as farming closer to water sources. When experts intervene to enforce the laws, they are told by the politicians to leave people to proceed with their activities because they are their voters. This suggests that the success of by-laws on ensuring land tenure security does not only rest on their existence but also enforcement, actors' willingness, and people's compliance. Weakly designed by-laws in terms of low fines, low education level among village leaders, and low motivation among village leaders, due to the inadequate budget for the implementation of strategies identified in the Community Action Plan (CAP), are other disincentives of enforcement. This implies that by-laws established to implement VLUP are rubber-stamped and a manifestation of the business-as-usual process.

9.4 INTEGRATING CUSTOMARY LAND REGISTRY AND CONFLICTS SETTLEMENTS INTO STATUTORY SYSTEMS

9.4.1 Village Land Registry and Statutory Land Tenure Systems

The village land registry is an important component of land administration systems, because important documents of land rights are stored. Setting of village and district land registries becomes obligatory under section 21(3) of Village Land Act 1999. The village land registry is supposed to be a simple record of intra-village customary ownership, as well as all internal land transactions and dispositions. At the village level, Village Executive Officers (VEO) supervise the village land registry. In this regard, the VEO record all land transactions, available information on land use and tenure in case of land disputes, and update the records of the latest landowners. The Village Land Regulations 2001 sections 52–59 state that VEOs are also responsible for feeding this information to the district land registry (URT, 1999, 2001). Due to these legal requirements, land registries were established in Babati and Iringa districts

The establishment of a land registry in rural villages integrates rural land rights with the national statutory systems. It is through these registries that CCROs, land use plans, and reports on land disputes are kept. Traditional information about land rights is legally stored through these registries. Nevertheless; efforts to enhance information access and reduce asymmetry in rural areas suffer from various malpractices. Firstly, enforcement of record-keeping is not properly undertaken, some land transactions are still being done without the involvement of the village government, and most land subdivisions and transactions proceed without official land transfers. This is contrary to the Village Land Act 1999 section 3 and the Village Land Regulations 2001 section 38 which state that "the records of all dispositions and transactions involving customary rights of occupancy and derivative rights, including any cautions entered on the register section 3 of the Title register" must be kept. The problem is also exacerbated by a lack of understanding among the VEOs and farmers on the land laws and regulations. According to Pedersen (2010), Lugoe (2015), and Fitzgerald (2017), the majority of village leaders have not received training on land policies, laws, and regulations; and frequent transfer of government officials at village level interfere with formalization efforts.

Secondly, the design and operation of the registry shows that the documents stored in this registry have a lower value. There are poor storage devices and they are unsecured, as presented in

FIGURE 9.5 Improper storage of CCROs in Ndiwili Village Land Registry. Source: Field Data, 2016.

FIGURE 9.6 Improper storage of CCROs in Ndiwili Village Land Registry. Source: Field Data, 2017.

Figures 9.5, 9.6 and 9.7. Some land registries have doors which are of a wooden design, and documents are kept on the floor and in boxes. This is contrary to what has been stated in the Village Land Act 1999 section 21 which requires the document should be kept in a dry and safe place.

Thirdly, the operation of the village land registry shows that documents are not seriously kept or stored, and not computerized (analogous to the system of filing). The village land registry is not linked to the national land management system. It is a village-isolated information management system. A lack of serious documentation suggests the land registry is not valued. The implications for the unvalued documentation of land information shows that the current practices and operation of land registry do not guarantee land owners their security of land rights. This occurs because one

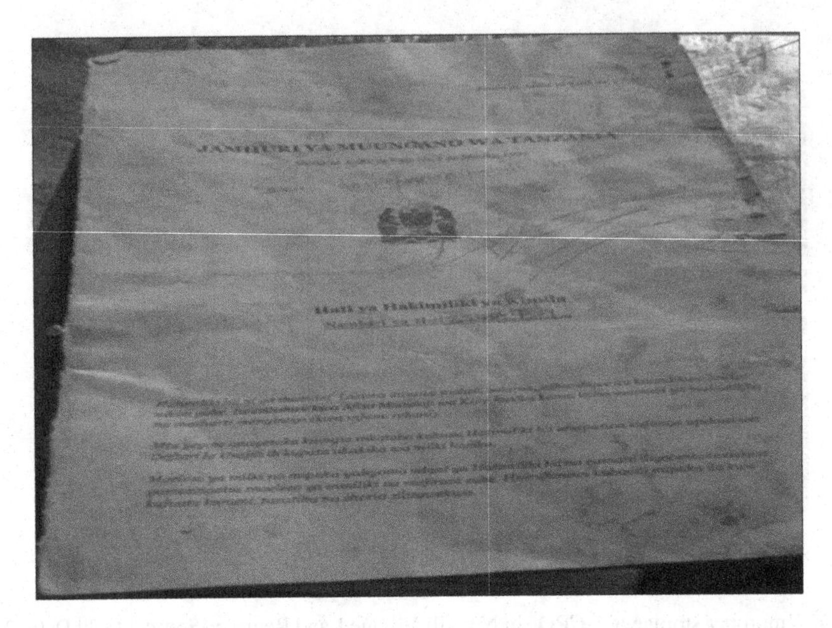

FIGURE 9.7 Improper storage of CCROs in Ndiwili Village Land Registry. Source: Field Data, 2018.

copy of a CCRO is kept at the village land registry, and in case it is stolen or village leaders conspire to change the ownership status, the buyers or sellers are threatened over their ownership.

9.4.2 Land Dispute Settlements as a Platform for Integrating Customary Land Tenure into Statutory Systems

Land disputes have been one among the challenges of land administration system in Africa for the past three decades. The challenges have been mitigating escalating conflicts, and handling their impacts (Benjaminsen, Maganga, and Abdallah, 2009). The dominating conflicts have been farmer–herder, boundary conflicts, especially in borders between villages and conservation or forest areas. This case may be highlighted by a conflict between Sangaiwe village and Tarangire national park in Babati district which persisted between 1990s and 2000s. The villagers claimed that the boundary boards in the mountain which separates Tarangire National Park and Sangaiwe village were put in the wrong place by the government during the village boundary survey, village conflicts, and institutional conflicts. Traditionally, these conflicts were dealt through negotiation, fines, penalties, traditional arrangement of land uses, and shared practices. Village, hamlets and family meetings were key platforms for conflict settlements. Post-conflicts and trauma have been worked through with traditional healings, taboos, and prayers (Massay, 2016). In all these practices, clan heads, the elderly, chiefs, and herbalists were involved in dispute settlements. In the context of land reforms, and formalization, statutory land disputes settlement platforms have been established. These are the village councils, ward tribunals, district land, and housing tribunals (DLHT), the high court (land division), and the court of appeal of Tanzania. These are supported by laws and regulations such as the Courts (Land Dispute Settlement) Act 2002 (No. 2 of 2002).

The enactment of these laws and regulations is an integration of customary practices of conflicts settlements and mediation. Village councils are legally recognized by the Land Dispute Settlement Act and the policy as a lawful platform for conflict settlement. However, the law has not recognized the hamlets, family, and clan meetings as appropriate platforms. This chapter considers this as a drawback because these are immediate platforms to initiate the process of conflict settlement. Lastly, this integration, although it established a legitimate procedure for conflict settlement, has undermined traditional procedures of fines and penalties which previously encouraged togetherness.

Traditional punishments were mostly in form of non-monetary value and compensation. For example, offenders paid crops, animals, and alcohol to the village and affected persons. The established punishments in terms of compensation money and recovery money have affected the traditional conflict settlements proceedings. This is because monetary punishments are accompanied by costs which discourage rural people. Formalization of DLHT analogous to formal courts acts as a disincentive for people who are afraid of the court due to their lack of education. Consequently, this integration disfavors efforts geared towards security of land tenure.

9.5 CCROS FOR INTEGRATING CUSTOMARY LAND REGISTRATION AND OWNERSHIP INTO STATUTORY SYSTEMS

9.5.1 Registration and Adjudication of Customary Rights

Registration and adjudication of customary rights are other strategies to integrate customary rights into statutory systems. Registration entails legal recognition and identifying the owner and appropriate land uses. As established by the Village Land Act 1999 sections 51–55, registration and adjudication take two forms: spot adjudication* and systematic† village adjudication. In Babati and Iringa districts, a Systematic Adjudication (SA) or carta procedure using satellite imagery was used to delineate the boundaries of a land parcel. The use of satellite imagery maps reduces costs and time needed to mark boundaries. Also, as the whole village is visible in one or a series of adjacent images, it is also credited for being more inclusive, as landowners are able to see their parcels before the registration process proceeds (Zein, 2017). This process of adjudication was done piecemeal, whereby the villages were divided into different hamlets (*Kitongoji*) to carry the process. In view of Kurwakumire (2013), the piecemeal process of adjudication is considered flexible, inclusive, participatory, affordable, reliable, attainable, and upgradable, can be applied within short time frame, and the involvement of the local communities can help guarantee the acceptance of the result at local level.

Systematic adjudication involved various steps aimed to ensure that the Participatory Land Use Management team (PLUM) and Land Adjudication Committee (LAC) hamlet chairpersons were involved, and landowners and all neighboring individuals had to be present during the identification and verification of parcel boundary records. All these structures are a means of integrating customary tenure system into statutory mechanisms. At the end, the landowners and neighbors sign on the systematic adjudication record form and landowners obtain a Parcel Identification Number (PIN). According to Zevenbergen (2002) and Lemmen et al. (2009), this process is likely to reduce future land disputes, evections, and dispossessions, since it enhances transparency and identifies parcel owners. Indeed, adjudication and registration culminate in provision of CCROs which entails entitlement of land rights. The extent to which CCROs integrate customary rights into statutory systems remain a subject of discussion.

9.5.2 Customary Certificate of Rights of Occupants (CCROs) and Its Statutory Integration

Registration of land rights and ownership are done through granting of CCROs (URT, 1999). Section 8 of the Village Land Act 1999 states that ownership of land and associated rights are evidenced

* Spot adjudication is applied in response to a demand from an individual applicant for CCRO. A pre-established adjudication team, together with the CCRO applicant and contiguous neighbors, will visit the applicant's land parcel and make records of the coordinates of the boundary details of the parcel (TFCG, 2015).

† Systematic adjudication involves multiple land holdings; this option is capital intensive (financial, skilled technical personnel, labor, time, and equipment) requires mobilization and participation of large number of community members, and requires application of effective participatory methods. This is common when a third party interest is involved and willing to finance the process, such as an NGO project or government operations, or the communities in the village implement the process with their own resources (TFCG, 2015).

through holding CCROs. This is a change from customary procedures where land rights were traditionally owned, and exclusion was done through general boundaries (Fairley, 2012). Instead of giving emphasis to individualization of land rights, customary arrangements fostered shared and communal rights. Clanship farming was among the pointers for shared rights. The introduction of CCROs has integrated the traditional practices of land ownership and use. Today, formalization has introduced individual, joint, and group CCROs. These may take diverse forms (Schreiber, 2017). The Village Land Act 1999 established six types of occupancy certificates:

- Men and women at least 18 years of age could register in their own names any land granted to them by their local government or land they had used for more than 12 years;
- The law also permitted joint registration, for example, a husband and wife could register as co-owners, and the certificate would include both of their names;
- In case of polygamous marriage, or when more than two family members jointly owned land, the title could include the names of all of the users under a group registration;
- When a dispute arose over who should inherit the land of a deceased landowner, the family could appoint an administrator to temporarily manage the land until the family reached an agreement on how to divide the property;
- In the case of a person younger than 18 years, the legal guardian could obtain a guardianship occupancy certificate, which could expire when the child turned 18;
- Finally, organizations like schools or dispensaries could obtain institutional occupancy certificates.

These typologies of CCROs amalgamated traditional land ownership and the entitlements of a legally recognized system. CCROs provide evidence of ownership, justification for dispute settlement, and instruments of compensation (Fitzgerald, 2017). Between 2009 and 2017, data from the district land registry showed that a total of 15,196 CCROs were granted from 27 villages out of 102 villages in Babati district, while in Iringa DC about 14,990 CCROs were granted from 2005 to 2016, from 63 villages out of 133 villages.

Despite such benefits associated with the integration of customary land tenure rights into statutory systems, CCROs suffer from their poorly produced quality, they are not connected with the national land management system, and there are no systems that integrate customary land transactions, CCROs issuance, and land disputes. CCROs issuance and monitoring have remained analogous to traditional practice. Moreover, although CCROs' integration into legal systems in Tanzania suggests a desired land administration milestone, the practice has not made village council autonomous and secure in their land ownership. This is because land in Tanzania is still under presidential custody; the president has power to revoke any CVL and acquire any plot. Hence, this integration has not liberated village ownership and registration from these challenges.

The process of adjudication demonstrates diverse facets of its implications on land tenure security among rural people. The theoretical gains of adjudication link it with reduced conflicts, and improved land rights and access to loans (Deininger, 2003; Byamugisha, 2013). On the contrary, the practices of land adjudication demonstrate a failed intervention. Instead of providing remedies for the existing land use conflict, the process of adjudication has been a cause of land use conflicts. This occurs due to some errors in surveying and measurements, allocation errors, and deficiencies in software. The process has not improved women's ownership and rights to land; this manifests through continuing use of family CCROs and group CCROs ensuring that customary ways of land ownership that work in favor of the men rather than women continue to prevail.

9.6 CONCLUSION AND RECOMMENDATIONS

This chapter mainly assessed the extent to which customary land tenure system has been integrated into the statutory system to enhance security of land tenure. Assessing how customary land tenure has been altered through this integration is important, especially in developing countries where statutory rights are handled in tandem with customary rights. Three research questions were analyzed. Firstly, how is this integration empowering village governments to plan, administer, and recognize the village land? As drawn from Babati and Iringa DC, the results have shown that formalization legally recognizes the existence of a village and empowered village council to register and allocated land into statutory systems. Despite such an achievement, the central government is still the supreme organ in surveying, registration, and approval of land rights.

Secondly, in what ways do integrating customary land rights into statutory systems reduce land disputes, improve information access, record-keeping, and registration? While integration of customary land rights into statutory systems manifests in terms of having an established village land registry, security land tenure is far from being realized. Enforcement of record-keeping is not properly undertaken, there are land transactions done without involving the village council, land subdivision and transfer are done without official involvement, and the village land registry is not linked to national land management system. The current practices of the village registry do not suggest improvements in information related to land, but has contributed to people continuing relying on the customary system of information, access, and land management. Land dispute cases are still high, and mitigation strategies are weak.

Thirdly, to what extent is customary land registration and ownership being integrated into statutory land tenure systems? Customary land rights have been integrated into statutory systems through the entitlement of CCROs, whereby between 2009 and 2017 a total of 15,196 CCROs and 14,990 CCROs were granted in Babati and Iringa, respectively. CCROs are assumed to provide evidence of ownership for those possessing them. Despite the associated benefits, the majority of people have not used CCROs for acquiring loans, CCROs have not significantly contributed to mitigate or reduce land use conflicts, and cases registered with the DLHT are still on high side. In light of these shortcomings of formalization, a need still exists for reorienting and transforming the practices of adjudication, recognition ownership, and registration. To enhance authenticity of ownership, CCROs need to be integrated into the national management systems. Limited establishment of CCROs in the national land management systems characterize CCROs as remaining customary in operation. Also, the nature of formalization should be treated as a universal approach following standard procedures which show greater support for traditional practices. Generally, formalization should not be treated as a replacement of the African history of land values, morals, and ownership. Lastly, sustainability of formalization should be built in terms of encouraging local capacities, should be participatory, and should be owned by communities themselves.

REFERENCES

Ampadu, R. A. 2013. *Finding the Middle Ground: Land Tenure Reform and Customary Claims Negotiability in Rural Ghana.* PhD., Programme of Ceres, Research School for Resource Studies for Development, Netherlands.

Babati District Land and Housing Tribunal. 2013, Cases Register Book Manyara Region, Tanzania.

Barrows, R., and Roth, M. 1990. Land Tenure and Investment in African Agriculture: Theory and Evidence. *The Journal of Modern African Studies* 28(2): 265–297.

Benjaminsen, T. A., Maganga, F. P. and Abdallah, J. M. 2009. The Kilosa Killings: Political Ecology of a Farmer-Herder Conflict in Tanzania. *Development and Change* 40(3): 423–445.

Binswanger, H. and Deininger, K. 2007. History of Land Concentration and Land Reforms. *Land Redistribution: Towards a Common Vision, Regional Course*, Southern Africa 913.

Bruce, J. 1993. The Variety of Reform: A Review of Recent Experience with Land Reform and the Reform of Land Tenure with Particular Reference to the African Experience. In: Marcussen, H. S. (ed.), *Institutional Issues in Natural Resource Management*, International Development Studies.

Bruce, J. W. and Migot-Adhola, S. E. 1994. *Searching for Land Security in Africa*, The World Bank, Washington, DC.

Byamugisha, F. 2013. *Securing Africa's Land for Shared Prosperity. A Programme to Scale Up Reforms and Investment*, The World Bank, Washington, DC.

Cotula, L. and Mathieu, P. 2008. *Legal Empowerment in Practice; Using Legal Tools to Secure Land Rights in Africa*, IIED, London.
Dale, P. and Baldwin, R. 2000. Emerging Land Markets in Central and Eastern Europe. In: Csaki and Lerman (eds.), *Structural Change in the Farming Sectors in Central and Eastern Europe*, WB Technical Paper, No. 465.
Deininger, H., Hilhors, T. and Songwe, V. 2014. *Identifying and Addressing Land Governance Constraints to Support Intensification and Land Market Operation: Evidence from 10 African Countries. Food Policy*, Elsevier.
Deininger, K. 2003. *Land Policies for Growth and Poverty Reduction*, World Bank. The World Bank, Washington, DC.
Fairley, E. C. 2012. Upholding Customary Land Rights Through Formalization: Evidence from Tanzania's Program of Land Reform. *A Paper Presented at World Bank annual Conference on Land and Poverty*, Washington, DC.
Fitzgerald, H. 2017. Wearing an Amulet: Land Titling and Tenure (in)Security in Tanzania. PhD dissertation, University of Maynooth, Ireland.
Hart, A., Tumsifu, E., Nguni, W., Malley, Z., Masha, R. and Buck, L. 2014. *Participatory Land Use Planning to Support Tanzanian Farmer and Pastoralist Investment. Experiences from Mbarali Districts, Mbeya Region, Tanzania*, International Land Coalition, Rome, Italy.
Huggins, C. 2016. Village Land Use Planning and Commercialization of Land in Tanzania. Land Governance for Equitable and Sustainable Development. *Research Brief* 01.
International Land Coalition. 2013. *Rangelands Village Land Use Planning in Rangelands in Tanzania: Good Practice and Lessons Learned*, International Land Coalition, Rome, Italy.
Kimbi, E., Madoffe, S., Materu, C. J., Mbwile, R. P., Mwambene, P. L., Mwaiganju, A. and Udo, H. F. 2014. Assessing Dynamics of Force Livestock Movement, Livelihoods and Future Development Options for Pastoralists/ Agro-Pastoralists in Ruvuma and Lindi Region in the Southern Tanzania.
Kironde, J. M. 1995. Access to Land by the Urban Poor in Tanzania: Some Findings from Dar es Salaam. *Environment and Urbanization* 7(1): 77–96.
Knight, R. S. 2010. Statutory Recognition of Customary Land Rights in Africa: An Investigation into Best Practices for Lawmaking and Implementation. *FAO Legislative Study.*
Kosyando, Ole. L. M. 2006. *MKURABITA and Implementation of the Village Land Law- Act No.5 of 1999: A Participation Report of the Pilot Project Handeni District, September 18-December 8*, TAPHGO, Arusha.
Krantz, L. 2015. Securing Customary Land Rights in Sub- Saharan Africa. Learning from New Approaches to Land Tenure Reform. *Working in Human Geography* 2015: 1. Department of Economy and society, Göteborgs Universitet.
Kurwakumire, E. 2013. Towards the Design of a Pro-Poor Land Adjudication Procedure for Communal Land South Africa Surveying+ Geomatics Indaba (SASGI). *Proceedings 2013.*
Kushoka, N. A. 2011. Land Use Plan and Farmers- Pastoralists Conflict in Mvomero District: Its Implications on Household Food Production. Master dissertation, Sokoine University of Agriculture, Tanzania.
Lemmen, C., Zevenbergen, J. A., Lengoiboni, M., Deininger, K. and Burn, T. 2009. *First Experiences with High Resolution Imagery Based Adjudication Approach in Ethiopia*, Delft University of Technology, Netherlands.
Lugalla, J. L. P. 2010. Why Planning Does Not Work? Land Use Planning and Residents Right in Tanzania. *Review of African Political Economy* 37(125): 381–386.
Luoge, F. 2015. The Performance of the Land Reform Sub-Component in the Context the of World Bank Project in Tanzania. Land Evaluation Project Evaluation Report 2006-2015.
Mango, G. and Kalenzi, D. 2011. *Report on the Study to Develop a Strategy for Establishing Cost Effective Land Use Plan in Iringa and Njombe Regional*, Government Printers, Dar es Salaam, Tanzania.
Manji, A. 2006. *The Politics of Land Reform in Africa: From Communal Tenure to Free Markets*, Zed Books Ltd, London.
Massay, G. E. 2016. *Tanzania's Village Land Act, 15 Years on Tanzania Natural Resource Forum*, Arusha, Tanzania.
Milder, J. C., Buck, L. E., Hart, A. K., Scherr, S. J. and Shames, S. A. 2013. *A Framework for Agriculture Green Growth: Green Print for the Southern Agriculture Growth Corridor of Tanzania (SAGCOT)*, Dar es Salaam Centre.
National Land Use Lanning Commission. 2011. *Guideline for Participatory Village Land Use Planning Administration and Management in Tanzania*, 2nd Edition, Ministry of Lands and Human Development, Dar es Salaam.

Pedersen, R. H. 2010. *Tanzania's Land Law Reform: The Implementation Challenges, Danish Institute for International Studies*, Working Paper 2010: 37.

Pedersen, R. H. 2015. A Less Gendered Access to Land? The Impact of Tanzania's New Wave of Land Reform. *Development Policy Review* 33(4): 415–432.

Peters, P. E. 2004. Inequality and Social Conflict Over Land in Africa. *Journal of Agrarian Change* 4(3): 269–314.

Platteau, J. P. 1996. The Evolutionary Theory of Land Rights as Applied to sub-Saharan Africa: A Critical Assessment. *Development and Change* 27(1), 29–86.

Sangeda, A. Z. and Malole, J. L. 2013. Tanzanian Rangeland in Changing Climate: Impacts, Adaptations and Mitigation. *Net Journals* 2(1): 1–10.

Schreiber, L. 2017. Registering Rural Rights: Village Land Titling in Tanzania, 2008–2017, Innovation for Successful Societies, Princeton; University, http://successfulsocieties.princeton.edu, accessed on 20th March 2019.

Smith, R. E. 2003. Land Tenure Reform in Africa: A Shift to the Defensive. *Progress in Development Studies* 3(3): 210–222.

TFCG. 2015. Securing Village Land Certificates and Acquisition of Certificates of Customary Right of Occupancy: A case study of 10 villages in Kilosa, Mpwapwa, Lindi and Rufiji Districts. TFCG Technical Report 50.

Ujamaa Community Resource Team. 2010. *Participatory Land Use Planning as a Tool for Community Empowerment in Northern Tanzania Gatekeeper Series No.147,* IIED, London.

United Republic of Tanzania. 1999. *Village Land Act No.5,* Government Printers, Dar es Salaam, Tanzania.

United Republic of Tanzania. 2001. *Village Land Act Regulation*, Government Printers, Dar es Salaam, Tanzania.

United Republic of Tanzania. 2004. *Environmental Management Act 2004*. Government Printers, Dar es Salaam, Tanzania.

United Republic of Tanzania. 2009. *Water Resource Management Act 2009*. Government Printers, Dar es Salaam, Tanzania.

USAID. 2011. *Performance Evaluation of Mobile Application to Secure Tenure (MAST) Pilot*. USAID Country Report: Property Rights and Resource Governance, Tanzania.

Wily, L. A. 2011. *Customary Land Tenure in the Modern World. Right to Resources in Crisis*. Working paper. Rights and Resources Initiative, Washington, DC. Series: Rights to Resources in Crisis: Reviewing the Fate of Customary Tenure in Africa, Brief #1 of 5.

World Bank. 2003. Land Policies for Growth and Poverty Reduction. *The World Development, Washington* 18(5), 659–671.

Zein, T. 2017. Mass Registration of Land Parcel Using FIT for Purpose Land Administration; Procedures and Methods. *Paper prepared for presentation at 2017 World Bank conference on land and poverty-Washington DC*, March 20–24, 2017.

Zevenbergen, J. A. 2002. *Systems of Land Registration Aspects and Effects NCG, Nederlandse Commissievoor Geodesie*. Netherlands Geodetic Commission, Delft, the Netherlands.

10 Urban Identities in the Context of Land Management

Evidence from Urban Development Processes in Kibuga in Kampala, Uganda

Pamela Durán-Díaz, Priscilla Tusiime, and Walter Dachaga

CONTENTS

10.1 INTRODUCTION

The rapid urban growth and urban population explosion experienced by world cities come with tremendous challenges, such as the loss of local cultural heritage in the struggle to attain the highest and best use of every square meter of urban land, resulting in an urban life devoid of identity.

> Culture is key to what makes cities attractive, creative and sustainable. History shows that culture is at the heart of urban development, evidenced through cultural landmarks, heritage and traditions. Without culture, cities as vibrant life-spaces do not exist; they are merely concrete and steel constructions, prone to social degradation and fracture. It is culture that makes the difference. It is culture that defines the city as what the ancient Romans called the *civitas*, a coherent social complex, the collective body of all citizens (UNESCO, 2016, p.17).

Davies, Whimster, and Clayton (2009) report that people who live in areas rich in historic buildings tend to have a stronger sense of place. In other words, a city's spatial and architectural fabric that manifests its intangible historical and cultural values fosters a sense of identity and place, which, according to Chigbu (2013), promotes development. Nevertheless, in a highly urbanizing world, cities preparing to host the urban populace pursue land value maximization through interventions such as urban planning, urban regeneration (renewal), modern (smart) cities, slum upgrades, and housing infrastructural development. Within this context, it is a common phenomenon for old cities and historic buildings to make way for modern buildings and urban infrastructure in the quest for urban development. Conversely, there is a growing concern about the decline or near-loss of local cultural identity in modern cities. While organizations such as UNESCO have spearheaded the safeguarding of cultural heritage especially in the cities, it is important to understand how cities' spatial structural growth leads to transformation of social structure and family kinship as well as the decline in cultural activity practice, norms, and values and the sense of belonging which this study collectively identifies as the urban intangibles. The design of the cities by professionals such as architects, engineers, and urban planners are usually influenced by factors such as funding, funders' (clients) desires, time for the project, different policies, and building regulations from municipal councils. Therefore, matters of how cities affect the native communities' social and cultural cohesion are hardly considered, and where they are considered, are challenging to manage. Hence, this chapter focuses on the state of urban intangibles amidst urban development processes by observing the progressive spatial and social changes of delimited areas of Kampala in Uganda throughout time. It examines how spatial transformation of the city is deteriorating the intangible heritage of the local people and leading to a loosely coherent society marked by individualism, secularization, disorganization, and lack of solidarity. This was achieved by asking three key questions: 1) how did the physical spatial structure of Kampala evolve in terms of morphological structure and road networks from 1858 to 2016? 2) How did the spatial structural growth and heterogeneity of the city transform local land use, cultural activities, sense of belonging, and values of the Baganda in Kibuga? 3) What are the challenges faced by built environment professionals in accounting for urban intangibles in the design of modern cities in pursuit of urban development? The answers to these questions provide the basis for proffering suggestions on sustaining urban intangibles in modern cities. Consequently, the morphological structure, the road networks, the land uses, the cultural activities, the sense of place/belonging, and the social values/norms were studied in two heritage sites of Kampala within a bounded timeframe, from 1858 to 2016, by collecting old maps and pictures and conducting interviews and focus group discussions. The chapter is organized as follows: the foregoing section, which is the introduction, is followed (Section 10.2) by a theoretical perspective to deconstruct the concept of urban intangibles. The next section consisted of the methodology that was used to examine the state of urban intangibles in Kampala. Next, findings of the case study were presented and discussed (Section 10.4), suggestions were outlined for safeguarding urban intangibles (Section 10.5), and conclusions drawn (Section 10.6).

10.2 UNDERSTANDING URBAN INTANGIBLES: THEORETICAL PERSPECTIVES

According to Scott and Storper (2015), urbanization touches many social, cultural, and political/administrative dimensions of human life; and as a result, it has powerful feedback effects not only on economic development, but also on society. The morphology of urban spaces in cities are the result of multiple layers of complexity, encompassing both the tangible built environment and intangible cultural values that together influence the living heritage that forms the spirit of place (Garagnani, Arteaga, and Bravo, n.d). To emphasize this, Taylor (2008) noted that one of the deepest needs of humans is a sense of identity and belonging, which makes urban landscapes cultural constructs in which our sense of place and memories inhere, and a repository of intangible values and human meanings that nurture our very existence. Conzen (2009) also argued that the morphological features of urban places at all scales can be reduced to a logical system of explanation, which can lead to an incisive and nuanced understanding of the relationship between urban communities and the physical fabric they create and recreate around them as social needs change over time. Drawing from this line of thinking, the urban intangibles concept is a pro-conservationist urban concept that advances cultural and social sensitive urban development. It uses the term urban intangibles to describe the aspects of urbanism that lack physical form but give cultural identity to urban space in urban design to differentiate cities from one another. The concept emphasizes the relation between everyday life activities, social norms, cultural activities, beliefs, local land uses of the urban dweller, and their natural and built environment. Ogaily (2015) argues that the vertical expansion (high-rise modern buildings) of Dubai city is dwarfing and phasing out the historical traditional Arabic fabric of the city together with the religious-cultural values, while the rapid urbanization is eliminating the transitional periods which historical towns enabled for culture to evolve and allow each generation to experience vibrant communities with a strong identity. Chigbu (2013) used the term sense of place to portray the intangible relationships people share with their place of habitation, usually based on their feelings about land-based practices, cultural patterns, values, attitudes, traditions, and customs they share in connection with their environment. Kubat and Kürkçüoğlu (2014) described urban spaces as social constructs and each area's culture and social relations have deep effects on urban structure. Orbasli (2002) described this relationship in his book *Tourists in Historic Towns*, in which he examines the relationship of culture, heritage, conservation, and tourism development in historic towns and urban centers, debating the impact of tourism on historic towns and the role tourism plays in conservation and urban continuity. Muratori (1959) expressed this idea in a type—urban tissue—organism—operative history relation. According to Muratori, a certain type of building could not be identified except within a particular application in the urban tissue (urban form or structure), the urban tissue could not be identified except in its involving context in the urban organism (the typology of urban networks that connect people and places), and the urban organism would only become real in its historical dimension, as part of a temporal construction that is always grounded on the conditions suggested by the past. Hence a relation between history and planning and architecture, in which cultural identity is paramount. Oliveira (2016) in a synthesis of images of cities posed two key questions that underscore how people relate with the city in non-physical ways: 1) what does the city's form mean to the people who live there?, and 2) can the city planner make the city's image more vivid and memorable to the city-dweller? Other terms have been used to express intangibles in urban context, including culture-led regeneration (Aykac, 2019), sense of community (Tan et al., 2018), musealization (Nelle, 2009), tourism-led regeneration (Ozden, 2008), and cultural landscapes (Taylor, 2008). In all the ideas expressed, the theme that runs through them is the conservation of intangible cultural heritage in the urban fabric that people identify with in urban development processes. These intangibles are what UNESCO (2003) describe as the practices, representations, expressions, knowledge, skills—as well as the instruments, objects, artefacts, and cultural spaces associated therewith—that communities, groups, and, in some cases, individuals recognize as part of their cultural heritage. Therefore, within our context, urban intangibles connote the social and cultural practices, representations, expressions, knowledge, and skills, that define the urban space and form of urban communities. Social and cultural identity with urban forms is necessary for community approval of urban development projects, as well as collaborative, participatory, and inclusive urban land use planning and decision-making (Chigbu, Alemayehu, and Dachaga, 2019). Viewed from a responsible land management perspective, the urban intangibles concept advances urban development as an intervention that relate to positive changes in socio-spatial relations, perceptions and beliefs, and behavior (de Vries and Chigbu, 2017). Therefore, an understanding of how urban expansion affects social values, norms, cultural identity, and local land use and practices is necessary for creating responsible cities.

10.3 METHODOLOGY

In line with the objectives of this chapter, a case study approach was adopted to undertake a qualitative study of Kibuga (meaning "city" in the Luganda language), the city for the Buganda Kingdom that was merged with and became part of the present-day Kampala city, in order to understand the transformation of its spatial structure and how that affects the intangible heritage of the Buganda Kingdom. It involved the collection of old maps, pictures, and any descriptive archived information on the spatial configuration of Kibuga, observation of the territorial dynamics and cultural heritage sites, and identifying the intangible assets through a focus group discussion. In addition, interviews were used to study variables such as effects of the spatial transformation on the memory, the collective imagination, the local identity, and the sense of belonging that were associated with the urban set-up. In-depth interviews were also conducted with selected key professionals of the built environment to understand the challenges of incorporating urban intangibles (for instance, the cultural heritage of local communities) into new urban planning and design.

Eight (8) persons were successfully interviewed—two (2) architects on the board of the Physical Planning Building Approval at Kampala City Council Authority (KCCA); three (3) from the Department of Physical Planning (at the Ministry of Lands, Housing and Urban Development (MLHUD)); two (2) assistants to the Managing Director of the Tourism Board from Mengo; one (1) Managing Director from the Historic Resources Conservation Initiatives (HRCI), and two (2) academics from University of Makerere from the spatial planning field.

Tracing the evolution of identity in the local community was achieved through focus group discussions, in which the selection criteria of the participants relied on their age and knowledge of intangible heritage. These were supported by site visits and observations.

The geographical scope was narrowed down from the whole of Kampala city to the Rubaga division, part of Kawempe, and the Makindye division; that is, the west of the city, since most of the historical events concerning the Buganda Kingdom and the city took place in this region.

10.4 FINDINGS

10.4.1 The Morphological Structure of the Buganda Kingdom before Formal Planning

Prior to formal planning, as can be seen in Figure 10.1, the area had nine (9) hills, separated by rivers and swamps in the valley. The river Bwaisa is located to the north, on the periphery of Kibuga, then the river Nabisasiro is located south of river Bwaisa. In between these rivers, from the west to the northeast are the hills of Lugala, Lubya, Kasubi, Makerere, and Mulago. Then to the south of the river Nabisasiro is the river Nalukolongo with its tributary river Wakaliga. In between these two rivers is a cluster of hills: Rubaga, Nakasero, Mengo, and Old Kampala. Old Kampala, also known as Kampala Hill, was the nucleus of Uganda's capital (Kampala), bordered by Makerere, Nakasero, Mengo, and Namirembe to the north, east, south, and west respectively. Kampala Hill, which used to be a favorite hunting ground of the King of Buganda, attained its name of Old Kampala when the city expanded to the neighboring hills. A swamp/wetland then makes a boundary for the Kibuga from the north, west, and south. Major landmarks such as religious institutions, cultural institutions, educational institutions, and health institutions are located on hilltops. Feeder roads were on the hillsides, with the settlements and major roads and highways going through the valleys.

In 1858, the first indication of spatial organization was from Kagwa's drawing (Gutkind, 1960). It marked the genesis of a more organized urban morphology. There was a main road beginning directly from the King's palace at the top of the hill and going down across the swamp. Other secondary and minor roads also sprouted from the hill top to the valley, and they had adjoining narrow roads as well. Between 1891 and 1899, the road network precisely showed roads going directly to and from the top of the hill and directly crossing the valleys (see Figure 10.2). The ring road was then at Mengo, going around the King's palace, and not at Kasubi. Rubaga and Namirembe Hills hosted the cathedrals, while a mosque was constructed at Kibuli Hill. Markets were built at the base of the hills at Namirembe and Old Kampala.

10.4.2 The Advent of Formal Planning and Urban Infrastructure in Kibuga

Implementation of planning schemes started by 1919, and this brought about changes in the urban tissue of several towns, including the old Kampala town, the new Kampala township, Namirembe, and Rubaga. From Figure 10.3, it can be seen that the road networks don't seem to follow a defined pattern. However, closer observation shows that feeder roads are concentrated on the hills above 1.106 m, but not below because settlements were mainly concentrated around the slopes of the hills

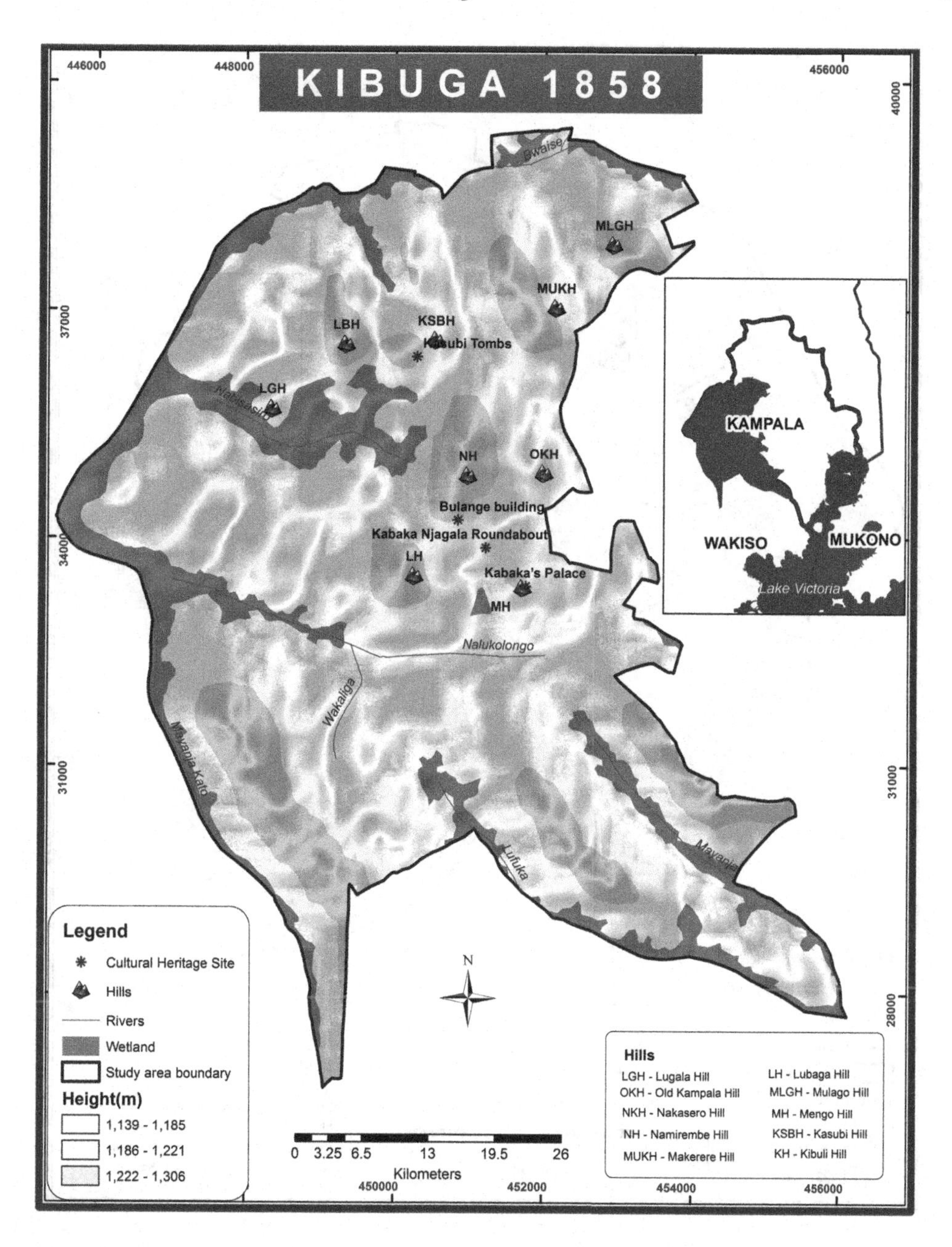

FIGURE 10.1 Morphology and different capitals of Buganda kings. Source: Tusiime, 2017.

while crop cultivation and animal rearing was done at the base of the hills, closer to the swamps. Thus, there was no need to create feeder roads. Some main roads and highways built by the government go through the valleys, as it is a cheaper option than building over the hill. Residential buildings were spread from the hillside into the valleys. A few individual houses were enclosed within a boundary and a gate. There sprang up a mixture of commercial complexes and/or buildings, some of which were mixed-use type, with shops on the ground floor and residential units on the upper floors. Educational faculties (schools and institutes), health centers, and hospitals were built, and industries were located in the valleys or in the swamps. According to the physical planner at MLHUD, parts of the swamps are drained, filled with soil, and then built over with residential structures and industries. However, the hilltops of Kasubi and Lubiri remained the green areas of present-day Kampala.

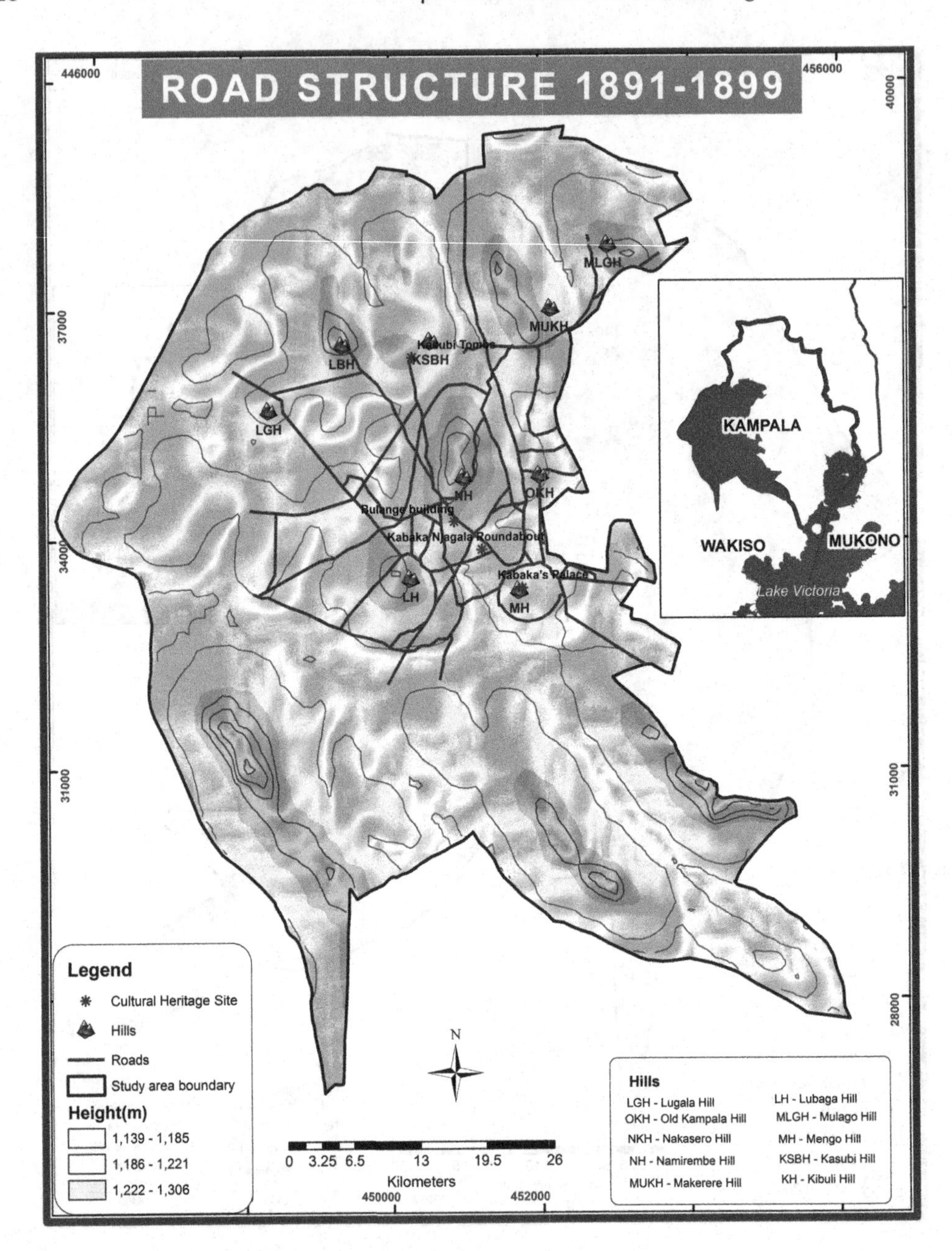

FIGURE 10.2 Road structure in Kibuga 1891–1899. Source: digitized by Tusiime, 2017.

All other hilltops accommodate institutions such as universities, hospitals, cathedrals, or mosques. These developments gave the area a new urban tissue and changed the spatial structure of the town. Consequently, the new urban form brought about changes, including changes in cultural land uses, cultural practices, and sense of belonging, as well as cultural norms and values.

10.4.3 Morphological Transformation and its Effects on Urban Intangibles

- **Fading historical land uses:** The Baganda people were predominantly agriculturalists. Farming was carried out on the hill slopes due to the fertile soils and proximity to the water in the valleys. Crop cultivation and animal rearing are still carried out in the backyards of

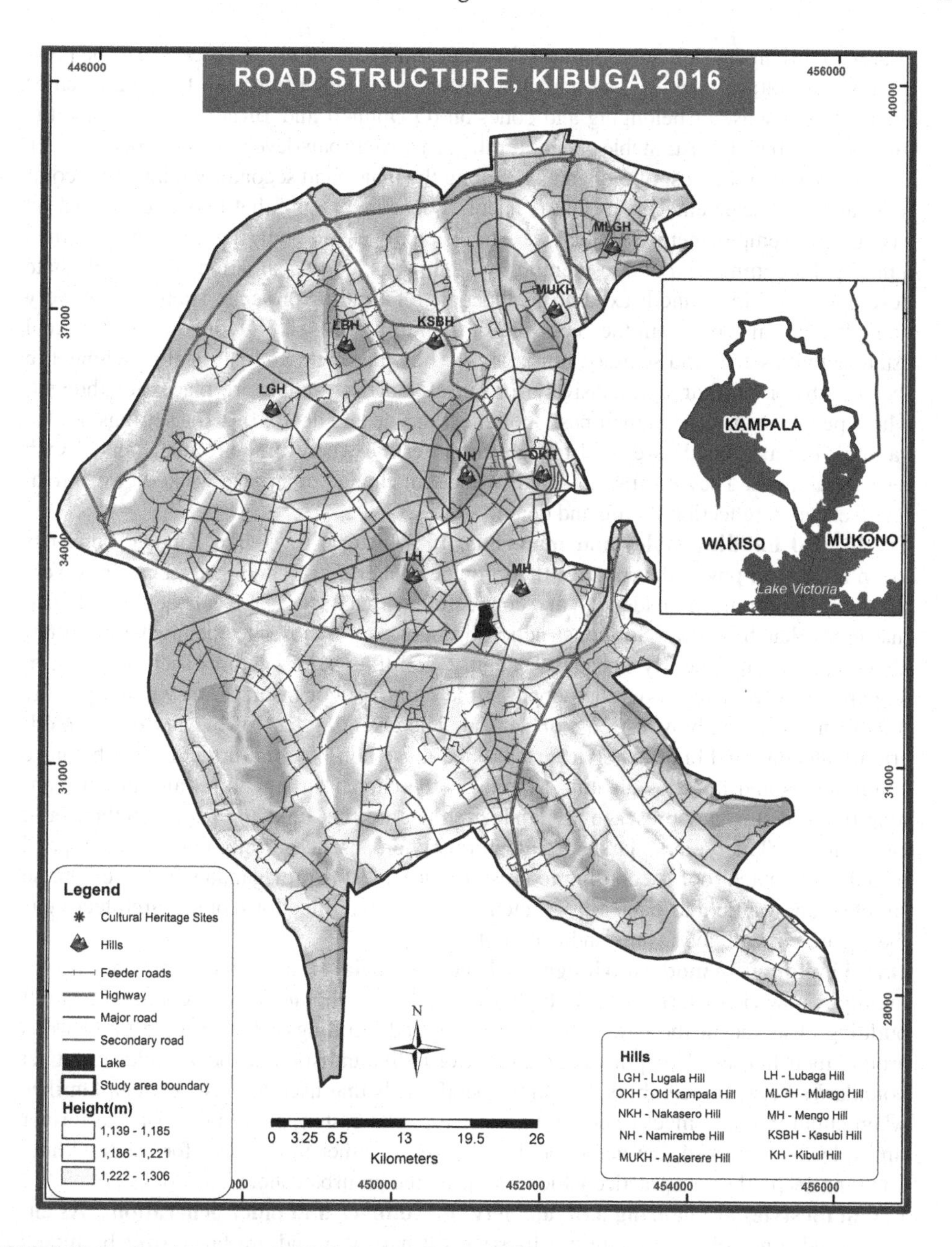

FIGURE 10.3 Road structure in 2016. Source: Tusiime, 2017.

some individual properties. Due to urban expansion, Mengo Hill and Kasubi Hill are the only non-built-up areas, as they are considered private property of the Buganda Kingdom. An analysis of land use in Kibuga showed that areas where subsistence farming is still taking place fall within the boundaries of the Kasubi Tombs and the Lubiri and a few vacant lands in the city. The rest of the area is represented by highly dense settlements which are gradually trespassing the cultural boundaries of the Kasubi Tombs and the Lubiri. The Tombs are an important spiritual and political site to the Baganda and tell both the history and culture of the people. Culture is recognized as an integral part of people's wellbeing, and local development and equity were linked with acknowledging diversity in cultural heritages and values (Duxbury, Hosagrahar, and Pascual, 2016). Within the context of urban development, cultural diversity is a driver of urban success and the preservation of cultural identity and activities fosters social inclusion and dialogue in cities by creating a sense of belonging and cohesion (O'Donnell and Turner, 2012). Hence the Tombs contribute to sustainable and culturally sensitive urban development (Nocca, 2017). It was observed that most of the land uses along the major and secondary roads were commercial, small-scale enterprises and residential apartments and that people depended on salaries from employment. Sites that were traditionally used as burial grounds have gained a market for commercial and residential uses. This has led to a situation where real estate developers and city councils exhume dead people for building projects. Traditionally every home had burial grounds in the back yard of the family property. Owing to booming real estate activities and land scarcity, families now buy land with no historical attachment in the suburbs or rural areas exclusively for burial. Burial grounds are places for thought, where people connect with their past and reflect on life and death. In some settings, graveyards or burial grounds are used to fight urban crime where they are interspersed with residential areas. They portray spirituality and can change otherwise wicked or criminal mindsets upon reflection on life and death.

- **Traditional building styles and materials:** In a historic town like Kibuga, buildings mean more than physical structures that provide shelter. For many locals, buildings were means of communication, identity, and connection with nature. The local people explained that using thatch or straw for construction brought nature closer to them. The buildings also communicated family lineage or identity, and the use of symbols such as a lion or leopard spoke volumes about the people in the building. For instance, they could portray inhabitants as being brave, strong, or wicked. The changing face of the city due to modern architecture and building styles has eroded these non-physical urban forms that gave communities and families identity. Moreover, the current architectural and urban planning trends distort any presence of Ganda architecture that links the people to their built environment. The typical Ganda architectural style consisted of circular houses, topped by a domed thatch roof. According to Otiso (2006), most rural Ugandan houses comprise simple round wattle-and-daub grass-thatched huts—a modified form of this architecture is the square hut with mud walls and iron roofs.
- **Gated buildings in modern Kibuga are breaking social ties:** A once homogenous community is now characterized by both physical and social boundaries. A common form of building observed in the new urban fabric is gated buildings where people build walls around their houses. This practice is a mark of individualism and social exclusion. It set boundaries between neighbors, breaking social bonds that used to exist between families when children could meet freely and socialize or share stories and tales. Unlike former times, locals remarked that neighbors have stayed together in one area for years without knowing each other—a practice which is considered an urban shock.
- **Urban lifestyles are causing a divide between younger and older generations:** As cities expand, lines of unique cultural lifestyle get narrower and are taken over by mixed lifestyles. This is what is happening and causing divide among young and old generations in Kibuga. While younger generations within the urban setting feel enlightened and see certain lifestyles as colloquial, the older generation are mindful of the social value associated with those lifestyles. They range from dressing styles, eating styles, and greetings, to moral conduct. The difference in ideologies is exacerbated by the urban experiences of young generations who learn new urban lifestyles and unlearn acceptable local lifestyles.
- **Loss of sense of place/community:** Communal living creates a sense of belonging and identity. However, the new urban fabric in Kibuga does not promote such belonging and identity. In previous times, what kept communities together were courtyards, open spaces, and cultural grounds. They acted as meeting points for entertainment, cultural education, and community development initiatives. They created an environment where everyone

felt part of a community and actively participating in community life. The new urban design coupled with individualism and urban stress makes people keep to themselves and this break communal ties. Again, mixed cultures and lifestyles due to city expansion make it difficult to define identifiable communities. This is because the culture and local lifestyle that bonded people is taken over by urban lifestyle of individualism. One source of a sense of belonging and community for the locals in Kibuga is the cultural heritage sites especially Kasubi Tombs, due to its unadulterated representation of the Ganda architecture as well being a burial ground for the last four kings of Buganda. There is a strong intangible value attached to the Tombs. According to Lugard (2013), regardless of the reduced autonomy of the Buganda Kingdom, solidarity is derived from the traditional administrative system in place and the reality of having a king. New urban designs and architecture are, however, degrading this sense of attachment associated with these traditional monuments. Also, people decried the breakdown of family bonds. There were complaints that relatives now live far from each other and hardly communicate or make visits. Also, the daily routine in the city leaves families with little time to spend with others. It was common in the pre-urban days for in-laws to reside in the same space or use the same furniture in a house with the family, which strengthened their bonds. Yet for others, speaking their native language defines, bonds, and set communities apart and makes people feel a part of it. All these are, however, things of the past due to a new urban tissue that arose as the city expand.

- **Highly auto-locomotive urban fabric discourages social interaction:** People build ties as they interact more; however, the new urban morphology does not encourage such interactions in Kibuga. Increased urban infrastructure has decreased walking and increased the use of automobiles. Most roads do not make provisions for pedestrians and this reduces human interaction within the urban setting.
- **New urban design discourages cultural activities and deteriorates values and norms:** According to Duxbury, Hosagrahar, and Pascual (2016), culture is an integral part of human development which constitutes the fabric for the dynamic construction of individual and collective identities. They noted that the active participation of people in local cultural activities (such as poetry, dance, sculpture, theatre, music, etc.) improves their quality of life and wellbeing and enhances life opportunities and options. This notwithstanding, parcellation and limited urban space discourages the practice of cultural activities especially those requiring open space. This has both social and spatial dimensions. Spatially the new urban fabric fails to meet the space requirements. Individuals living on properties under 60 m^2 cannot afford the complexities of cultural rituals. Only those that can be carried out in confines of small private spaces are still done. Socially, people shy away from practicing these activities for fear of mockery from people who consider themselves urbanized and enlightened. On the other hand, coupled with religious roots in the country, people feel ashamed to be identified with certain cultural activities in the urban center, and sometimes find such practices heathen and repulsive. For such people, especially young generations, such cultural activities have lost their relevance. The natural landscape/environment, such as earth, grass, trees, fire, or wind that favor these cultural activities are replaced by concrete landscapes. Cultural values and norms are still upheld by most of the older generation due to their conservative nature and the long practice of these cultural norms and values in their lives. Therefore, regardless of living in the city, they still had memories of the past. Nevertheless, busy and stressful urban lifestyles inhibit parents from passing down these norms and values to their young ones.

In sum, the morphological transformations transformed traditional land uses, led to loss of indigenous architectural building styles, loss of sense of place, breakdown of social ties and interaction, decline in cultural values and activities, and generational divide between the young and old.

10.4.4 The Challenge of Fostering Urban Intangibles in a Changing Urban Morphology

Fostering urban intangibles is a question of maintaining the right balance between modernization in urban design and conservation of intangible social and cultural heritage. To what extent modernity is enough, and how much intangible socio-cultural heritage should be conserved, is a question that is debatable. This notwithstanding, fostering urban intangibles has setbacks that are not only peculiar to Kibuga but also to many historical towns. First of all, built environment professionals blame the westernized skyline of Kampala on durability and easy access to foreign building materials, compared to less durable materials. Secondly, the government policy for building regulations doesn't foster culture. It also doesn't include mandates such as Cultural Impact Assessment reports for building projects, and thus implementation on the part of the local authorities is difficult. Furthermore, building projects also look at reducing cost and maximizing commercial gain, which poses a threat to cultural heritage sites as they are seen as not promising substantial economic returns, especially in a context where the tourism market is not vibrant. Finally, the architectural curriculum does not focus on local architecture and designs. This is one reason why typical Bugandan culture is fading in the new urban fabric of Kibuga.

10.5 SUGGESTIONS FOR FOSTERING URBAN INTANGIBLES IN NEW URBAN DESIGN

Advancing urban intangibles is an urban sustainability approach which strikes a balance between economic development and maintaining social structures and environmental elements. Achieving this balance is not usually easy, as economic gains often offset social and environmental considerations. Beset with negative consequences of urban development, this chapter gives some suggestions that can set the right environment for conserving urban intangibles:

a. Adopt the cultural-mapping methodology in order to skillfully include culture in the planning process of the city by inventorying, recording, and making use of GIS and cartography in the indigenous communities and historic towns.
b. Engage participatory urban planning and design approaches to meet local social and cultural needs. This approach helps in needs identification, creates a sense of belonging, and fosters community endorsement of urban development interventions.
c. It is impossible to stop modernization, but we can find a way to pass on culture in an interesting manner to the younger generation. Make provisions for creativity and incorporation of culture in the modern environment, thus fostering the continuity and diversity of cultural identity. With the increasing rates of migration and cities becoming more heterogeneous in nature, it is important that individuals are not stripped of their identity, but encourage social inclusion as advocated for by the new principles of the Urban Paradigm under the UN-HABITAT program. These principles stipulate that the cities we need are socially inclusive and engaging; affordable, accessible, and equitable; economically vibrant and inclusive; collectively managed and democratically governed; foster cohesive territorial development; regenerative and resilient; have shared identities and a sense of place; well-planned, walkable, and transit-friendly; safe, healthy, and promote wellbeing; and learn and innovate (UN-HABITAT, 2016).
d. Making the intangible tangible by integrating new digital augmented reality technology with physical innovative solutions of place activation. The idea is to unearth the latent artistic cultural layers and interpretation of a site. To do this would require a commitment by developers, councils, and design professionals to engage artists and cultural planners as an integral part of creating responsive urban designs.

e. Formulate and implement culturally sensitive urban development policies and regulations, including giving incentives for culturally sensitive designs and requiring cultural impact assessments for urban renewal and expansion projects.
f. Considering the issue of westernized building trends, professional architects, and urban designers should strive towards working with traditional architecture in the modern era. A starting point would be an introduction of local or culture-based architectural curriculums in universities, colleges, vocational, and technical institutes.
g. Finally, substantial amount of finances should be geared toward the construction of traditional cultural heritage facilities to mimic cultural settings, in a bid to promote the continued practice of cultural activities, arts, and norms. This action would not only foster creativity in the younger generation, but would advance the local economy's sustainability based on the use of native resources and knowledge, with a positive impact on the natural, intangible, and built environment.

10.6 CONCLUSION

This chapter sought to reveal how the spatial structure of the city evolved from 1858 to 2016 and how this evolution brought about the deterioration of what is described as urban intangibles—cultural values and norms, cultural activities, and sense of belonging. It is noteworthy that there is an identity relation between the built environment and the people who inhabit it. This relation is expressed in how people identify and interact with their environments in both physical and non-physical forms. While urban development is a physical thing, it has social dimensions that are reflected in how the changing urban face affects social relations and cultural values and norms. Understanding this relation is necessary for rethinking urban regeneration and historical urban patterns. Hence, the insights offered in this chapter advance the pursuit of urban transformation in a manner that transfers intangible social and cultural values to the present.

REFERENCES

Aykaç, P. 2019. Musealisation as a strategy for the reconstruction of an idealised Ottoman past: Istanbul's Sultanahmet district as a 'museum-quarter'. *International Journal of Heritage Studies* 25(2): 160–177.
Chigbu, U.E. 2013. Fostering rural sense of place: The missing piece in Uturu, Nigeria. *Development in Practice* 23(2): 264–277.
Chigbu, U.E., Z. Alemayehu, and W. Dachaga. 2019. Uncovering land tenure insecurities: Tips for tenure responsive land-use planning in Ethiopia. *Development in Practice* 29(3): 371–383.
Conzen, M.P. 2009. Conzen MRG 1960: Alnwick, Northumberland response. *Progress in Human Geography* 33: 862–864.
Davies, J., R. Whimster, and L. Clayton. 2009. *Heritage Count 2009*. London: English Heritage.
de Vries, W.T., and U.E. Chigbu. 2017. Responsible land management-concept and application in a territorial rural context. *fub. Flächenmanagement und Bodenordnung* 79(2): 65–73.
Duxbury, N., J. Hosagrahar, and J. Pascual. 2016. Why must culture be at the heart of sustainable urban development? *Agenda 21 for Culture*.
Garagnani, S., J. Arteaga, and L. Bravo. n.d. Understanding intangible cultural landscapes. *Parametricism vs. Materialism* 431.
Gutkind, P.C.W. 1960. Notes on the Kibuga of Buganda. *Uganda Journal* 24(1): 29–43.
Kubat, A.S., and E. Kürkçüoğlu. 2014. Morphological evolution of urban form components in the historical peninsula of Istanbul. *Plenary Sessions* 25: 170.
Lugard, L.F.J.D. 2013. *The Rise of Our East African Empire (1893): Early Efforts in Nyasaland and Uganda (volume 2, of 2 vols)*. Abingdon, Oxon: Routledge.
Muratori, S. 1959. Studies for an operating urban history of Venice. *Palladio*: 1–113.
Nelle, A.B. 2009. Museality in the urban context: An investigation of museality and musealisation processes in three Spanish-colonial world heritage towns. *Urban Design International* 14(3): 152–171.
Nocca, F. 2017. The role of cultural heritage in sustainable development: Multidimensional indicators as decision-making tool. *Sustainability* 9(10): 1882.

O'Donnell, P.M., and M. Turner. 2012. The historic urban landscape recommendation: A new UNESCO tool for a sustainable future. In: *Meeting of IFLA*, Cape Town.

Ogaily, A. 2015. Urban planning in Dubai; cultural and human scale context. In: *Annual Meeting for the Society of Council on Tall Buildings and Urban Habitat*, New York, October, 26–30.

Oliveira, V. 2016. *Urban Morphology: An Introduction to the Study of the Physical Form of Cities*. Switzerland: Springer International Publishing.

Orbasli, A. 2002. *Tourists in Historic Towns: Urban Conservation and Heritage Management*. London: Taylor & Francis.

Otiso, K.M. 2006. *Culture and Customs of Uganda*. London: Greenwood Publishing Group.

Ozden, P. 2008. An opportunity missed in tourism-led regeneration: Sulukule. In: *WIT Transactions on Ecology and the Environment*. WIT Press, London, UK.

Tan, S.K., S.H. Tan, Y.S. Kok, and S.W. Choon. 2018. Sense of place and sustainability of intangible cultural heritage–the case of George Town and Melaka. *Tourism Management* 67: 376–387.

Taylor, K. 2008. Landscape and memory: Cultural landscapes, intangible values and some thoughts on Asia. *ICOMOS Open Archive*: 1–14. Available at: http://openarchive.icomos.org/139/1/77-wrVW-272.pdf.

Tusiime, P. 2018. The influence of the built environment on Cultural Identity. MSc thesis, Technical University Munich (TUM), 63.

Scott, A.J., and M. Storper. 2015. The nature of cities: The scope and limits of urban theory. *International Journal of Urban and Regional Research* 39(1): 1–15.

UNESCO. 2003. *Convention for the Safeguarding of the Intangible Cultural Heritage*. Paris: UNESCO.

UNESCO. 2016. *Culture: Urban Future–Global Report on Culture for Sustainable Urban Development*. Paris: UNESCO.

UN-HABITAT. 2016. *The City we Need 2.0: Towards a New Urban Paradigm*. Nairobi: UN-HABITAT.

11 Authenticating Deeds/ Organizing Society

Considerations for Blockchain-Based Land Registries

Gianluca Miscione, Christine Richter, and Rafael Ziolkowski

CONTENTS

11.1 INTRODUCTION

For years now the hype surrounding blockchain as a new and more decentralized architecture to facilitate transactions has led people to search for cases of application for this emerging technology. Before thematizing the hype that has been propelling blockchain on the global stage, it is useful to mention the main novelty that blockchain brings: authentication of uniqueness of data in digital settings or "native authentication," i.e. without relying on an external/non-digital intermediary. Without this property, it could not be used for digital currency, which, of course, must be unforgeable. Other relevant, and related properties are: decentralization (even if no unique guarantor is often substituted by clusters of miners), immutability (records cannot be deleted), auditability (mathematically provable), and pseudonymity (via key-pair identification), according to Maurer and DuPont (2015).

Specifically regarding land administration, two salient aspects of blockchains motivate the interest for it: immutability of records, which promises to reduce corruption to make land management more responsible, and so-called smart-contracts, which can facilitate transactions involving multiple parties (Griggs et al., 2017).* Since digital innovation has not produced general theories that can direct the development of new technologies in organizations and societies, the innovation process remains idiosyncratic, i.e. based on people trying out things all around the world. Here, we discuss the prospects of blockchain for land registries from different theoretically informed angles. Those theories do not aim at predicting future successes. Rather, they provide distinct lenses to highlight different aspects and prospects of this technology in practice. Those insights may turn out to be useful for practitioners operating in very diverse settings. Although blockchain has the features of other global technologies like the Web, our aim is to make our views sensitive to the African context and developing economies more broadly.

In developing countries, it is not uncommon to have weak and often inconsistent systems to keep track of land rights. For instance, customary land ownership may be family- or clan-based rather than individual, like in Western cadastres and land registers. When disasters strike, legal land registries may be lost, as happened with the Haiti earthquake that left farmers fighting among each other. De Soto (2000), and international organizations, also stress the lost opportunity due to unregistered land, which cannot be used as collateral to grant people access to financial services. Other positive side effects of consistent land ownership registration have been put forward. For example, in Peru, clear registration of land ownership reduced the racketeering that farmers were exposed to in areas where coca leaves were cultivated.

So, what are the gridlocks that obstruct the consolidation of basic tools of land management like land registries? Of course, there is no exhaustive answer to that. What is worth highlighting here is that corruption, vote of exchange, speculation, and vested interests all undermine the basic agreements needed to establish a common land registry. In the domain of land registration, the promise that blockchain brings about is that of inherent immutability and transparency of the distributed ledger, which leaves less space for "maneuvering." Before introducing the cornerstones of blockchain as a technology in the Blockchain theoretical section (Section 11.3.1), we are setting the scene for this work, starting with the high expectations that surround blockchains.

11.2 SETTING THE SCENE

Why blockchain and why now? Agre (2003) states that the narratives in which new technologies always come wrapped do not explain technologies themselves, but the energy that propels them into societies. Blockchain, like other information technologies, can be considered a statement of intent towards a target audience, designed to initiate and guide action (Bekkers and Homburg, 2007), so they can also be seen as a myth (Mosco, 2004). Myths are tales providing shared frames of reference that enable individuals, groups, and organizations to deal with, or overlook, contradictions that cannot be fully resolved. From this perspective, the legitimation of blockchain applications derives from their persuasive power: whether the allure of blockchain mobilizes necessary resources from decision-makers, investors, developers, and users (Miscione, 2015). Homburg and Georgiadou (2009) explored how the narrative wrapping and legitimizing of spatial data infrastructures (SDI), a previous wave of technological innovation, traveled from North America to Europe, then to other continents since the 1990s. The kind of analysis they propose is discursive: SDI are conceptualized as a myth, able to organize human action and mobilize resources independently from the hard facts, often claimed to be the foundation of those projects.

The narrative of blockchain technology is recurrent and well-crafted: it originated during a global financial crisis and was built to overcome distrusted intermediaries like states and banks. Its most well-known application, which proved its functionality, has been for Bitcoin, the first and largest crypto-currency to date. Nakamoto (2008), the unknown and mysterious author of the Bitcoin

* In a nutshell, blockchains are databases whose integrity is guaranteed by nodes. Their independence limits the possibilities of tampering with the records. More technological details are provided in the 11.3.1 Blockchain section (Section 11.3.1).

seminal paper, together with others from the same crypto-libertarian circles, gave paramount importance to decentralization to circumvent intermediaries. A provocative paper by Atzori (2015) makes this attitude clear right from his article's title: "Blockchain technology and decentralized governance: Is the state still necessary?" The hype surrounding blockchain systems seemed endless until its peak: the crypto-currency market capitalized close to US$800 billion in winter 2017, not counting the initiation of a plethora of projects which explored the application of blockchain technology in domains other than finance. All these prove the forces that a convincing narrative can mobilize.

Bitcoin recently had its ten-year anniversary, establishing it as the first successful, apparently independent global financial infrastructure. If Bitcoin had not increased in value so dramatically, it is very likely it would have had neither the global outreach nor the security it nowadays has. So its hyped narrative had material, undeniable effects. One of the keys of its success is the incentive for miners to partake (1) in the distributed authentication process to seal a rightful transaction—e.g. avoiding double-spending/forgery—but also (2) in the maintenance of the ledger—e.g. continuing the longest, thus most reliable, chain of transactions. However, Bitcoin's governance problems are undeniable as well. The never-ending conflicts about its block size, with consequences also to the incentives for miners, show the short-sightedness of thinking that such a financial system would organize spontaneously, without requiring governance.

Moving beyond its origin in financial applications and start-ups of all sorts, cryptocurrencies have put blockchains on the agenda of many multinational companies as well as governments and international organizations, which explore potential applications of blockchain technology that would not have been tried otherwise. Land registries are one of those cases. So, the hype propelling blockchain is real, not because it predicts if this technology will be successful and for what purposes, but because it has real consequences for people and organizations trying to apply it to the most diverse domains and countries, including land administration in Africa.

Before moving our attention to three distinctive theoretically informed views on blockchain prospects for land registries, it is worth noting the common roots of De Soto's focus on land property and smart-contracts* as documented by DuPont (2019). De Soto started influencing the Peruvian economic system from the 1980s and 1990s, especially establishing land titles, which in turn helped in formalizing the Peruvian black economy, then gave impetus to economic development. Similar approaches were later embraced by international organizations like the International Monetary Fund and pushed to other developing countries. More recently, De Soto himself turned his attention to blockchain for land registration with a project in Georgia. In theory, pairing blockchain architecture with smart-contracts would facilitate a more dynamic economy. On the side of smart-contracts, it is remarkable to note that Szabo, a prominent figure in the crypto-libertarian scene, refers to land registration as prone to being forged if not maintained in a decentralized manner (DuPont, 2019). So the alliance of property rights and decentralized authentications are the cornerstones of his envisioned mode of organizing society.

In this chapter we do not speculate on the possible consequences of these views. The aim is to identify the basic principles of, and ideas behind, blockchain in dialogue with essential characteristics of an empirical context (i.e. land governance in Ghana) through the use of different theoretical lenses. The final aim is then to distill future research directions and questions that are both empirically relevant and theoretically informed.

The structure of this chapter reflects its theoretical foci and continues as follows: three theoretical angles are used to shed light on different prospects of blockchain for land registries in Ghana and countries with comparable land governance in Africa. Those theoretical cornerstones are: neo-institutionalism, structuration theory, and actor-network theory. After a brief introduction to each

* Self-enforcing code when defined criteria are met. The term "contract" is misleading; it rather regards encoded logic, which only in rare cases resembles contracts. Prominent use cases of smart-contracts are blockchain-based crowdfunding (Initial Coin Offerings), exchange of values, auctioning, or property rights management.

theoretical lens, each of them is considered sequentially in relation to: 1) blockchain technology, 2) Ghana's land governance context, 3) prospects for further research.

11.3 THEORETICAL ANGLE: NEOINSTITUTIONALISM

North (1990) defines institutions as the formal and informal rules that shape social interactions. In other words, they are akin to the rules of a game. In terms of neoinstitutional theory, institutions are defined as accepted social models informing human behavior. Institutional theory, compared to structuration theory or actor-network theory, is less sensitive to action, and emphasizes more the social models that explain and constrain patterns of individual and collective action. Neoinstitutionalism, especially the work of Powell and DiMaggio (1991), introduced the concept of an organizational field to draw a boundary around organizations engaged in the same kind of activities (e.g. projects for registering land, in our case). Within a field, it is common to observe the phenomena of isomorphism, even when rational choice models would predict differentiation processes based on outcome maximization efforts. Although institutions are often mentioned in research on land registries, actual use of institutional theory is quite rare in a field that is more sensitive to envisioned developments than to persistence; for an example of such work, see de Vries, Bennett, and Zevenbergen (2015).

From the cognate domain of SDI, Silva (2007) discusses how institutionalization of technology does not occur by decree. For the case of a land administration system in a Central American country, he highlights and discusses the institutional constraints for roll-out of a spatial information system. His qualitative study describes in detail the diverging approaches of the locals and information system promoters regarding the rationality of institutions, the link between the information system and work tasks, and historical resistance to change in the institutions that regulate land ownership. Prominently from this account, one sees how interorganizational cooperation is not created by an information system; rather it is a condition for the roll-out of the system.

To explain and manage blockchains for land registration in diverse settings like Ghana and other African countries, neoinstitutional theory has two strengths. Firstly, it can account for a variety of groups, across which technology use spans. Secondly, neoinstitutionalism has a peculiar explanatory capacity for phenomena that happen within an organizational field, even when no direct interaction between people in that field takes place. For example, different countries opt for similar modes of land registration under the pressures of imitation, rather than independent analyses. Indeed, isomorphism is already detectable to the extent that the idea of using blockchain in land registration was embraced by international organizations, tested in Georgia, and now tried out elsewhere across the same organizational field.

11.3.1 Blockchain

It is important to highlight some fundamental characteristics of blockchains to figure out how they may relate to land administration in a variety of institutional settings. Peters and Panayi (2015) classify blockchain systems along (1) whether the access to transactions is public (for everyone to see) or private (for selected parties only), and (2) whether validation of transactions is permissionless (any participating party can validate transactions) or permissioned (selected parties validate transactions), as summarized in Table 11.1. The parties are called "nodes" in a blockchain system.

Consider Bitcoin, the best-known and most widely spread blockchain system to date. Bitcoin was initiated at the same time as the beginning of the 2008 financial crisis as a system sidelining the centralization of power and avoiding external influence, such as corrupted and untrustworthy intermediaries. This is constitutive of its design; as shown in Table 11.1, Bitcoin's transactions are publicly accessible, and every party can partake in transaction validation by providing their computational power. From a user's perspective, money transfers via Bitcoin can be faster and cheaper, especially when international payments are considered.

TABLE 11.1
Classification of Blockchain Types

Access to Transactions		Access to Transaction Validation: *Permissioned*	*Permissionless*
	Public	All nodes can read/submit transactions; authorized nodes validate transactions.	All nodes can read, submit, and validate transactions.
	Private	Only authorized nodes can read, submit, and validate transactions.	N/A

(Adapted from Peters and Panayi, 2015).

What differentiates blockchain systems from previous systems for land registration is the provision of uniqueness of data (native authentication) without relying on any guarantor who mediates transactions. Technically, the blockchain can reliably keep records of transactions without having to rely on cadastres and/or notary services. Of course, those transactions may not have legal standing, but at the very least, they can mathematically prove each transaction that occurred. So, blockchain systems rely on decentralization (no unique guarantor), immutability (records cannot be tampered with or deleted), auditability (mathematically provable), and pseudonymity (via key-pair identification) (Maurer and DuPont, 2015). It is important to stress that blockchains themselves do not guarantee land data quality, which depends on who enters them, but only data immutability. Nonetheless, the persistence of data may act as a deterrent to entering false data because fraudulent entries can be traced back and not deleted.

From an institutional perspective, land management through a blockchain system is peculiar for many reasons. For one, the intermediary's function is thereby substituted by distributed consensus maintenance (mining), introducing rivalry in digital settings (Miscione et al., 2018; Ziolkowski, Miscione, and Schwabe, 2018). Validating parties thereby compete for a reward in finding the next block by solving a complex mathematical puzzle, which is labeled the consensus algorithm. In short, Bitcoin's immutability rests on computational power: as long as 51% of the hashing capacity* remains honest (complies with the code), the system continues to work as expected. The quantity of computational power devoted to the consensus algorithm and its related rewarding scheme, hence, correlates with system security.

For a second reason, considering Bitcoin again, the rules for the system to work were formulated and enacted early on by its initiator (Nakamoto, 2008). While changes to the system are possible, with varying impact levels (soft vs hard forks), every change requires consensus among the majority of nodes and parties. More than with free and open-source software (FOSS), every change of a software's functionality may leave those who favored the previous version behind (no backward compatibility). Something similar happened to Bitcoin and also to other blockchain systems. The history of disagreement (ending up in forks) within the Bitcoin community is well-documented (De Filippi and Loveluck, 2016). Forks have endangered a system which capitalized hundreds of billions of USD at its peak (Campbell-Verduyn, 2017), and showed to researchers and practitioners the consequences of the collision of incompatible social models and values hard-coded into IT.

11.3.2 Specificities in Ghana: Institutional Plurality and the Plurality of Documentation

As in many African countries, land governance in Ghana is characterized by fluid boundaries between formal and informal institutions (Benjaminsen and Lund, 2002; Hydén, 2006; de Herdt,

* The sum of computational power of all validating nodes at a given time. Solving the mathematical puzzle relies on searching for the right "hash." This process is hereby referred to as "hashing."

2015; de Sardan, 2015) and a plurality of norms and laws that govern land access, uses, and transfers. The Land Administration Project of the World Bank (2003–2008) consolidated various organizations under the Lands Commission (LC) and created the Customary Land Secretariats (CLS) with the aim "to help customary authorities to improve and develop customary land administration" (Arko-Adjei, 2011, p.81). The LC holds the mandate to register land rights and maintain land tenure records as per statutory law. The customary sector includes the chieftaincy institution, namely, the "stool" chieftaincy in the south and the "skin" chieftaincy in the north, the latter being accompanied in customary offices by earth priests in some northern regions (Lund, 2008). Many of the statutory laws concerning land rights and associated administrative agencies originated in the time of Ghana's independence. The overall move towards formalization of land administration across various types of land sectors (Oberdorf, 2017) plays a role in defining the institutional landscape around land rights documentation, and various types of institutions and associations may vie for recognition as public authorities. Therefore, we can speak of a kind of isomorphism that follows the insignia of the state, an isomorphism where formal norms (de Sardan, 2015) carry across different organizations and associations of actors and practices. At the same time, the "plurality of institutions produces … ambiguous practical meanings of law and property" (Lund, 2008, p.6). Procedures and practices of land rights' registration are not uniform across the country, but involve different actors and different proofs of evidence in the chain of validation that lead to an eventual registration, mostly of leasehold titles, with the Lands Commission (Abubakari, Richter, and Zevenbergen, 2018). In sum, despite isomorphic trends towards formalization and uniformization of registration processes, the institutional plurality and local politics over land also produce a plurality in methods to establish what counts as evidence for land claims with (contesting) oral knowledge and witness accounts playing an important role in land claim-making processes (Berry, 2001). In conclusion, even if the original values of blockchain and Ghana land registrations are unrelated, they might still converge in as far as the lack of a unique intermediary authority can be circumvented by an unalterable record of transactions which can allow unrelated parties to trade.

11.3.3 Theoretical Highlights and Possible Research Questions

Institutional theory in any of its variations considers especially the legitimate social models, thus the socially accepted courses of action that people have in front of them when dealing with the many aspects of land management and use. As is strikingly clear from the previous two sections, the blockchain mode of organizing and land registration in Ghana show little to no overlapping. However, two considerations may avoid discouragement: firstly, the origins of technology do not exhaustively predict where it ends up. The internet was invented to reduce the consequences of a Soviet nuclear attack on the US (Abbate, 2000), and ended being used to share pictures of cats and vacations. Secondly, even if the institutional logics remained different, there is no reason to exclude a priori that different logics could co-exist side by side. For example, even if nuclear power came from military purposes, and is still managed in a highly hierarchical way, it serves civilian purposes no less.

So, we should not get carried away by an overemphasis on decentralized vs centralized registrations and different courses of actions that institutionalism highlights. The assessment of blockchain for land management should not be based on the fit between blockchain consensus and "statutory institutions" or "customary institutions." Rather, what is likely to happen is that different systems (paper-based, oral, blockchain-signed …) will interplay with each other. Institutional theory invites us to transcend the idiosyncrasies of individuals and consider social models to manage these interplays. With these points in mind, institutional theory can prompt further action and research on blockchain and land registries by focusing on these central concepts: immutability, auditability, incentive scheme, legal system vs "code is law." Immutability over long periods of time is obviously central for any reliable land registry. Uncertain longevity of records undermines their validity from their inception. Immutability is also the core of blockchain technology, which relies on it to diminish the influence of third parties. Thus, to the extent third parties are a threat to records (corrupt officials, speculation …), blockchain and land registries can converge in practice. Still, the actual longevity of decentralized authenticated tokens* over decades has to be proved. Also, their legal standing in courts awaits confirmation.

Accordingly, research questions could be: which land management practices would embrace or avoid blockchain-based registries; and for what reasons? Then, which existing authorities would gain or lose from the progressive introduction of those records? Finally, would the institutional landscape in Ghana see an increase or decrease of competing land registration practices?

Partly descending from immutability, auditability (allowing reading rights to non-validating parties, e.g. not-mining users) holds some promise of streamlining land management because it makes easier to check who did what and when.

Moving to the more technical aspects of blockchain, an innovation that proved viable at scale is its incentive scheme that keeps actors compliant with a consensus algorithm set up a priori. Ideally, this makes it conceivable to move away from current cost-recovery strategies (fees, general taxation, etc.) and hard-code the incentives into the blockchain ledger itself. How to do that would be a design research question with specific answers in specific contexts.

Lastly, institutionalism considers the legal system as one of the regulatory forces that shape organizations. Famously, Lessig (1999) stated that "code is law." Therefore, code, especially if paired with suitable incentive systems, can help to enforce law or, alternatively, to consolidate the legal system in Ghana or elsewhere. More on this line of thinking can be found at the end of the next section.

Empirically, institutionalism invites collecting data where different social models encounter. An obvious starting point for such an inquiry is the existing registration procedures. Another one is in statutory and/or customary courts, where disputes are to be settled. Also, the definition of the consensus algorithm should not be overlooked when studying land registration, because its regulatory function is likely to be far-reaching, due to the immutability it enforces.

Institutional approaches have been criticized for treating people as "cultural dopes" (i.e. defined by their circumstances). So, besides the prospects presented so far, it is useful to consider other theoretical angles that help where institutionalism runs the risk of downplaying human agency.

11.4 THEORETICAL ANGLE: STRUCTURATION THEORY

Orlikowski (2000) is usually associated with the translation of structuration theory into information system research (see also Orlikowski and Barley, 2001). This stance, which moves the traditional dichotomy between structures and agencies to an analytical (rather than empirical) level, can help in understanding if and how a land management system reproduces existing structures by facilitating established courses of action, or, conversely, if and how new patterns of action become possible. Conceptualizing technology in use as a process of enactment opens up a better understanding of how practices change technology through their adoption.

Puri's (2006) concept, relying on Orlikowski and Gash (1994), of technological frames as sensemaking devices, puts stakeholders' frames at the center of attention. Puri argues that an overemphasis on technological and economic resources tends to overlook organizational and institutional dimensions. With the intention of accounting for the relation between structures and agencies, the case of the Indian National SDI is depicted through a variety of stakeholders' perspectives, with a specific emphasis on the implementation side. The account provides a rich picture of how SDI is socially constructed, and also of cases of failure. For example, he reports an excerpt of the design/reality gap from an interview with a public officer and SDI expected user:

> We have no idea what NSDI is all about! No one has consulted us. Probably, like in the past, they [scientific institutions concerned] would design something inappropriate to our needs, and then we would be

---

* A digital, intangible, unique representation of *something* such as a crypto-currency. Native to a blockchain system.

> asked to use it effectively. This has been going on. I am not sure how large volumes would be accessible online given the rather poor status of data communication infrastructure outside metros. (Puri, 2006)

Structuration theory is a relevant lens to look at blockchain and land registries, because it allows us to see how social structures are reproduced, and how they may harmonize or clash when they enter in interplay with new land registries. Symmetrically, this theoretical stance is sensitive to how agencies may change social structures.

11.4.1 Blockchain

Bitcoin served as a glaring example of the risks blockchain systems would face without functioning governance modes, at least when the need arises. For instance, never-ending conflicts about block size made it evident that technical consensus about transactions does not suffice to regulate the whole network. Organizational consensus is necessary to avoid forks when disputes cannot be reconciled. A first attempt to formalize organizational consensus "on-chain" has been carried out by TheDAO (DuPont, 2017). A DAO (decentralized autonomous organization) aims to predefine its operation in code, thus, code becomes the equivalent of law, supposedly. TheDAO, while generating a great deal of enthusiasm and mobilizing hundreds of millions USD in funding, became a victim of its own code-is-law dogma when an alleged hacker stole part of its funds through a bug in the TheDAO's software (Siegel, 2016). The TheDAO community found itself in a dilemma: preserving the dogma of immutability, and hence losing the funds in question, or creating a precedent by reversing the transaction, thus breaking the dogma of immutability. After heated debates, the TheDAO community, also influenced by the leading figure Vitalik Buterin, opted for the latter, forking the underlying Ethereum blockchain while leaving the dissidents to continue running the original chain, which is currently called Ethereum Classic.

From an agent-structure perspective, "code is law" entails the agent's adherence to whatever outcome is predefined by the code, which keeps the interplay between the agent and structure constricted; the agent's influence on the structure, hence, is supposed to end after the design and development phase. As became apparent to Bitcoin and even more with TheDAO, this line of thinking (which assumes the possibility of complete/self-contained contracts) cannot handle unforeseen circumstances. Thus, there must be a means for rules to change the rules, as one cannot anticipate the unknown. "Code is law," hence, can turn into "Code is Constitution," i.e. rules to change the rules. And, in fact, this is what happened in more recent developments: in the time that followed, on-chain governance gained in importance with large and heavily-funded projects such as Tezos (Crunchbase, 2019), Aragon (Cuende, 2017), and D-finity (Crunchbase, 2019). These aim to preserve the benefits of immutability while allowing for changes to the system upon consensus; and by that, facilitating the interplay between agents and structures in front of unknown circumstances.

11.4.2 Specificities in Ghana: Structures of Opportunity and Strategic Documentation

At a simplified level, the diverse geography of land documentation and records in Ghana derives from the dynamic tension between two forces: the force towards an ideal Weberian-type state structure to govern Ghana, on the one hand, and the forces of decentralization and fragmentation driven by the "politico-legal institutions that compete for political authority [and which] operate to legitimize their undertakings partly through territorial strategies" (Lund, 2013, p.16) on the other. This dynamic gives rise not only to legal stipulations for the registration of land rights to be implemented variously across the country (Abubakari, Richter, and Zevenbergen, 2018), but also provides what Lund (2008, p.4) calls "structures of opportunity for the negotiation of rights and distribution of resources." For example, in decision-making over land use and transfers, individuals and groups of agents may act as what Moore (2000) refers to as semiautonomous fields drawing on different rationales to legitimize claims to land. The choices and decisions to document land claims as a land right (of whichever nature this may be) is contingent upon the anticipated effects of documentation on the social and economic positions of individuals and groups. Berry's work (2001) in the southern Asante region of Ghana, for instance, illustrates that claims to land are made and legitimized through processes of re-narration and revisions of constructed histories of people–land relations. Thus, while land itself may form the immutable entity that provides family or clan lineages' identities through time, the documentation of land may be used to break with, reshuffle, or comply with customary social structures depending on circumstances, need, and aims. The overall outcome then of a given process of negotiation and contestation may be likened to the consensus building in Blockchain systems, insofar as there is no central node with final verification and legitimation power. However, in Ghana, this production of records does not run according to programmable, mathematical rules, but entails political contestations and differs according to the problem's context. Neither law nor code is law. What is strategically and politically deployed is the "mutability of records."

11.4.3 Theoretical Highlights and Possible Research Questions

Also, in this case, a common theoretical angle does not hide the wide differences of structuration processes in relation to blockchain on the one hand, and Ghanaian land registries on the other. Nonetheless, some common central concepts can again help us in action and research: the shift from "code is law" to "code is constitution" is a salient one. Since algorithms—it does not matter how sophisticated—do not always achieve putting on track the unpredictable future, consensus-building has to move one level down: rather than rules (i.e. structures) to regulate people (i.e. agents), the consensus needs to be about how to change rules when the need arises. The clear analogy here is to constitutional laws that regulate how laws can be changed. This can be seen as the product of tensions between structures and agencies. Indeed, since agencies and structures cannot be always reconciled, the need to have a higher level of appeal may arise.

In relation to land registries in Ghana, this theoretical angle highlights the misalignment and frictions between rules and actors. When operationalizing structuration theory into a concrete research design, it can be useful to keep in mind that rules/actors do not necessarily correspond to algorithms/people. Rules can be humans and actors can be software agents, especially in the context of on-chain governance, which intends to enable DAOs.*

Another recommendation for further action and research about blockchain and governance is to consider that, contrary to previous IT architectures, blockchain are relatively inflexible because of their promise to guarantee immutability. Therefore, consortia designing and developing blockchain-based land registries have to agree quite precisely on details of how their registry is going to work early on in the process. Later changes risk to undermine the credibility of the whole record, or at least to generate inconsistencies (Ziolkowski, Miscione, and Schwabe, 2018).

Finally, an interesting terrain for practitioners and researchers is to figure out if and how it is possible to design incentive schemes that make this distributed ledger scalable and self-sustainable. Indeed, making the unknowns known would certainly help in figuring out how to balance what goes into the code and what remains a matter of people's discretion.

Possible research questions are: if the overall scene is one of validation of evidence as instruments in politics, how would blockchains play out in this context? Whose agency would they support in Ghana? What new structures would they create or which structures would they legitimize?

Despite its strengths, structuration theory tends to be weak in accounting for difference and radical transformations, as both of them are not necessarily produced by incremental change. The last section on actor-network theory may help in that direction.

* In principle, land governance actors in Ghana could be receptive to blockchain conceived of as another form of database to legitimize something already existing (e.g. statutory or a chieftaincy registry), but they would be reluctant regarding the immutability trait of blockchain. Quite likely, they would seek to build-in what happened in TheDAO's case: rules to change the rules. Then, the question is: what form would such a constitution take in an instance of implementation in Ghana? Would it embed practical norms implicitly? Who would be the author of such a blockchain constitution?

11.5 THEORETICAL ANGLE: EMPHASIS ON ACTORS AND OBJECTS

Related to the direct and indirect influence of actor-network theory (ANT), actors, artefacts, and objects have been coming to the foreground in a number of disciplines. The concept of a boundary object was first introduced by Star and Griesemer (1989) to explain the case of the foundation of a museum, which was possible because different stakeholders did not need to negotiate a common understanding about the museum, nor common goals. Rather, it was enough for each of them to act according to their own social world; the boundary object was, at the same time, the product of those actions and the mediator across them. The basic idea of boundary objects has been developed and specified in several directions: for instance, intermediary objects (Boujut and Blanco, 2003) and boundary negotiating artefacts (Lee, 2007) emphasize different stages of evolution of a boundary object during product development. Bechky (2003) relates boundary objects to the boundary between professional belonging and status.

Harvey and Chrisman (1998) introduced the concept of a boundary object in the geographic community by arguing how geographic information systems (GIS) act as boundary objects through mediating between different groups who do not share common understandings. Through a microscale study on GIS and wetlands, they show how it worked despite "wetland" having different meanings for different actors: "The agreement is only paper-thin. The boundary object serves to solve jurisdictional and administrative battles while it conceals continued geographic ambiguity" Harvey and Chrisman, 1998). Harvey (2006) discusses the necessarily elastic relation between the cadastre and land tenure in Poland. Local practices, also inherited over generations, have to find working accommodations to relate to Polish state and European Union regulations. Hence, this information infrastructure is tense between diverging civil and political interests. So, it acts as a historically grounded boundary object.

These kinds of objects are a suitable sense-making device to approach land registries and blockchain when the points of convergence between stakeholders are a determinant aspect to account for. For instance, when a land registry spans across areas regulated by different property regimes, conceptualizing a registry as a boundary object highlights how commonalities and differences converge on a possible solution.

11.5.1 BLOCKCHAIN

Many of the public and permissionless blockchains (Bitcoin included) have publicly available source codes, free for everyone to fork and to set up their own blockchains. This is similar to FOSS projects such as Linux or Firefox, whose mode of governance has been labeled "bazaar" by Raymond (1999) and Demil and Lecocq (2006): this mode of governance is characterized by openness and fairness, and an open license contract of the object in question. Further, there are fairly limited ways to exercise control over actors (to use a particular version of software), and an actors' motivation to contribute with new code is rather low unless reputation is pivotal.

According to Miscione et al. (2018), blockchains differ from FOSS in many ways: blockchains bring rivalry among actors, e.g. a token always belongs to one entity at a time; their value is rooted in their uniqueness. Having conflicting versions of token ownership would undermine the ledger's integrity and by that, affect the token's value. Forks are undesirable, because users of a blockchain have a mutually dependent interest in the integrity of data, which, in contrast to the "bazaar," leads to a stronger common interest. What's more, if a majority of actors running a particular blockchain instantiation changes its consensus mechanisms, this change can be enforced on other users, which is not a feature of FOSS.

Comparing FOSS and blockchain systems with an emphasis on artefacts offers two perspectives. FOSS can certainly be seen as a boundary object, where communities of actors with widely diverging interests contribute a variety of solutions; however, there is common ground with every release, patch, or update of a software. The elasticity of FOSS is exemplified in low switching costs from one version to another, so an arbitrary version (even a self-programmed version) of software can be instantiated upon one's liking. Does the same apply for blockchain systems? Even if a blockchain system can be seen as the result of a sense-making process across different actors or groups, likely with diverging interests, the elasticity of a blockchain system is far more constrained: blockchain systems are defined by the consensus protocol, which determines everyone's mode of participation.

11.5.2 Specificities in Ghana: Land Rights Documents and Fees as Boundary Objects

Within the context of land governance, land itself forms an important boundary object, across which different communities and institutions associate and disassociate (not unusually, violently). While land creates boundaries through various modes of exclusion, it also forms the object around which "webs of interests" develop and sustain (see e.g. Meinzen-Dick and Mwangi, 2009) through a variety of uses with corresponding bundles of land rights (Davy, 2018). Within this context, technologies to support or change processes of land documentation and recordation also function as boundary objects. The socio-technological assemblages include not only the survey equipment of certified land surveyors or the GPS-enabled mobile phones of "fit-for-purpose mappers" (Lengoiboni, Richter, and Zevenbergen, 2019), but also the digital data and the land rights documents that are being printed and issued to landholders, as well as the associated fees for various signatures and stamps on a series of documents that have come to verify a land claim or transaction. The associations of actors that emerge through time may stabilize into larger actors that endow the process with legitimacy, and at the same time bind together people who previously held different meanings regarding land, property, and documents themselves. For example, in the process of leasehold registration, we see that the technologies and templates for formal recording of the Lands Commission (LC) form boundary objects that negotiate with LC external processes and actors. In this meeting, meanings and norms of the bureaucratic arena become adjusted in response to social norms and administration external actors, which in turn also change. What emerges is a space of "practical norms" that can be quite literally visited and observed at the encounters between registrants, intermediaries, and officials in front of the Customer Service Access Units of the LC (Abubakari, Richter, and Zevenbergen, 2018). Fees paid for a variety of intermediary documents, services, and signatures constitute an important, if not constitutive, element in the creation of such networks across differing institutional contexts. Meridia, a for-profit organization that recently entered the market of land rights documentation in Ghana, actually flipped the logic insofar as it created different "documentation packages" according to the actors that need to be involved given the institutional scene that they encountered (Salifu, 2018). The latter in turn shape the content and format of a given type of documentation package, and the fees associated with various signatories create the glue in the emergence of action nets.

The examples above indicate that Ghana's context of institutional plurality is likely to give space to a new system, and it is imaginable that blockchain proponents, alongside their technology, may come to act as boundary objects in the emergence of new actor networks with their own normative frameworks and new shared meanings of land, property, and related recording technology. However, insofar as such endeavors would succeed amidst the contestations over legitimacy, the question as to whose claims are valid for entering on the blockchain's own sphere of "code as law" and, as recent controversies over block sizes have shown, "lawlessness of code," remains open. As such, it is as much a question of data governance as it is of land governance.

11.5.3 Theoretical Highlights and Possible Research Questions

Land rights documents in Ghana, generally speaking, play at least two important roles that evidence their value as boundary objects. Firstly, for the purpose of enrolment of communities, initiatives to map land rights are able to attract interest especially in communities where demand for (some kind of) land rights documents exists, e.g. as part of specific development projects, such as service provision (see, for instance, Lengoiboni, Richter, and Zevenbergen, 2019). Secondly, the production of documents and decisions on their form and content enroll further actors, for example, customary authorities, who sign in return for negotiated fees, and statutory actors to sanction the content of the documents. This kind of multi-party relation is exemplified by Meridia's initiative. In this case, partially overlapping actor networks emerge in conjunction with the development of a whole set of so-called "documentation packages" for different purposes and regions (Salifu, 2018).

Actors entering the land governance scene of Ghana often recognize the central role that the maintenance of digital land rights data plays in becoming "inevitable partners" (Oberdorf, 2017). In this sense, analogue or digital documents play a similar function, as both boundary objects, but also as devices in the processes of interessement, enrollment, and mobilization (Callon, 1986). In a study on blockchain implementation in Ghana by BitLand and Benben, Oberdorf (2017) observed that "although not the core of the functioning of the Blockchain, the process of digitization repeatedly returned in the interviews with the companies and public actors" (p.44), indicating both the hopes that blockchain can make documents in their digital form indisputable as well as searchable, and indicating the greatest perceived risks: the moment of entry of a document into the database and who should be the "single party" to add the initial data (Oberdorf, 2017). So, data quality and immutability are properties that offer affordance for blockchains acting as boundary objects across traditional and statutory regulations. Moving to more empirical entry points and question, what an emphasis on artefacts suggests asking is if, and how, blockchain is actually used in Ghana, for what functions exactly and at what scale; and what actors and associated normative frameworks become enrolled in the process? How is it different from what we see elsewhere, and why, for instance, what specific forms and uses of blockchain may function as boundary objects and what changes in meaning would we see then at the crossroads of different normative frameworks in land governance? If stabilizing into actor networks, how would these new patterns of action interplay over legitimacy with others; and to what outcome? (Table 11.2).

11.6 CONCLUSIONS

Henssen (2010) outlines a set of generic legal principles underlying reliable land registration systems, namely the booking or register principle, the consent principle, the principle of publicity, and the principle of specialty. These principles also inform recent discussions and research on the potential of blockchain-based land registries (e.g. Vos, 2016; Griggs et al., 2017). Since only the principle of publicity interplays directly with blockchain, one might hurriedly conclude that land registration and blockchains are unrelated. Any of the three theoretical stances discussed above helps in avoiding such a shortcut. Technology and its encounter with practices cannot be deduced from principles without well-documented risks of gross reductionism. So, empirical work cannot be supplanted by logical reasoning, but fostered by a suitable theoretical angle. The same applies to predictions of how the three main cadastral processes—e.g. adjudication of land rights, land transfer, and subdivision/consolidation (Zevenbergen, 2002)—would be affected by blockchain. For example, to be considered legally binding, transactions must take place in an unambiguously identifiable manner. Pseudonymity of blockchains may rule this out. We claim that unintended consequences are important, for example, traceability of pseudonyms may have side-effects, and thus are worth proper consideration. Still, it is important to bear in mind that blockchain does not validate data quality (from traders, surveyors, notaries, etc.) but only its immutability. Data quality is beyond blockchain, which has no native connection to real world assets. Rather, it crystallizes "garbage in, garbage out ...". Its only effect can be deterrence: people may think twice before entering the wrong data, because they are traceable as never before.

Looking at the dictionary, "deed" has two quite distinct meanings: "something that is done, performed, or accomplished; an act," and, in the vocabulary of law, it is "a writing or document executed under seal and delivered to effect a conveyance, especially of real estate." These two meanings of "deed" get exceptionally close in the context of blockchain for land registries in Ghana,

TABLE 11.2
Overview on Applied Theoretical Lenses, Blockchain Technology, Specificities in Ghana, and Theoretical Highlights

Theoretical angle	Blockchain	Specificities in Ghana	Theoretical highlights and possible research questions
Institutionalism: commonly accepted social models shape courses of action and legitimize them. Neoinstitutionalism: the institutional field (comprising actors engaged in the same kind of activity) tends towards isomorphism.	Originally, the distributed consensus that is at the foundation of blockchain as an architecture was intended to avoid existing social models and organizations. Algorithmically certified consensus, sealed by miners, substitutes other norms and modes of regulation. BC ensures honesty in the narrow sense of compliance with the incentive scheme.	A plurality of institutions entail differentiated processes and procedures for recording and non-recording of land rights and transfers. Isomorphism is evident, especially in processes of formalization of land governance institutions.	Highlight: theory invites us to transcend the idiosyncrasies of individuals and consider social models to explain and manage the interplays of different systems (paper, oral, BC-signed, etc.) Questions: where do/would different institutions encounter the registration court cases later on, in the definition of the consensus algorithm? Which existing authorities would gain or lose from the progressive introduction of those records; and would the institutional landscape see an increase or decrease of competing land registration practices? Design research question: how to move from current cost recovery strategies (fees, general taxation, etc.) and hard-code the incentives into the blockchain ledger itself ? Theoretical precaution: avoid treating people as "cultural dopes" (i.e. defined by their circumstances).
Structuration: agents and structures are in constant interplay and tend to align over time.	Most BCers, and the sector as a whole, have realized that a too-inflexible system does not work well when adjustments are needed due to unforeseeable circumstances (e.g. BTC size, DAO …) On-chain governance (rules to change the rules/"Code is constitution") may better fit the constant interplay between agents and structures.	Administrative structures provide opportunities for strategic documentation practices, and over time the emergence of "practical norms" at the interface of social and bureaucratic norms.	Highlight: theory highlights the misalignment and frictions between rules and actors. Questions: are consortia adopting or at least considering on-chain governance? If so, how concretely, and for what? Whose agency would BC-based registries support in Ghana; and what new structures would they create or which structures would they legitimize? What is kept out of agents' reach (in fixed structures)? Design research questions: if, and how, it is possible to design incentive schemes that make a distributed ledger scalable and self-sustainable. How to agree quite precisely on details of the registry's functioning early on in the process to avoid later changes to undermine the credibility of the whole record? Theoretical precaution: keep in mind that rules/actors do not necessarily correspond to algorithms/people.

(Continued)

TABLE 11.2 (CONTINUED)
Overview on Applied Theoretical Lenses, Blockchain Technology, Specificities in Ghana, and Theoretical Highlights

Theoretical angle	Blockchain	Specificities in Ghana	Theoretical highlights and possible research questions
ANT/objects/socio-materiality: emphasis on how human and non-human actors perform together. The distinction is not explanatory; better to focus on what those "actants" do and what cannot be done otherwise.	BC generated its own form of sociality that is not captured by bazaar, etc.	Land rights documents (on paper and digital data) and related technologies function as boundary objects across initiatives to document land rights and associating actors from different institutional settings creating ANTs tied into a "fee economy" that surrounds documents.	Highlight: approach shifts emphasis on artefacts and how they negotiate the boundaries between different norms, meanings, and methods. Questions (also for design research): if, and how, blockchain is actually used in Ghana, for what functions exactly and at what scale; and what actors and associated normative frameworks become enrolled in the process? How is it different from what we see elsewhere, and why; for instance, what specific forms and uses of blockchain may function as boundary objects and what changes in meaning would we see then at the crossroads of different normative frameworks in land governance? If stabilizing into actor networks, how would new patterns of action interplay with others; and to what outcome, e.g. legitimacy gains/losses? Theoretical precaution: avoid overemphasizing the possibilities of "becoming" and downplaying the constraints people and organizations live by.

where statutory regulation is minoritarian, and both customary regulation and blockchain derive their authority of authenticity from continued performance (or from tribal ties and mining, respectively), rather than legal principles.

Some have tried to outline the state of the art in relation to land property and blockchain, for example, Graglia and Mellon (2018) and Vos (2016). The challenges in front of practice and research are remarkable, and impossible to list here. So, we limit ourselves to the key issues:

- Relevance of transparency and immutability in low-trust settings;
- Longevity of records;
- Cross-jurisdictional issues;
- Disjunction between quality of data entry and immutability;
- Role of state authorities in data quality and legal standing;
- Prospects for multi-chain records to leverage different properties of different chains;
- Effects of increased visibility of records on bureaucratic functioning (role or auditors and civil society.

While it is certainly possible that some of the expectations from blockchain for land management are tall tales (Bennett, Pickering, and Sargent, 2019), we think that the three theoretical angles discussed above offer a toolkit to approach challenges of research and practice of land registries and blockchain. None of them is exhaustive; none of them can provide an ultimate guide to action and research in this domain. However, those quite distinctive theoretical views promise to account for the challenges and prospects of designing and implementing blockchain-based land registries in Ghana and beyond, especially where no one single regulatory regime is dominant.

Even though the early stage of this application domain prevented us from relying on sound empirical materials, existing studies (Pelizza and Kuhlmann, 2017; Ziolkowski et al., 2018; Ziolkowski et al., 2019) and our initial explorations made it apparent that, while blockchains strove to substitute human discretion with algorithmic authentication, humans and blockchains are more likely to re-adjust the division of labor. The outcomes of those adjustments may have far-reaching consequences in terms of "smart" land management, to the extent the rigidities of "code is law" allow for automations that would be impossible otherwise. On the side of "responsible" land management, technology by itself has little ethical agency. Nonetheless, the functioning of blockchains and the distributed and transparent mode of authentication hold the promise of a greater accountability even though, it has to be stressed, the lack of authorities in charge may turn up sour when problems arise.

REFERENCES

Abbate, Janet. 2000. *Inventing the Internet.* Cambridge, MA: MIT Press.

Abubakari, Zaid, Christine Richter, and Jaap Zevenbergen. 2018. "Exploring the 'implementation gap' in land registration: How it happens that Ghana's official registry contains mainly leaseholds." *Land Use Policy* 78:539–554. doi:10.1016/j.landusepol.2018.07.011.

Agre, Philip E. 2003. "P2p and the promise of internet equality." *Communications of the ACM* 46(2):39–42.

Arko-Adjei, Anthony. 2011. "Adapting land administration to the institutional framework of customary tenure - The case of peri-urban Ghana." PhD, Faculty for Geo-Information Science & Earth Observation, University of Twente.

Atzori, Marcella. 2015. "Blockchain technology and decentralized governance: Is the state still necessary?" Available at: https://papers.ssrn.com/sol3/papers.cfm?abstract_id=2709713.

Bechky, B.A. 2003. "Object lessons: Workplace artifacts as representations of occupational jurisdiction." *American Journal of Sociology* 109(3):720–752.

Bekkers, V., and V. Homburg. 2007. "The myths of e-government: Looking beyond the assumptions of a new and better government." *The Information Society* 23(5):373–382.

Benjaminsen, Tor A., and Christian Lund. 2002. "Formalisation and Informalisation of land and water rights in Africa: An introduction." *The European Journal of Development Research* 14(2):1–10. doi:10.1080/714000420.

Bennett, Rohan Mark, M. Pickering, and J. Sargent. 2019. "Transformations, transitions, or tall tales? A global review of the uptake and impact of NoSQL, blockchain, and big data analytics on the land administration sector." *Land Use Policy* 83:435–448.

Berry, Sara S. 2001. *Chiefs Know Their Boundaries: Essays on Property, Power, and the past in Asante, 1896-1996*. Portsmouth, Oxford and Cape Town: Heinemann, James Currey: David Philip.

Boujut, J.F., and E. Blanco. 2003. "Intermediary objects as a means to foster co-operation in engineering design." *Computer Supported Cooperative Work (CSCW)* 12(2):205–219.

Callon, Michel. 1986. "Some elements of a sociology of translation: Domestication of the scallops and the fishermen of St Brieuc Bay." In: *Power, Action and Belief*, edited by John Law, 196–233. London, Boston and Henley: Routledge & Kegan Paul.

Campbell-Verduyn, Malcolm. 2017. *Bitcoin and beyond: Cryptocurrencies, Blockchains, and Global Governance*. Routledge, London, UK.

Crunchbase. 2019. "Tezos." *Crunchbase.*

Cuende, Luis. 2017. "The Aragon token sale: The numbers: Aragon".

Davy, Benjamin. 2018. "After form: The credibility thesis meets property theory." *Land Use Policy* 79, 854–862. doi:10.106/j.landusepol.2017.02.036.

De Filippi, Primavera, and Benjamin Loveluck. 2016. "The invisible politics of bitcoin: Governance crisis of a decentralized infrastructure.". *Internet Policy Review*, 5(4): 32.

de Herdt, Tom. 2015. "Hybrid orders and practical norms - A Weberian view." In: *Real Governance and Practical Norms in Sub-Saharan Africa : The Game of the Rules*, edited by Tom de Herdt and Olivier de Sardan, 95–120. London: Routledge.

de Sardan, Olivier. 2015. "Practical norms: Informal regulations within public bureaucracies (in Africa and beyond)." In: *Real Governance and Practical Norms in Sub-Saharan Africa : The Game of the Rules*, edited by Tom de Herdt and Olivier de Sardan, 19–62. London: Routledge.

De Soto, Hernando. 2000. *The Mystery of Capital: Why Capitalism Triumphs in the West and Fails Everywhere Else*. New York, NY: Basic Civitas Books.

de Vries, Walter T., Rohan M. Bennett, and Jaap A. Zevenbergen. 2015. "Neo-cadastres: Innovative solution for land users without state based land rights, or just reflections of institutional isomorphism?" *Survey Review* 47(342):220–229.

Demil, B., and X. Lecocq. 2006. "Neither market nor hierarchy nor network: The emergence of bazaar governance." *Organization Studies* 27(10):1447–1466.

DuPont, Quinn. 2017. "Experiments in algorithmic governance: A history and ethnography of "The DAO," a failed decentralized autonomous organization." In: *Bitcoin and Beyond (Open Access)*, 157–177. Routledge.

DuPont, Quinn. 2019. *Cryptocurrencies and Blockchains*. John Wiley & Sons.

Graglia, J. Michael, and Christopher Mellon. 2018. "Blockchain and property in 2018: At the end of the beginning." *Innovations: Technology, Governance, Globalization* 12(1–2):90–116.

Griggs, Lynden, Rod Thomas, Rouhshi Low, and James Scheibner. 2017. "Blockchains, trust and Land Administration–the return of historical provenance." *Trust and Land Administration–The Return of Historical Provenance* 6. Available at: https://papers.ssrn.com/sol3/papers.cfm?abstract_id=3325558.

Harvey, Francis. 2006. "Elasticity between the cadastre and land tenure: Balancing civil and political society interests in Poland." *Information Technology for Development* 12(4):291–310.

Harvey, Francis, and Nick Chrisman. 1998. "Boundary objects and the social construction of GIS technology." *Environment and Planning. Part A* 30(9):1683–1694.

Henssen, J. 2010. *Land Registration and Cadastre Systems: Principles and Related Issues*. Technische Universität München, Germany.

Homburg, Vincent, and Yola Georgiadou. 2009. "A tale of two trajectories: How spatial data infrastructures travel in time and space." *The Information Society: An International Journal* 25(5):303–314. doi:10.1080/01972240903212524.

Hydén, Göran. 2006. *African Politics in Comparative Perspective*. Cambridge, UK and New York: Cambridge University Press.

Lee, C.P. 2007. "Boundary negotiating artifacts: Unbinding the routine of boundary objects and embracing chaos in collaborative work." *Computer Supported Cooperative Work (CSCW)* 16(3):307–339.

Lengoiboni, Monica, Christine Richter, and Jaap A. Zevenbergen. 2019. "Cross-cutting challenges to innovation in land tenure documentation." *Land Use Policy* 85: 21–32.

Lessig, Lawrence. 1999. *Code and Other Laws of Cyberspace*. New York, NY: Basic Books.

Lund, Christian. 2008. *Local Politics and the Dynamics of Property in Africa*. New York: Cambridge University Press.

Lund, Christian. 2013. "The past and space: On arguments in African land control." *Africa* 83(1):14–35. doi:10.1017/S0001972012000691.
Maurer, W.M., and Q.I. DuPont. 2015. *Ledgers and Law in the Blockchain*. Retrieved from https://escholarship.org/uc/item/6k65w4h3.
Meinzen-Dick, Ruth, and Esther Mwangi. 2009. "Cutting the web of interests: Pitfalls of formalizing property rights." *Land Use Policy* 26(1):36–43. doi:10.1016/j.landusepol.2007.06.003.
Miscione, G. 2015. "Myth - Management of the unknown." *Culture and Organisation* 22(1):67–87.
Miscione, G., R. Ziolkowski, L. Zavolokina, and G. Schwabe. 2018. "Tribal governance: The business of blockchain authentication." 2018/01/03/.
Moore, Sally Falk. 2000. *Law as Process: An Anthropological Approach*. Münster: LIT Verlag.
Mosco, V. 2004. *The Digital Sublime: Myth, Power, and Cyberspace*. Cambridge, MA: MIT Press.
Nakamoto, S. 2008. "Bitcoin: A peer-to-peer electronic cash system." *Consulted* 1:2012.
North, D.C. 1990. *Institutions, Institutional Change and Performance*. Cambridge: CUP.
Oberdorf, Vincent. 2017. "Building blocks for land administration - The potential impact of Blockchain-based land administration platforms in Ghana." Medicine Science International Development Studies, Department of Human Geography and Planning, Utrecht University.
Orlikowski, Wanda J. 2000. "Using technology and constituting structures: A practice lens for studying technology in organizations." *Organization Science* 11(4):404–428.
Orlikowski, W.J., and S.R. Barley. 2001. "Technology and institutions: What can research on information technology and research on organizations learn from each other?" *MIS Quarterly: Management Information Systems* 25(2):145–165.
Orlikowski, W.J., and D.C. Gash. 1994. "Technological frames - Making sense of information technology in organizations." *ACM Transactions on Information Systems* 12(2):174–207.
Pelizza, A., and S. Kuhlmann 2017. "Mining governance mechanisms: Innovation policy, practice and theory facing algorithmic decision-making." *Handbook of Cyber-Development, Cyber-Democracy, and Cyber-Defense*.
Peters, Gareth William, and Efstathios Panayi. 2015. "Understanding modern banking ledgers through blockchain technologies: Future of transaction processing and smart contracts on the internet of money." arXiv:1511.05740 *[cs]*.
Powell, Walter W., and Paul J. DiMaggio, eds. 1991. *The New Institutionalism in Organizational Analysis*. Chicago and London: The University of Chicago Press.
Puri, Satish K. 2006. "Technological frames of stakeholders shaping the SDI implementation: A case study from India." *Information Technology for Development* 12(4):311–331.
Raymond, Eric. 1999. "The cathedral and the bazaar." *Knowledge, Technology and Policy* 12(3):23–49.
Salifu, Fuseini Waah. 2018. "Innovative approaches to land tenure documentation in Ghana: An institutional perspective." MSc Geographic Information Science, Department of Urban and Regional Planning and Geo-Information Management. University of Twente.
Siegel, David. 2016. "Understanding the DAO attack." *CoinDesk*. Available at: https://www.coindesk.com/understanding-dao-hack-journalis.
Silva, L. 2007. "Institutionalization does not occur by decree: Institutional obstacles in implementing a land administration system in a developing country." *Information Technology for Development* 13(1):27–48.
Star, Susan Leigh, and James R. Griesemer. 1989. "Institutional ecology, 'translations' and boundary objects: Amateurs and professionals in Berkeley's Museum of Vertebrate Zoology, 1907–39." *Social Studies of Science* 19(3):387–420.
Vos, Jacques. 2016. "Blockchain-based land registry: Panacea illusion or something in between?" IPRA/CINDER Congress, Dubai.
Zevenbergen, Jaap. 2002. "Systems of land registration aspects and effects." *Publications on Geodesy* 51.
Ziolkowski, Rafael, Gianluca Miscione, and Gerhard Schwabe. 2018. "Consensus through blockchains: Exploring governance across inter-organizational settings." International Conference on Information Systems (ICIS), 2018, 2018/12/13/.
Ziolkowski, R., G. Parangi, G. Miscione, and G. Schwabe 2019. "Examining gentle rivalry: Decision-making in blockchain systems." Zurich Open Repository and Archive. https://www.zora.uzh.ch/id/eprint/160377/1/Blockchain_Governance.pdf

Section IV

Norms and Goals in Land Management Practice

12 The Need for Land Use Planning and Governance in Suleja Niger State, Nigeria

What It Is and What It Ought to Be

Gbenga Morenikeji, Bamiji Adeleye, Ekundayo A. Adesina, and Joseph O. Odumosu

CONTENTS

12.1 INTRODUCTION

A system of land governance should perceive land as a natural resource that must be sustainably developed and used. Land managers and administrators must operate within specific technical frameworks, as land governance is concerned with social, legal, cultural, and economic entities (Otubu, 2018). Land planning and governance is not negotiable amidst the population explosions in both urban and peri-urban areas. Lack of strategic planning or implementation of plans by constituted authorities, especially in sub-Saharan Africa, including Nigeria, exacerbates various land issues. The problem of population growth is linked with the urbanization that has created many negative effects on urban and peri-urban land, though the effect varies within communities. After the passive response of the government to solving the issues of unplanned use and growing misuse of land in Nigerian cities, a number of laws and policies were formulated by the Nigerian government which are applicable to all levels. These laws include the Land Use Act of 1978, the Urban Development Policy of 1992, the Urban and Regional Planning Decree of 1992, and the Housing and Urban Development Policy of 2002. It is sad to note that despite the enactment of these laws and policies, the anarchic system of land use activities still exists in Nigerian cities (Yahyaha and Ishiak, 2013). This calls for strategic synergy among all actors in land planning and development towards sustainable land governance.

Land in Suleja is becoming scarce due to high demand and immense settlements expansion as a result of population influx to Suleja due to its strategic location. Suleja is experiencing expansion, growth, and developmental activities due to its proximity to Abuja, the Federal Capital Territory (FCT) of Nigeria. The movement of the FCT from Lagos in 1991 to Abuja contributed to the massive movement of people from other parts of the country to settle in Suleja, and also to the high cost of accommodation within the Abuja City that many low- and middle-income earners could not afford. Another factor responsible for the influx of men and women into Suleja was the demolition of illegal structures within the FCT and people from the northeastern part of Nigeria who left their state due to the activities of insurgents.

Suleja is about 20 km north of Abuja, the Federal Capital of Nigeria, and about 100 km northeast of Minna, the State Capital of Niger State (Buba, Makwin, and Ogalla, 2016). Suleja has about ten (10) wards within the Local Government Area. The existing land use of the study area includes residential, commercial, recreational, industrial, educational, and probably minor agricultural land uses. These land use types create spatial imbalance due to human needs. The ethnic culture of the people in Suleja has influenced the types, pattern, and design of housing construction.

Over the years, there have been signs of environmental stress, such as loss of vegetation and valuable land, erection of substandard and illegal houses, overcrowding, unplanned settlements, and slums with unpleasant living environments due to rapid population increase and migration leading to settlement growth, with people competing for limited available land and other resources which could lead to indelible damage to the study area. The concern of this work is centered on how to harmonize population pressure, urbanization, and gradual loss of valuable land in Suleja.

Against this background, there is the need to clearly understand the relationship between population growth within urban areas and its effect on peri-urban areas. In the light of the above, this study sets out to examine the dynamics of urban growth in relation to land consumption and also to forecast the future land risk in Suleja for sustainable planning and governance system (Figure 12.1).

12.2 LITERATURE REVIEW

12.2.1 Land Use Planning

All around Africa, and in the world at large, land use impacts, occupation, and allocation of land rights are changing; the impact of these changes include increasing land scarcity, rapid urbanization, and growing hazards which put larger numbers of people at risk (de Vries, 2018). Why is land use planning important in the modern world? Land use planning is regarded as one of the most sensitive issues across countries; it is politically connected as it affects people's livelihoods and community's essential needs (Chigbu et al., 2016). Planning and governance in land use is inevitable. Land use is connected with individual wellbeing, and is also a support to environmental sustainability, to a country's economic growth, and social inclusion (OECD, 2017). However, if that is what land use does, then it has to be planned. Allocating available land resources to different land activities such as agricultural, residential, and industrial must be judiciously planned (Haseeb, 2018).

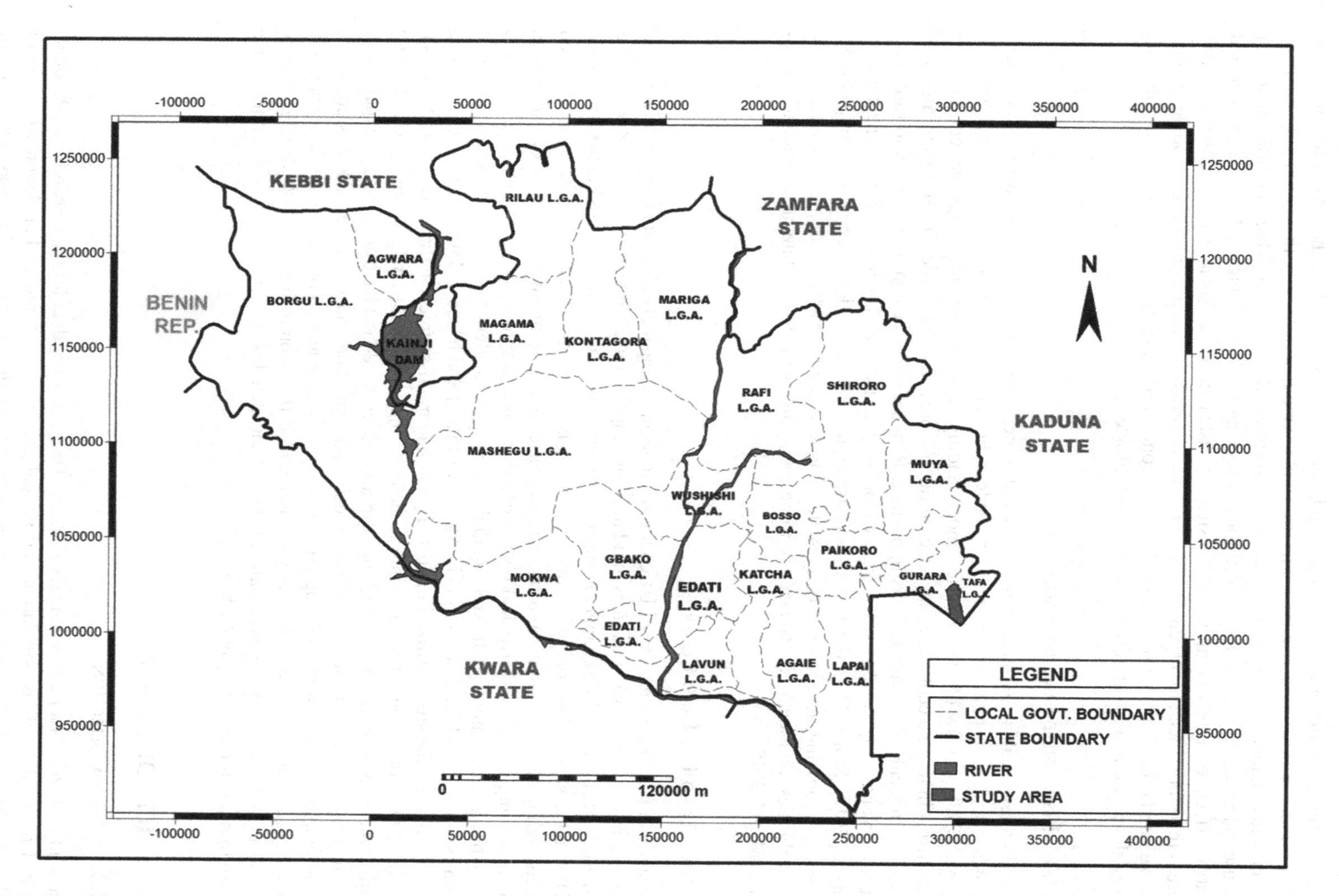

FIGURE 12.1 Map of the study area. Source: Department of Urban and Regional Planning, Federal University of Technology, Minna (2018).

12.2.2 Need for Land Governance

Land governance is the process by which decisions are made regarding the access to and use of land, the manner in which those decisions are implemented, and the way that conflicting interests are reconciled (Global Tool Network). According to Hernandez (2017), reliable land information for land use planning, zoning, and administration is lacking in most countries. This lack has created negative effects on urban planning and design, infrastructure, and socio-economic development. There is therefore a need for urgent land governance as pressures keep mounting on land due to factors like rapid urbanization, growing population, economic development, food insecurity, water and energy shortages, and the effects of conflicts and disasters.

12.2.3 Responsible Land Management

To attain a responsible land management in rural or peri-urban areas, it has to possess some characteristics, such as responsive, respected, reliable, robust, reflexive, and recognizable (de Vries and Chigbu, 2017). In view of that, a better understanding of the interrelationships and coordination mechanisms in linking ecological, social, cultural, political, and economic dimensions by all stakeholders from local to national levels is important. Joint action at the community level towards participatory planning approaches will serve as a tool for the sustainability of land development efforts (Chigbu, 2013). Similarly, in developing peri-urban areas, preconditions like peri-urban planning, visioning, improving accessibility of growth centers, citizens' participation, decentralization of responsibilities, and local governance are to be the focus of the local, state, and national government, (Magel, 2015).

12.2.4 Land Use Land Cover

The study of land use/land cover (LULC) has become an increasingly important, especially when it comes to monitoring urban dynamics and land consumption rate. Mohajane et al. (2018) described land use/land cover (LULC) changes as one of the most important applications of Earth Observation (EO) satellite sensor data; one of its main functions is that it provides a comprehensive and good understanding of ecosystem monitoring, and responses to environmental factors. In the same vein, remote sensing techniques have also been recognized as a powerful means to obtain information on Earth's surface features at different spatial and temporal scales. It was used to assess the rate of urban expansion and loss of vegetation in Akure North and South Local Government Areas of Ondo State of Nigeria; the study utilized multi-temporal and multi-source satellite imageries of Landsat data for 1991, 2002, and 2016. The study concluded that substantial land use/land cover (LULC) changes have taken place and the built-up land and agricultural land have continued to expand over the study period; while the forest land, bare rock, and water body have decreased. It was noted that the development of the urban built-up areas has resulted in reduction of the land under agriculture and other natural vegetation. The study recommended that monitoring of LULC through remote sensing and GIS should be institutionalized at local and state levels in order to provide coordination in environmental monitoring at all levels.

12.3 METHODOLOGY

12.3.1 Data Collection

The study utilized data from secondary sources. The secondary data for the research was the documented materials such as population and annual growth rate (National Population Commission (NPC) 2006), historical records, and a topographic map of the study area which was collected from the Ministry of Lands and Survey, Minna, Niger State. The satellite imagery of the study (Enhance

TABLE 12.1
Land Use and Land-Cover Classification Scheme

S/N	Classification	Description
1	Built-up area	All residential, commercial, and industrial areas, village settlement and transportation infrastructure.
2	Bare surface	Cropland and pasture, orchards, groves, vineyards, nurseries, and ornamental, horticultural areas, confirmed feeding operations.
3	Vegetation	Trees, shrubland, and semi-natural vegetation, deciduous, coniferous, and mixed forests, palms, orchids, herbs, gardens, and grasslands.

(Authors' classification, 2019)

Thematic Mapper (ETM) 2000, 2005, 2010, and 2018) over a period of 18 years were obtained from the National Remote Sensing Centre, Jos, Plateau State, Nigeria and analysed to determine land use trends. To determine the population figure for 2000 and 2005, the population figure of 1991, 108,561, from the National Population Commission was projected to obtain the figures for the years 2000 and 2005 (Table 12.1).

12.3.2 Image Processing Techniques (Classification)

The bands 3, 2, 1 satellite imagery was used to form the false color composite for the study. On the bands 3, 2, 1, false color composite, vegetation appears as red, built-up areas appear in cyan color, and bare surfaces/degraded lands appears in white color. The area of interest (Suleja) was clipped out from the four satellite images acquired (2000, 2005, 2010, and 2018), and sample sets (built-up areas, vegetation, and bare surfaces) were created for the respective years under study. A sample set stores locations of sampled pixels and the assigned class names. The sample sets created were then subjected to a supervised maximum likelihood classification on ILWIS 3.3 Academic software for the four satellite images used.

12.3.3 Overlay Process

Image overlay operation is the geospatial process or procedure prior to the determination of spatial topological relationships. The aim of an image overlay operation is to determine "what spatial feature is on top of what." An overlay operation is much more than mere merging of points, lines, and polygonal features, but it involves all the attributes of the features taking part in the overlay operations. For this study, the feature overlay analysis was carried out. In carrying out the feature overlay analysis, all the classified imageries were transformed from raster to polygon using ILWIS 3.3 Academic. The polygons for built-up areas were created on each of the four satellite images classified (2000, 2005, 2010, and 2018), and the layers of the polygons were overlaid with the polygon of built-up area 2018 serving as the base year (See Figure 12.2). The overlay of the built-up area shows how urban developments within Suleja have displaced or encroached upon other land uses/land cover.

12.4 RESULTS

12.4.1 Examination of the Dynamics of Urban Growth in Suleja Local Government Area

A population figure of 144,149 and 169,154 was obtained for the years 2000 and 2005 respectively, while the 2006 population figure for Suleja (216,578) was used for the projection of the population

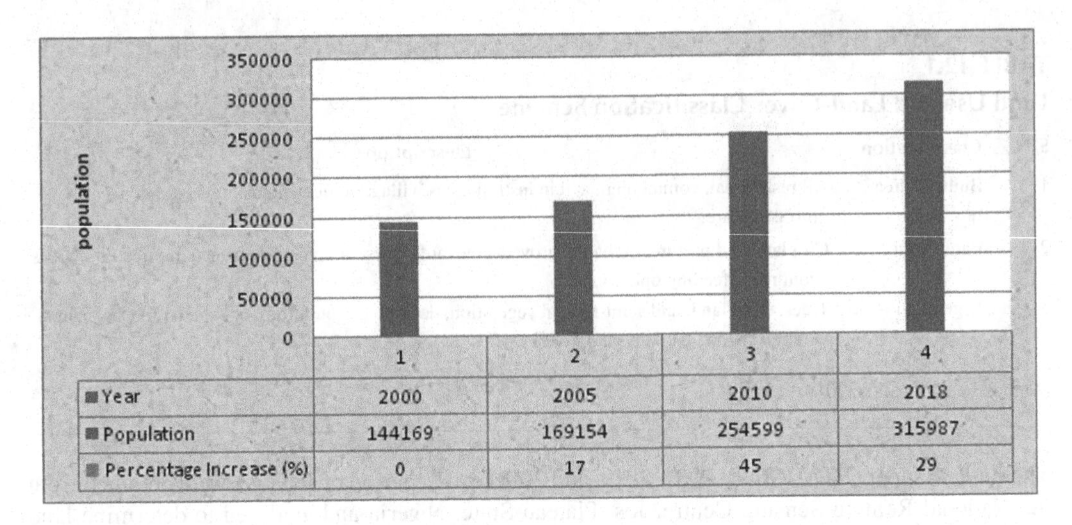

FIGURE 12.2 Percentage increase in population in Suleja between 2000 and 2018. (Authors' computation, 2018.)

figure of the year 2010 and 2018. Population figures of 245,599 and 315,987 were computed for years 2010 and 2018. The population figure of the years under study was calculated by Equation (12.1):

$$Pt = Po\left(1 + r/100\right) n \tag{12.1}$$

Po = the population figure for 1991, i.e. 108,561
Pt = projected population
r = annual growth rate (3.2) of Suleja
n = time lag between the base year and the target year

Figure 12.2 shows that there was a percentage increase of 17% in the population of Suleja between 2000 and 2005, between 2005 and 2010, there was 45% increase, while between 2010 and 2018, there was an increase of 29% in the population figure. The percentage increase was calculated by subtracting the previous population from the original or current population, then the total was divided by the previous population and multiplied by 100.

12.4.2 Examination of Land Consumption Rates in Suleja

12.4.2.1 Change in Land Cover of Suleja between 2000 and 2018

Figure 12.3 shows the land cover of Suleja in the year 2000. The bare surface covered land area of 9.23 km^2 (4.4%), and the built-up area and vegetation covered 16.32 km^2 (7.7%) and 186.08 km^2 (87.92%) respectively.

Figure 12.4 shows the land cover of Suleja in the year 2005. In the year 2005, there was a significant change in the land cover of Suleja. Also, it revealed that bare surface covered a land area of 5.37 km^2. The built-up area and vegetation covered 21.87 km^2 and 184.85 km^2 respectively.

Figure 12.5 revealed the land cover of Suleja in the year 2010; it shows that bare surface covers a land area of 0.27 km^2, the built-up area covers 27.10 km^2, and vegetation covers 184.0 km^2.

Figure 12.6 shows the extent of land cover of Suleja in the year 2018. Bare surface covers a land area of 1.72 km^2, the built-up area covers 45.74 km^2, and vegetation covers 289.4 km^2.

12.4.3 Magnitude of Change between the Years 2000 and 2005

In Table 12.2, the magnitude of change in Suleja (C) between the year 2000 and 2005 was calculated by subtracting the annual land use frequency A from B (land use in 2000 from 2005). The

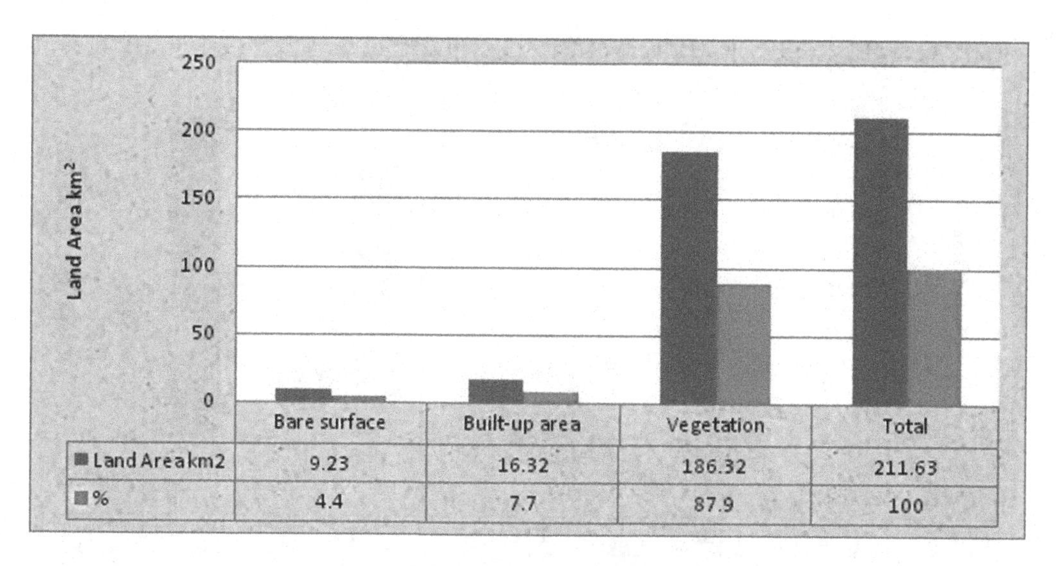

	Bare surface	Built-up area	Vegetation	Total
Land Area km2	9.23	16.32	186.32	211.63
%	4.4	7.7	87.9	100

FIGURE 12.3 Land cover of Suleja in 2000. (Authors' computation, 2018.)

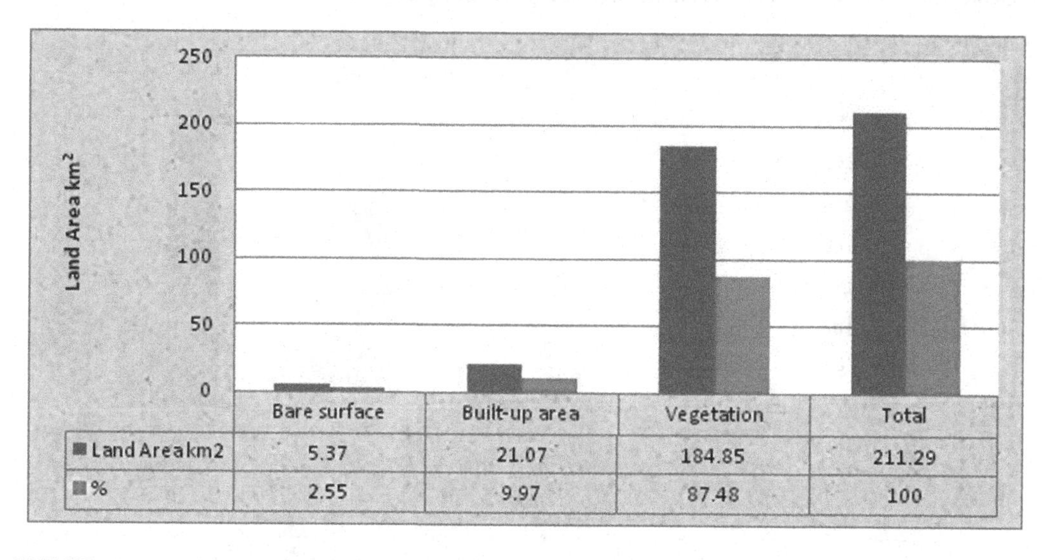

	Bare surface	Built-up area	Vegetation	Total
Land Area km2	5.37	21.07	184.85	211.29
%	2.55	9.97	87.48	100

FIGURE 12.4 Land cover of Suleja in 2005. (Authors' computation, 2018.)

annual frequency of change (D) is derived by dividing the magnitude of change of each land use by five years (reference year). The percentages of change (E) were calculated by dividing the magnitude of change of each land use by A and multiplying by 100. Table 12.2 shows that the built-up area has been growing at 0.95 km^2 yearly and bare surface has suffered a loss of –0.77 km^2 yearly.

12.4.4 Magnitude of Change between the Years 2005 and 2010

In Table 12.3, the magnitude of change in Suleja (C) between 2005 and 2010 was arrived at by subtracting A from B (land use in 2005 from 2010). The annual frequency of change (D) was arrived at by dividing the magnitude of change of each land use by five years (reference year). The percentages of change (E) were calculated by dividing the magnitude of change of each land use by A and multiplying by 100. Table 12.3 shows the built-up area has been growing at 1.21 km^2 yearly, and bare surface growing at 1.02 km^2 yearly.

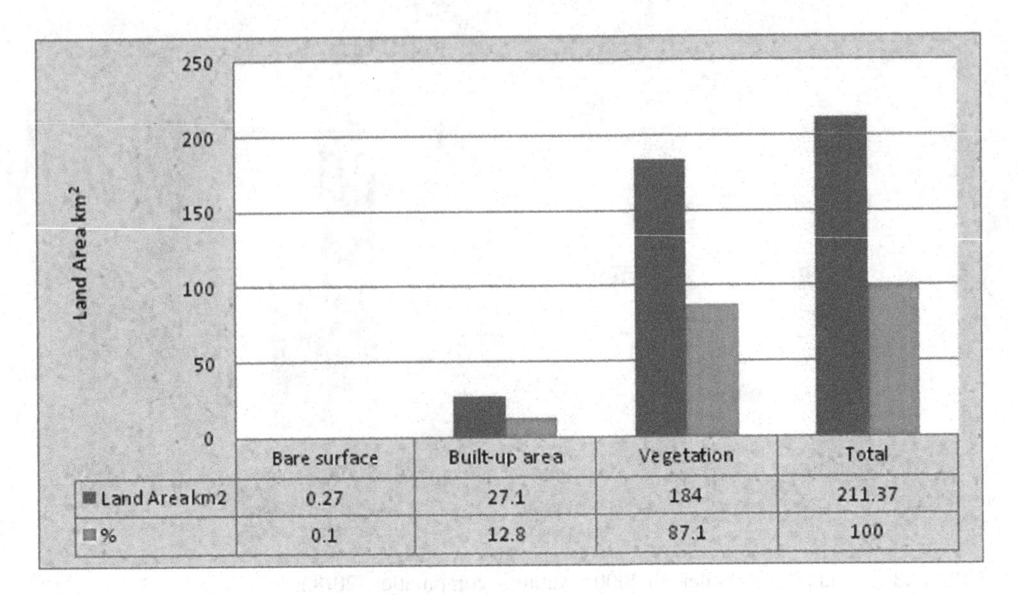

FIGURE 12.5 Land cover of Suleja in 2010. (Authors' computation, 2018.)

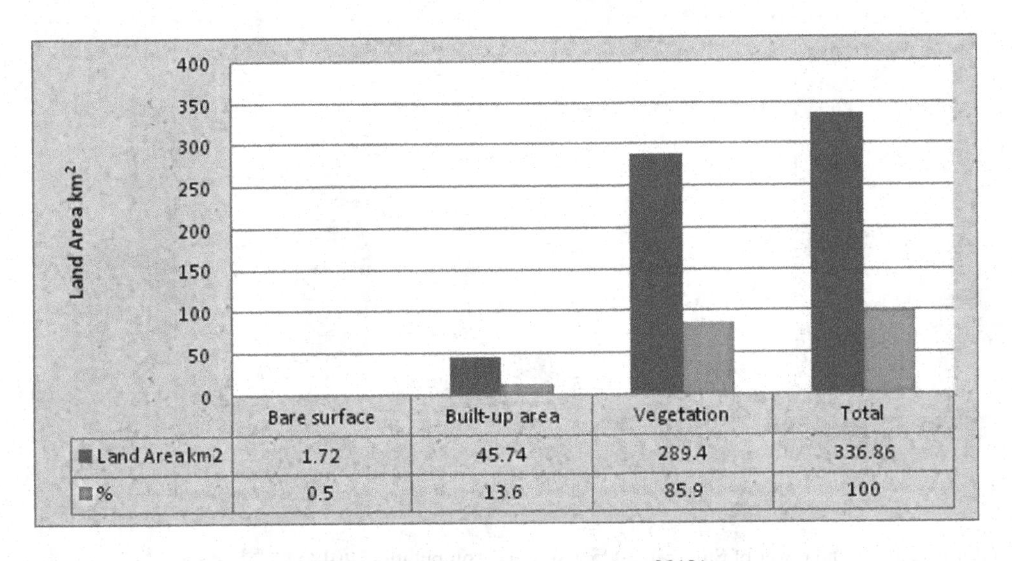

FIGURE 12.6 Land cover of Suleja in 2018. (Authors' computation, 2018.)

12.4.5 Magnitude of Change between the Years 2010 and 2018

In Table 12.4, the magnitude of change in Suleja (C) between 2010 and 2018 was calculated by subtracting A from B (land use in 2010 from 2018). The annual frequency of change (D) was derived by dividing the magnitude of change of each land use by eight years (reference year). The percentages of change (E) were calculated by dividing the magnitude of change of each land use by A and multiplying by 100. Table 12.4 shows the built-up area has been growing at 2.33 km^2 yearly, but the bare surface reduced by 0.18 km^2 yearly compared to 2005 and 2010.

12.4.6 Land Use Pattern/Cover of Suleja

Figure 12.7 shows the image characteristics of Suleja for the year 2000. The bare surface areas cover 9.23 km^2 (4.4%). The built-up area is 16.32 km^2 (7.7%). Vegetation has a total area of 186.08 km^2

TABLE 12.2
Magnitude and Percentage of Change in Land Cover between 2000 and 2005

CLASSES	A 2000	B 2005	C MAGNITUDE OF CHANGE (B–A)	D ANNUAL FREQUENCY OF CHANGE C/5	E PERCENTAGE OF CHANGE C/A * 100
Bare surface	9.23	5.37	−3.86	−0.77	−41.82
Built-up area	16.32	21.07	4.75	0.95	29.11
Vegetation	186.08	184.85	−1.23	−0.25	−0.66
Total	211.63	211.29	−0.34	0.70	−13.37

(Authors' computation, 2018)

TABLE 12.3
Magnitude and Percentage of Change in Land Cover between 2005 and 2010

CLASSES	A 2005	B 2010	C MAGNITUDE OF CHANGE (B–A)	D ANNUAL FREQUENCY OF CHANGE C/5	E PERCENTAGE OF CHANGE C/A * 100
Bare surface	5.37	0.27	−5.1	1.02	95.0
Built-up area	21.07	27.10	6.03	1.21	28.62
Vegetation	184.85	184.00	−0.85	−0.17	0.50
Total	211.29	211.37	0.08	2.06	124.12

(Authors' computation, 2018)

TABLE 12.4
Magnitude and Percentage of Change in Land Cover between 2010 and 2018

CLASSES	A 2010	B 2018	C MAGNITUDE OF CHANGE (B–A)	D ANNUAL FREQUENCY OF CHANGE C/8	E PERCENTAGE OF CHANGE C/A * 100
Bare surface	0.27	1.72	1.45	0.18	537.03
Built-up area	27.10	45.74	18.64	2.33	68.78
Vegetation	184.00	289.4	105.4	13.175	57.28
Total	211.37	211.37	125.50	15.70	663.1

(Authors' computation, 2018)

(87.92%). Figure 12.2 shows the image characteristics of Suleja in 2005. The bare surface area has an area of 5.37 km^2 (2.5%). The built-up area covers an area of 21.07 km^2 (10%), and vegetation has an area of 184.85 km^2 (87.5%).

Figure 12.2 depicts the spread of Suleja in 2010; the bare surface covers an area of 0.27 km^2 (0.1%). The built-up area covers an area of 27.10 km^2 (12.8%). Vegetation covers an area of 184.00 km^2 (87.1%) , and Figure 12.2 also shows the spread of Suleja in 2018; a bare surface covers an area of 1.72 km^2 (0.5%). The built-up area covers an area of 45.74 km^2 (13.6%). Vegetation has an area of 289.4 km^2 (85.9%).

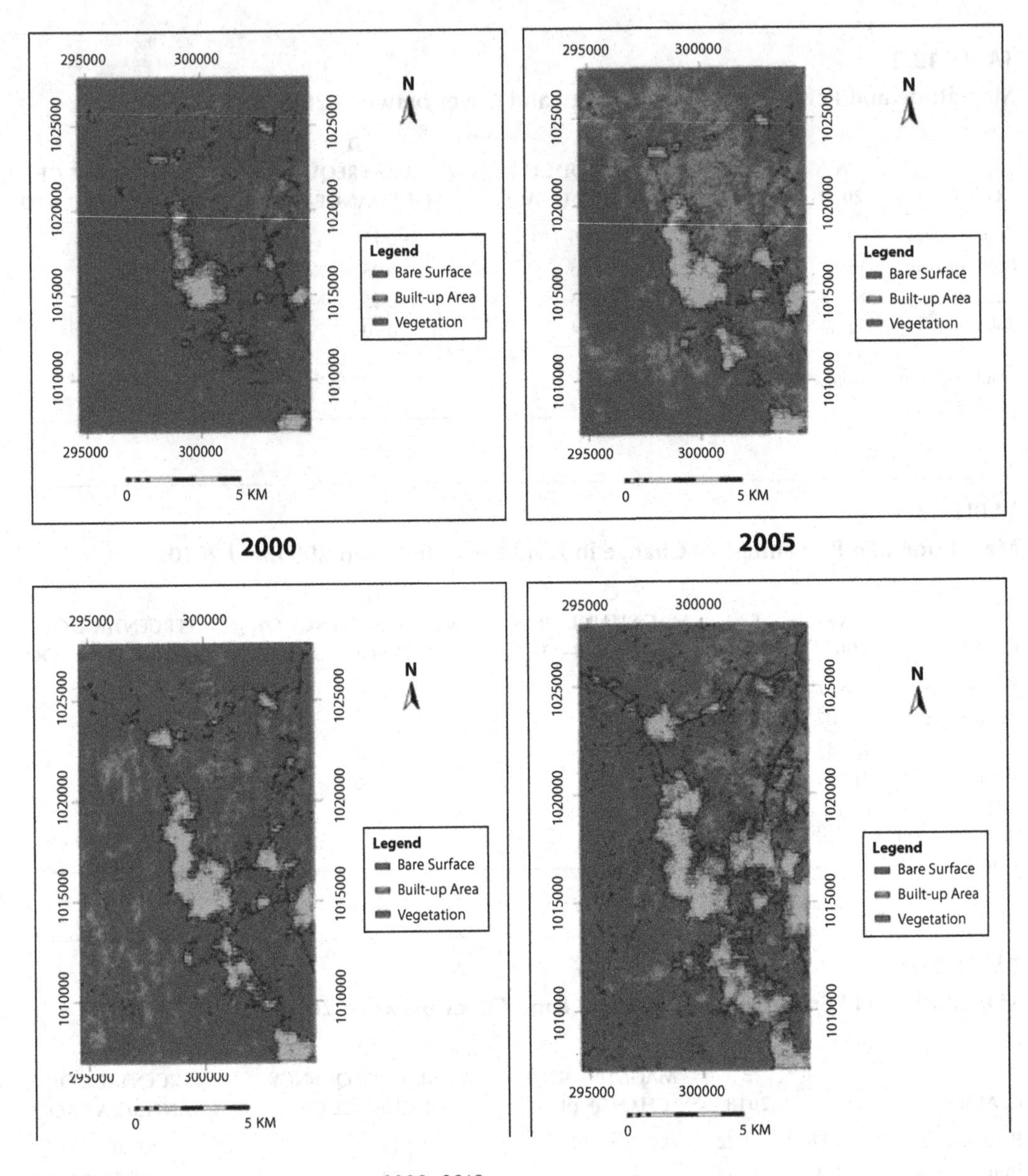

FIGURE 12.7 Changes in land cover 2000–2018.

12.4.7 Forecasting the Future Land Risk in Suleja

Figure 12.7 shows that Suleja is growing toward the southeast of the city which is the main express road linking to the Federal Capital Territory of Nigeria. Many scholars have identified a failure to evolve master plan provision, the general lack of development control, environmental inconvenience, a debilitating state of disorder, chaos, and high crime rate, a massive backlog of infrastructure neglect, and a shortage of qualitative housing dwellings as part of Suleja's problems emanating from incessant growth; these problems will double in ten years' time. The economic situation of the country has forced more people to leave the Federal Capital Territory Abuja of Nigeria to search for cheaper accommodation and a cheaper life that is available in the study area. Likewise, people along this axis are vulnerable to flooding and epidemics.

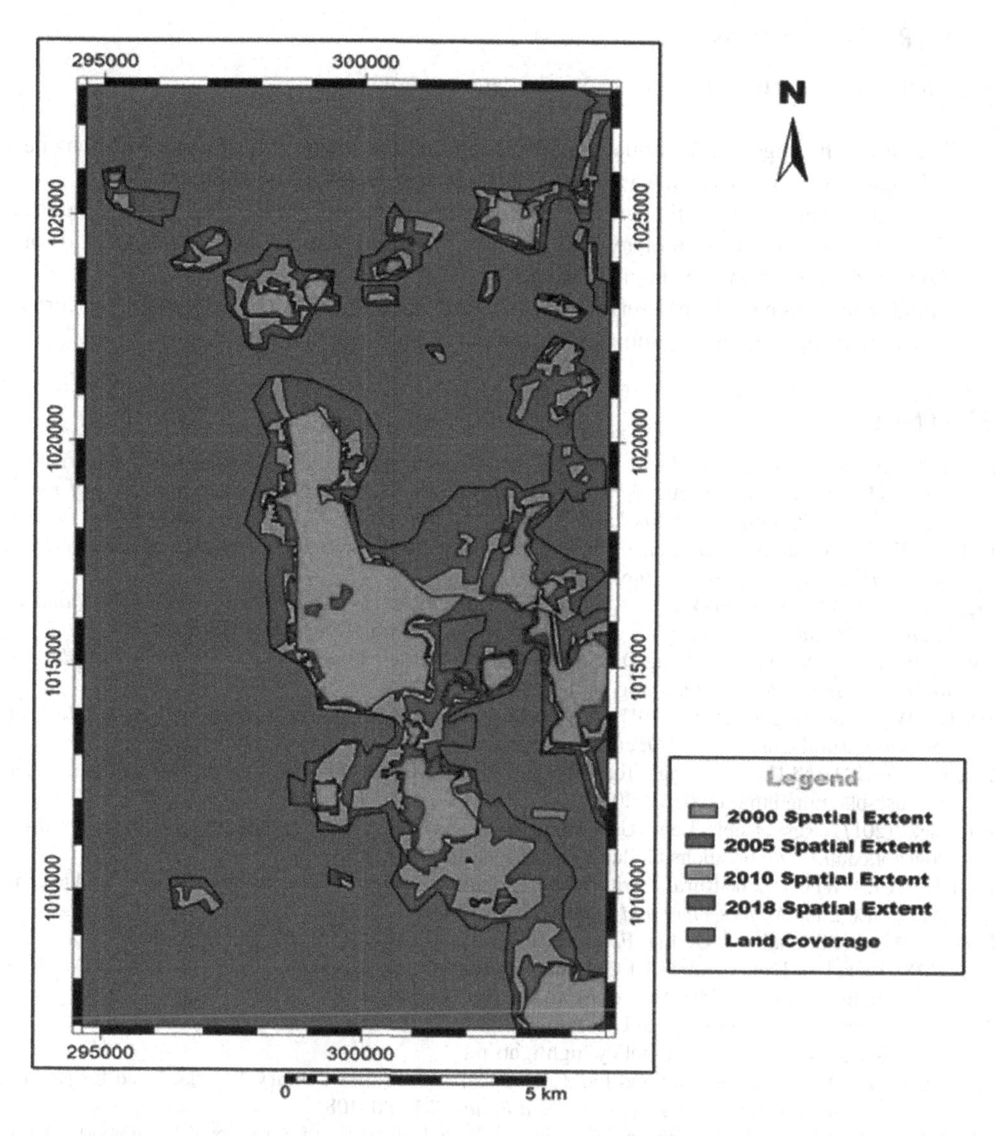

FIGURE 12.8 Changes in spatial extent 2000–2018.

12.5 CONCLUSION

This study has shown the pattern and direction of growth of Suleja between year 2000 and 2018 using remote sensing and GIS tools. The results show that the city is growing at a rapid rate; there was a percentage increase of 17% in the population of Suleja between 2000 and 2005 and between 2005 and 2010, there was a 45% increase, while between 2010 and 2018, there was an increase of 29%. The population figure is also growing toward the southeastern direction mainly along the main transportation routes (see Figure 12.8). The growth pattern is largely influenced by factors such as low rent and cheap household commodities. The growth in Suleja as observed from the prediction may lead to a high cost of housing, decline in agricultural lands, high cost of living, chronic slum generation, increased crime rate patterns, and by extension, a high rate of unemployment.

In the next 10 to 15 years, many more areas not fit for habitation will be occupied, and many more will lose land rights through demolition as they continue to build towards the FCT.

12.5.1 Recommendations

Based on this research, the following recommendations were made:

1. The research suggests that valuable plots of land on the major and strategic locations be redesigned as fit for purpose so as to regulate and stop leapfrog development;
2. The state government should partner with the National Centre for Remote Sensing and the Federal University of Technology, Minna for training and capacity-building of the staff of the Niger State Urban Development Board;
3. Immediate update of the existing master plan and adoption of computerized land records (GIS) for proper land registration and titling in the Suleja should commence.

REFERENCES

Buba, Y. A., Makwin, U. G., & Ogalla, M. (2016). Urban Growth and Landuse Cover Change in Nigeria Using GIS and Remote Sensing Applications. Case Study of Suleja L.G.A., Niger State. *International Journal of Engineering Research & Technology (IJERT)*, *5*(8), 124–138.

Chigbu, U. E. (2013). Rurality as a Choice: Towards Ruralising Rural Areas in sub-Saharan African Countries. *Development Southern Africa*, *30*(6), 812–825.

Chigbu, U. E., Haub, O., Mabikke, S., Antonio, D., & Espinoza, J. (2016). *Tenure Responsive Land Use Planning: A Guide for Country Level Implementation*. Nairobi, Kenya: UN-Habitat.

de Vries, W. T. (2018). Potential of Big Data for Pro-Active Participatory Land Use Planning. *Geoplanning: Journal of Geomatics and Planning*, *5*(2), 205–214.

de Vries, W. T., & Chigbu, U. E. (2017). Responsible Land Management-Concept and Application in a Territorial Rural Context. *fub. Flächenmanagement und Bodenordnung, 79*(2), 65–73.

Haseeb, J. (2018). Land Use Planning - Techniques, Classification & Objectives. https://www.aboutcivil.org/land-use-planning.html, on 03/02/2019.

Hernandez. (2017). Responsible Land Governance in Urban Areas. https://equalrights4womenworldwide.blogspot.com/2017/04/responsible-land-governance-in-urban.html.

Magel, H. (2015). Where Is the Rural Territorial Development Going? Reflections on the Theory and Practice. *Geomatics, Land Management and Landscape*, *1*, 55–67.

Mohajane, M., Essahlaoui, A., Oudija, F., El Hafyani, M., Hmaidi, A. E., El Ouali, A., & Teodoro, A. C. (2018). Land Use/Land Cover (LULC) Using Landsat Data Series (MSS, TM, ETM+ and OLI) in Azrou Forest, in the Central Middle Atlas of Morocco. *Environments*, *5*(12), 131.

OECD. (2017). The Governance of Land Use; Policy Highlight. (February). https://www.oecd.org/cfe/regional-policy/governance-of-land-use-policy-highlights.pdf.

Otubu, A. (2018). The Land Use Act and Land Administration in 21st Century Nigeria: Need for Reforms. *Journal of Sustainable Development Law and Policy*, *9*(1), 80–108.

Wehrmann, B. (2017). Land Governance: A Review and Analysis of Key International Frameworks, United Nations Human Settlements Programme (UN-Habitat).

Yahyaha, O. Y., & Ishiak, Y. (2013). Effective Urban Land Use Planning in Nigeria: Issues and Constraints. *Journal of Environmental Management and Safety*, *4*(2), 103–114.

13 Inner City Development in Ghana

Nature and Extent of City Expansion in Wa Municipality

E.D. Kuusaana, E.A. Kosoe, Ninminga-Beka, and A.R. Ahmed

CONTENTS

13.1 INTRODUCTION

The growth of cities in the Global South is probably one of the most distinctively modern facts of urbanization and a product of history of human mobility. In 2006, for the first time in human history, the majority of the world's population lived in urban areas (UN-Habitat, 2006). Indeed, the world is increasingly becoming urbanized, and the rate at which city populations grow and countries urbanize shows the leap of social and economic change (Donk, 2006). Thus, urbanization is both a mirror and an instrument of broad socio-economic changes in society.

In Africa, projections indicate that by 2030 the continent's population will exceed the combined population of Europe, North America, and South America (UN-Habitat, 2014). Related to general demographic transition in the continent, the growing population of African cities has been unprecedented. Much of the population growth is expected to take place in cities, as well as in new urban developments. In the last two decades, African cities have experienced a 3.5% growth rate per year and this is expected to increase until 2050 (Lall et al., 2017). Estimates show that densities in cities are expected to increase from 34 to 79 persons per square kilometer by 2050 (UN-Habitat, 2014). The confluence of high fertility rates and economic development are noted to be the key drivers of population growth in African cities for decades to come. The urban population of Africa is expected to increase from 400 million to 1.2 billion by 2050, with over 58% of the African population expected to be in urban areas (UN-Habitat, 2014).

Amidst all the difficulties of economic growth and development, housing supply and access to social services have not kept pace with rates of urbanization (UN-Habitat, 2014). As a result, different aspects of urbanization such as urban poverty, urban sprawl, and the emergence of slums are increasingly being discussed as major consequences of urbanization in Africa (Baker, 2008; Tacoli et al., 2015; UN-Habitat, 2016). However, consensus in urban history, especially from the West has increasingly favored urban renewal as a measure of addressing problems of both the inner city and the peripheries (Segrue, 2014). In the Global South, and particularly in China and Singapore, inner city redevelopment has been used in addressing issues of sprawl, gentrification, and the emergence of slums (Leaf, 1995; Zhang et al., 2014). However, there are limited perspectives regarding the dynamics of inner-city development, particularly in Africa where tenure arrangements can limit state access and control over land. There is less emphasis on over-concentration and excessive infilling within inner cities, even though new studies are showing that recent form of urbanization in Africa is characterized by redevelopment of land within the urban core rather than in the periphery (e.g. Linard et al., 2013).

In Ghana, the population tripled between 1960 and 2010, with over 51% of the people living in urban areas (Ghana Statistical Service, 2012). The urban population grew from 8.3 million in 2000 to 12.5 million in 2010, indicating a 4.2% annual growth rate (Ghana Statistical Service, 2012). Even though the actual growth rate of urban slum population has witnessed a decline since the 1990s, the proportion of urban slum population stood at 37.9% as of 2014 (see Figure 13.1 for further details of urban population size and growth pattern). What is more significant is that this figure remains high and these slum areas are very noticeable in the largest cities of Ghana such as Accra, Kumasi, and Sekondi-Takoradi, among others, with growing incidences in the secondary cities as well (Owusu and Afutu-Kotey, 2010).

Most secondary cities and many small towns in Ghana including Wa have a mixed character of both radial and sectorial developments. Hence, developments emerge out of autochthonous settlements in the inner cities, and over decades, over-concentration at the inner cities has been the trend. The implication of this is that buildings become compact and dense with limited access to basic amenities like adequate sanitation and water. Notwithstanding the important role that inner cities play in the characterization of the city, much of the literature on urban development has focused excessively on peri-urbanization (see Farvacque-Vitkovic et al., 2008; White et al., 2007; Cohen, 2005; Lerise et al., 2004; Fodor, 1999) to the neglect of the dynamics in the inner cities. This chapter therefore draws on the concept of spatial justice as an analytical lens to explore the nature and extent of inner-city developments in Ghana, with the Wa Municipality as a case study.

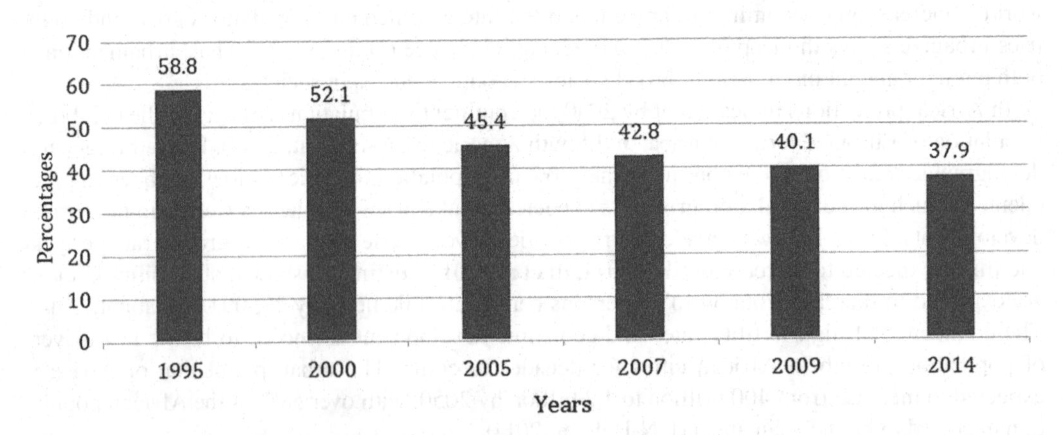

FIGURE 13.1 Urban slum population growth pattern in Ghana. (UN-Habitat, retrieved from the United Nation's Millennium Development Goals database. Data are available at: http://mdgs.un.org/.)

13.2 SPATIAL JUSTICE: ANALYTICAL LENS

Spatial processes of redeveloping urban areas arise out of the need to reorganize cities and areas that have not developed according to modern principles of spatial planning. Such processes sometimes include upgrading of slums, development of public amenities (Zheng et al., 2016), development of new zoning rules, and the conversion of agricultural lands into other land use-types, like commercial and residential (Uwayezu and de Vries, 2018). These processes, when embarked on from political viewpoints with emphasis on economic growth, often result in social injustices (Uwayezu and de Vries, 2018).

Social justice is a concept that originates from the conceptualization of social justice into space. Social justice concerns itself with environmental and social discrimination while levitating political ecology from planning systems (Hafeznia and Hajat, 2016). While there is no universally agreed definition of spatial justice, authors such as Harvey (1998), Soja (2010), Fainstein (2010), and Lévy (2012, 2013) have been debating the connections between spatial thoughts and theories of justice. Edward Soja, a key proponent of the concept, while leaving the definition of spatial justice open contends that, however, it might be defined, spatial justice "has a consequential geography, and a spatial expression that is more than just a background reflection or a set of physical attributes to be descriptively mapped" (Soja, 2010, p.1).

Spatial justice therefore refers to justice in the physical space. It is the combination of space, politics and social justice.

> It is an application of spatial or geographical aspect of justice, fair and equitable distribution of resources and wealthy opportunities in the society which also can be considered as output or a process of geographical patterns or distribution which are fair or unfair or process which produce these outputs (Soja, 2008, p.4).

Social justice entails the recognition and respect of fundamental human rights for people in a geographic space and the allocation of resources through the principle of equity. It calls for inclusive spatial development in an attempt to reduce economic inequalities and social schisms caused by urban (re)development (Soja, 2009).

Spatial justice therefore becomes a useful analytical lens for understanding the dynamics of inner city (re)development, because it offers a platform for unpacking the underlying factors shaping access, control, and use of space within the inner city, as well as the socio-economic inequalities that come with it. Spatial justice as an analytical framework that argues the role of space (a set of material and ideological relations) is formed by social relations that can (re)produce justice or injustice. As a normative concept, spatial justice is concerned with questions such as what kind of justice can be deployed in analyzing spatial arrangements, if spatial justice is about the distribution of social justice, and what spatial justice can do that environmental justice or social justice cannot (Williams, 2013; Connelly and Bradley, 2004). Although spatial justice literature contends that geographic space is an important component in producing justice relations, most ideal questions (as outlined above) are underdeveloped.

13.3 MATERIAL AND METHODS

13.3.1 Study Area

Wa is the capital of the Wa Municipality, which shares administrative boundaries with Nadowli District to the north, Wa East District to the east, and to the west and the south Wa-West District. It lies within latitudes 1°40'N to 2°45'N and longitudes 9°32'W to 10°20'W. Wa Municipality has Wa as its capital, which also serves as the regional capital of the Upper West Region. It has a land area of approximately 579.86 square kilometers, which is about 6.4% of the Region (Ghana Statistical Service, 2014). The Wa Municipality has a total population of 107,214 and forms 15.3% of the population of the Upper West Region. The most dominant type of dwelling unit in the Municipality and Wa, the capital town, is compound houses (Ghana Statistical Service, 2014). The increase in population in Wa has put pressure on existing dwellings; hence the expansion of the city due to the proliferation of construction. Paradoxically, the expansion does not only take the form of sprawl, but also expansion within the city centers. While in most cases settlers and migrants build outside the city center, the indigenes are developing the limited spaces in the city centers with new residential buildings. This situation has resulted in the development of slums within the Wa Township. The study area is shown in Figure 13.2.

The study collected data from three major inner-city residential communities: Wa Central residential area, Zongo-Kabanye residential area, and Dondoli-Limanyiri residential area. These neighborhoods were selected because they lie within the Central Business District and in close proximity to Wa Central market. Also, they are characterized by neighborhoods of original settlements and some settler communities. For example, Puohuyiri, Daanaayiri, Sokpayiri, and Suuriyiri are native settlements of Wa, while Zongo, Limanyiri, Tagrayiri, Nayiri, and Wapaani are old settler communities that provided various essential services to the city and its native settlers. These areas also serve as major enclaves for Islamic education and clerical activities. This explains why the central mosques of the Ahmadiyyah and orthodox Muslim groups are both located in this area. These three selected study areas are depicted in Figure 13.3. The shaded parcels represent the randomly selected parcels for this study in the different residential areas in the inner city.

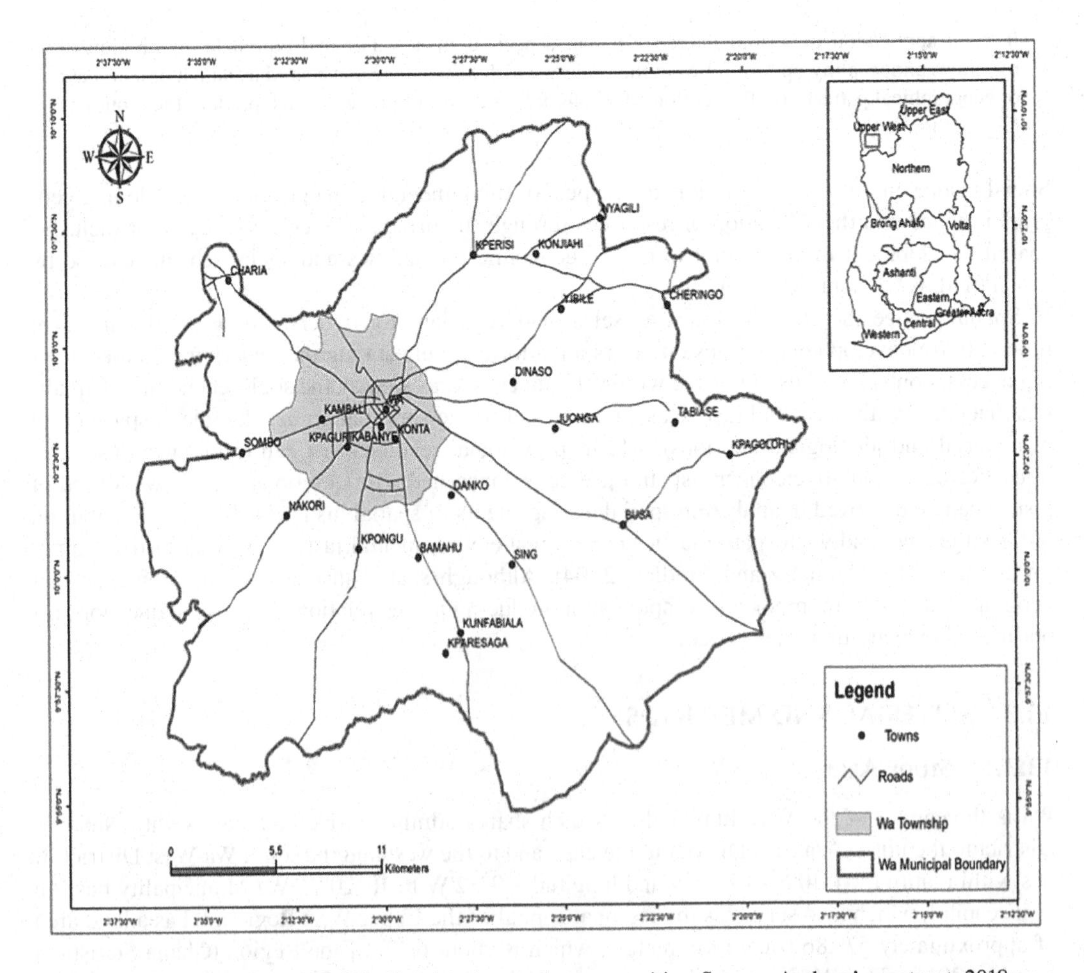

FIGURE 13.2 Map of Wa Municipality showing study communities. Source: Authors' construct, 2018.

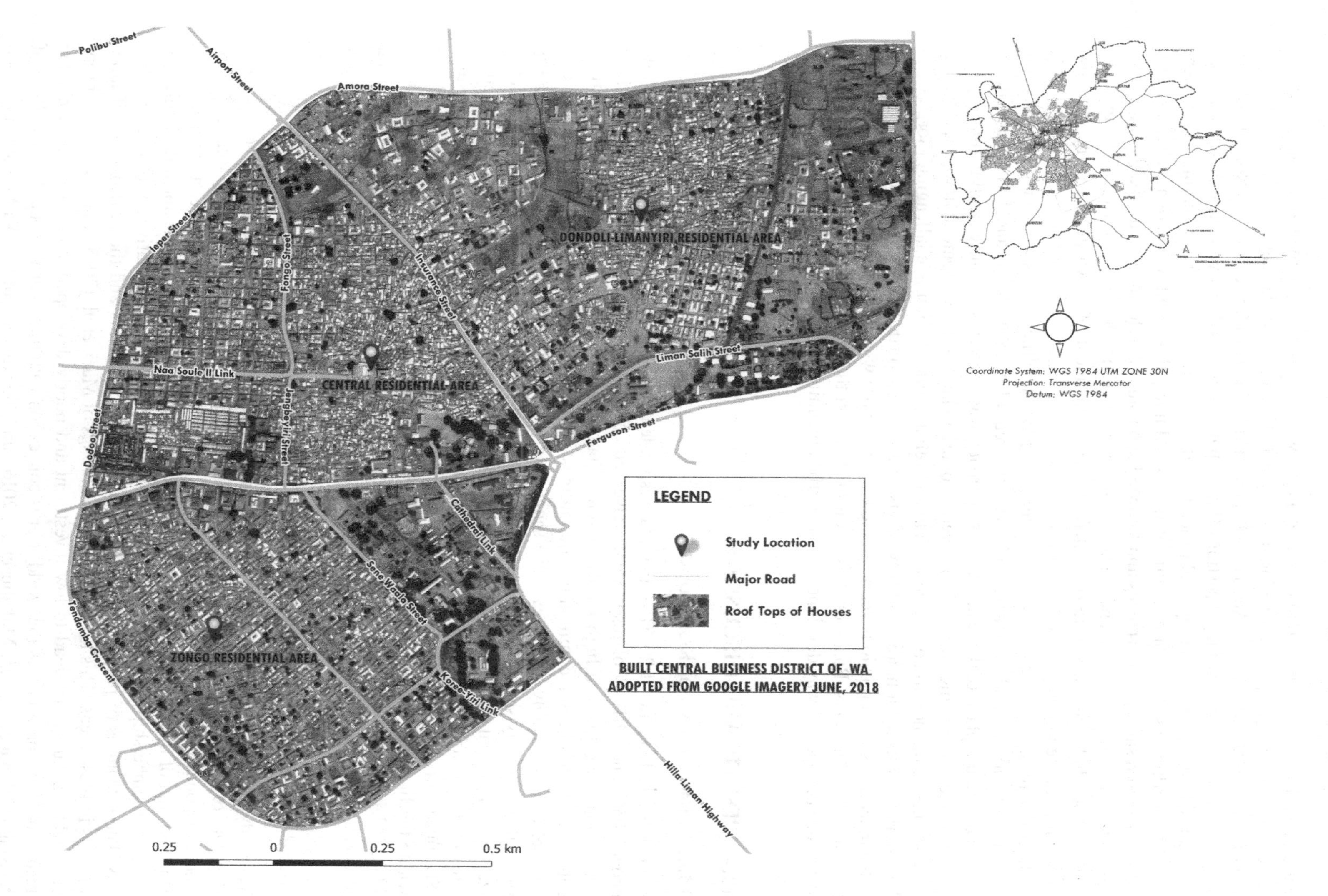

FIGURE 13.3 Three selected study residential areas showing selected residential parcels. Source: Authors' construct, 2018.

13.3.2 Data Collection and Analysis

Data was collected using questionnaires and interview guides to gather both quantitative and qualitative data. The study employs an exploratory sequential mixed method design using cross-sectional research data collected between August and September 2018. The choice of this design enables the study to ask precise and probing questions like "what," "how," and "why" of respondents including tenants, house owners, and regional officials of the Land Use and Spatial Planning Department (LUSPD) working as the planning unit of the Wa Municipal Assembly. Qualitatively, data was collected from residents' experiences and knowledge on slum development and urban sprawl. Data was collected on the processes and causes of urban sprawl/slum development from 135 randomly selected households and a key informant from the Department of Town and Country Planning Department at the municipal level.

To enable the analyses of the nature of inner city of Wa, satellite images were analyzed and overlaid on neighborhood land use maps, to observe the areas of intensifications and deviations with the proposed land use plans of those areas. This approach was necessary because satellite imagery has the advantage of providing the physical coverage of urban land and development; however, choosing the appropriate method to collect up-to-date and reliable information is challenging (Sori, 2012). The overlay presented a more vivid way of visualizing both vertical and horizontal developments in the inner cities from previous years, and to examine those with and without building permits for these new developments. Based on these examinations on the maps, it became possible to further synthesize the reasons why developers violate urban planning regulations.

13.4 RESULTS AND DISCUSSION

13.4.1 Background of Respondents

From Table 13.1, a total of 20.7% of the total respondents had no form of formal education. While about 24.4% and 14.1%% of respondents have had secondary education and junior high school completion respectively, a high number (34.8%) had either completed or were pursuing tertiary education. As has been a common practice in most slums area, many people start to take care of themselves immediately after they leave senior and junior high schools. The literature generally shows that inner city slums are hubs of people with low education, cheaper housing, and characterized by an informal economy (UN-Habitat, 2003). However, in the case of the three sites, there is a reasonable number of educated dwellers and students in tertiary schools, as well as civil servants (See Table 13.1). This is a bit different from the general characteristics of slums in the Global South within the literature (see UN-OHRLLS, 2016). The reasons for this can be diverse (See Sections 13.4.3–13.4.5) and can be sources of different levels of power which have implications on spatial justice, by (re)production of justice and injustice.

13.4.2 Property Ownership and Housing Characteristics

The study found that many of the respondents were natives with various degrees of property ownership. Out of the total respondents, only 20% had ownership of the properties they were living in. These respondents acquired their properties through self-construction, purchase, and inheritance. However, out of the respondents who did not own the houses, about 93.5% occupied the houses by virtue of membership of their families, while others, especially the non-natives, rented the properties. These features of property rights have facilitated investments and the development of labor markets in these areas. This is contrary to the findings of Marx et al. (2013), that ambiguous property rights in slums may lead to low investment and poorly developed labor markets. A notable feature of slum communities the world over is congestion with high densities and low standards of housing (UN-Habitat, 2003; Mahabir et al., 2016), and the situation in the Wa Municipality is no

TABLE 13.1
Summary of Respondents' Characteristics

Variable	Frequency (n = 135)	Percentage (%)
Gender		
Male	103	76.3
Female	32	23.7
Age		
20–29	14	30.4
30–39	27	20.0
40–49	24	17.8
50–59	23	17.0
60+	20	14.8
Educational Level		
No schooling	28	20.7
Primary	8	5.9
Middle/JHS	19	14.1
Secondary/SHS	33	24.4
Tertiary	47	34.8
Marital Status		
Single	50	37.0
Married	76	56.3
Divorced	4	3.0
Widowed	5	3.7
Occupation		
Student	24	17.8
Trader	28	20.7
Artisan	27	20.0
Farmer	14	10.4
Public servant	28	20.8
Unemployed	14	10.3
Religion		
Islam	129	95.6
Christianity	6	4.4

Source: Field survey, 2018

different. The study found large household numbers across all slum communities over the years. More than 46% of the slum houses have between 1–5 households; 33% between 6–10 households; 12% between 11–15 households; 3% between 16–20 households; while 6% of the households are made up of more than 20 households. The average number of households per house in the slum communities is 8.23. This is, however, does not compare favorably with the Wa municipality average household per house of 2.0 (Ghana Statistical Service, 2012). On the number of rooms per house, 51% of houses have between 11–20 rooms; 40% have between 1–10 rooms; while 10% of houses have more than 20 rooms, giving an average number of rooms per house of 29. Consequently, the study found high house occupancy rates in the slum areas. The study showed that the households consisted of extended families with the majority (79%) of the living in single rooms, 11% of respondents living in two rooms, 6% living in three rooms, and 4% in four rooms.

As a result of the high densities, the study found the majority of the housing units are compound houses with the capacity to accommodate many households. In almost all the slum areas, there were a few temporal structures housing people, which were in the form of kiosks and iron sheets and other wooden materials. This is in line with Turner's (1970) concept of "self-help initiatives," where residents have their own initiatives and coping strategies put in place to supplement the housing deficit. However, such self-help initiatives have serious implications for access, control, and use of land, as well as issues of territoriality, as the siting of temporary structures and their extensions go beyond locally agreed boundaries, including encroaching on public spaces. The everyday lived reality is that injustices are therefore apparent in such "self-help initiatives," although they are seen as means of coping with the challenges of the inner-city struggle.

Like in most slums in Ghana's most populated areas, most of the housing structures and units were in deplorable conditions as well as high densities. The study showed that the majority (74.1%) were more than 40 years old (Figure 13.4). In fact, some of the buildings were more than five centuries old. However, such structures are increasingly being replaced. The study revealed that only 6.7% of the buildings were ten years old or less, and such developments are highly contested, as the replacement of old structures come with new facilities that have implications on the encroachment on public space.

13.4.3 Inner City Densification: Characterization of Inner Cities

Over half (52%) of the slum areas were built with mud clay and grass roofs, especially houses that are more than 20 years old. About 48% of the houses in the communities were built with sandcrete blocks roofed with aluminium sheets, and considered to be durable and quality materials for housing structures. Such houses are generally those built within the last two decades.

One key feature of inner-city congestion and densification is the expansion of original housing units. About 78% of respondents indicated that there have been changes to the building which they occupy. Respondents indicated the changes were made to add new rooms to the original structures and the reasons behind such changes are displayed in Figure 13.5. This finding reinforces Abebe's (2011) classification of informal settlements: from infancy through consolidation to saturation. According to Abebe (2011), at the saturation stage further construction is continued through horizontal densification and this is accounted for by the need to find more rooms for increasing household sizes. However, making changes to existing structure is highly contested for a number of reasons.

Figure 13.6 shows the different reasons that are highly contested among dwellers in terms of their inabilities to modify existing structures. Access to land remains the most important factor in the modification of existing structures for a number of reasons. Both private and public spaces within the slums are massively encroached upon, especially by the growing middle class who can

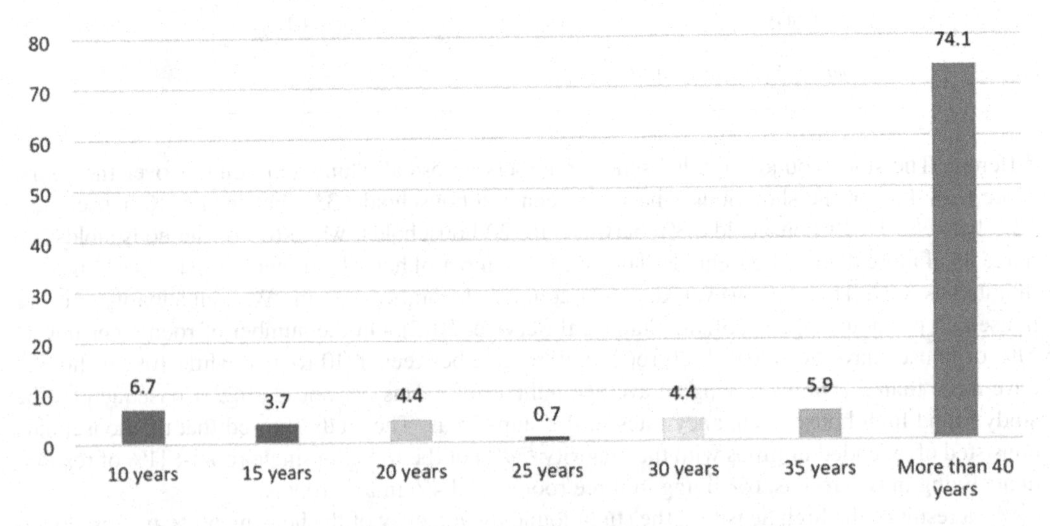

FIGURE 13.4 Number of years building has existed. Source: Field survey, 2018.

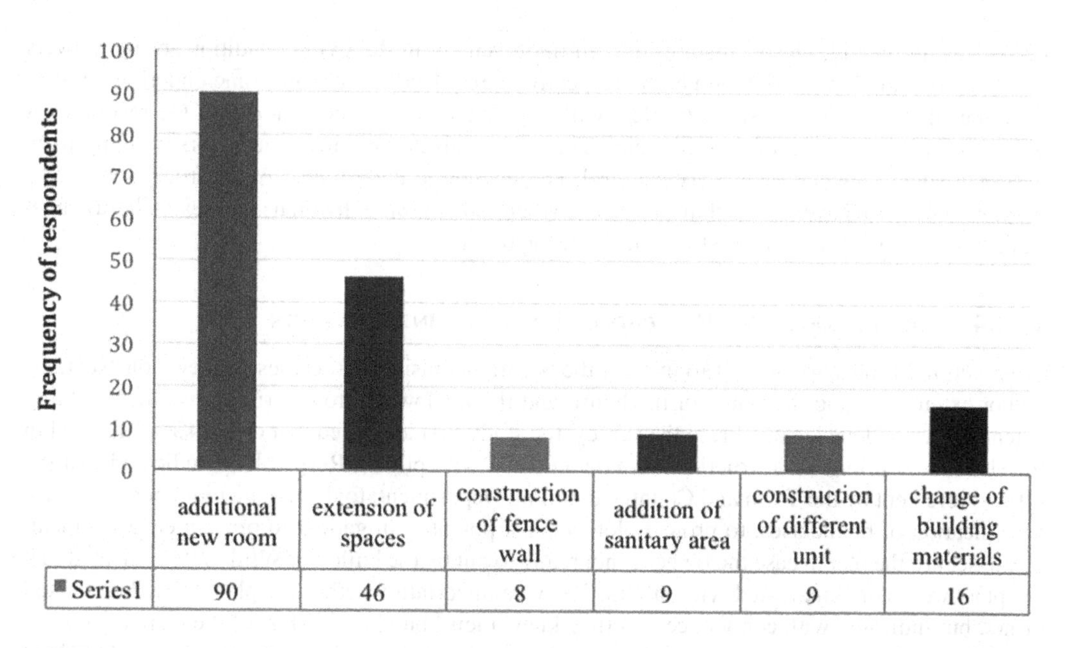

FIGURE 13.5 Reasons why house owners make changes to existing houses. Source: Field survey, 2018.

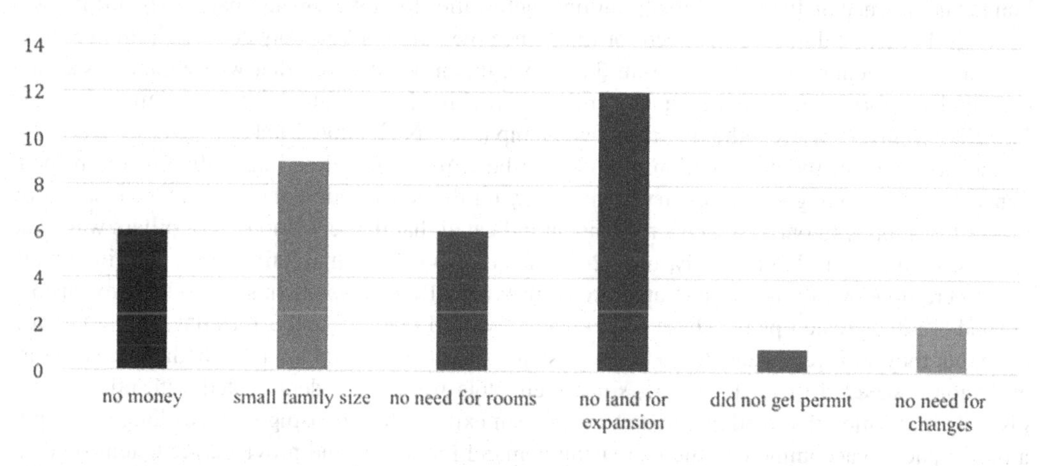

FIGURE 13.6 Reasons why house owners have failed to make changes to existing buildings. Source: Field survey, 2018.

build faster. As a result, injustices are apparent, especially when a poor household requires the smallest amount of space for basic necessities of life such as building a toilet and water supply. The study found that the nature of housing in the slum areas makes it difficult to connect water to all the houses in the neighborhoods due to limited public space and the unwillingness of dwellers to have their private land trespassed upon. The overprotection of private property rights leaves little space for major public infrastructure that even the most efficient self-help initiatives do not provide.

As a result, instead of expanding the existing housing units, some respondents want to acquire new plots (i.e. even through loans and borrowing) of land at different places and move out of the slum areas. This can be argued from two perspectives. Firstly, it can be seen that they are being coerced into leaving the place to which they have a socio-cultural attachment. This can be a gradual form of pushing people out of the inner city, leading to urban decay and creating new demand for utilities in new settlements (see Keil and Kipfer, 2003). Secondly, the movement out of the slum can be seen as an upgrade in social status and an improvement in the living conditions of the movers (see Giliani et al., 2017). Although both perspectives are relevant, there are some underlying power dynamics that determine if slum dwellers will stay or move out. The majority of the respondents did not want to move out of the inner city. They are motivated by different factors, ranging from personal, social, collective security, cultural, and economic undertones. Aside from land needs, those who want to move out cited instances of robbery attacks on individuals and sexual harassment against young girls that were a bit too much for them to take.

13.4.4 Permit Acquisition: Processes of Inner City Intensification

This section of the study sought to unravel the permit acquisition processes for new constructions and/or extensions, and the status of the house and the landowners to determine whether building extensions were done legally. From the survey, the study first identified that only 24% of the land in the slum areas prior to construction were demarcated with pillars. Respondents indicated that the pillars were kept by the Town and Country Planning Department: to demarcate plot boundaries; for identification of boundaries; to protect plots against potential litigation; to prevent encroachment; and to secure the plot in case the owner is not ready to put up the building. While 71% indicated that the plots were not demarcated with pillars, 5% were uncertain whether the plots were demarcated or not, but indicated with confidence that they knew their boundaries. Most of the respondents did not indicate why their plots were not delineated with pillars but mainly insisted they owned the land originally and did not need to limit that ownership with boundaries in the neighborhoods. Others blamed negligence on the part of the traditional authorities for not ensuring that the right thing was done. Still others did not see the need for that, since they did not have any counterclaim over their land, and their tenure was secure. From the above, it can be observed that where tenure is secure even under informal landholding arrangements, parcel holders in autochthonous communities are unwilling to invest in strengthening their ownership (see UN-Habitat, 2004).

Further, most of the demarcation pillars have been repositioned and according to the respondents, the repositioning was necessary to make way for the construction of roads and the installation of a utility supply. In one instance a respondent indicated that the repositioning of pillars was done because of litigation. Interestingly, only 26% of the respondents maintained that building extensions were done within the demarcated areas. In total, 41% of respondents made their extensions beyond the demarcated plots. These respondents revealed that city authorities have come to warn them, but they still went ahead to do the extensions. About 33% could not tell whether the structure extensions exceeded their plot limits, since their plots were never demarcated with pillars. This gives the indication that land and house owners can expand their housing units, as long as there is a bit of space to accommodate the increasing demand for shelter and provided the extension does not encroach on another person's private property. This revelation presupposes that legitimacy in building extensions are only curtailed when the extension trespasses onto another's private land; so long as it is into a public space, even the neighbors do not raise objections. It also reinforces the need to strengthen the oversight responsibility of the authorities to control and manage land use within city centers, not merely focus on emerging trends in the urban peripheries. In line with Bradshaw (2000), city authorities turn a blind eye to such unauthorized developments city.

Buttressing the case for the oversight duties of city development officials, the study found that only 20% of the respondents actually acquired building permits for extensions on their original structures, while the remaining 80% never applied for development or building permits. In an attempt to understand the motivations behind the erection of these illegal redevelopments, Figure 13.7 gives reasons for this observed trend. From Figure 13.7, 61% of the respondents indicated that it was required by law to acquire the permit; 22% indicated that they did not want their structures to be demolished by authorities; while 17% wanted to secure their lands and buildings from encroachment and litigation. Out of the 84% that did not acquire a building permit for expansion, 37% of them revealed that the buildings were self-owned, so they did not need any permit to make changes

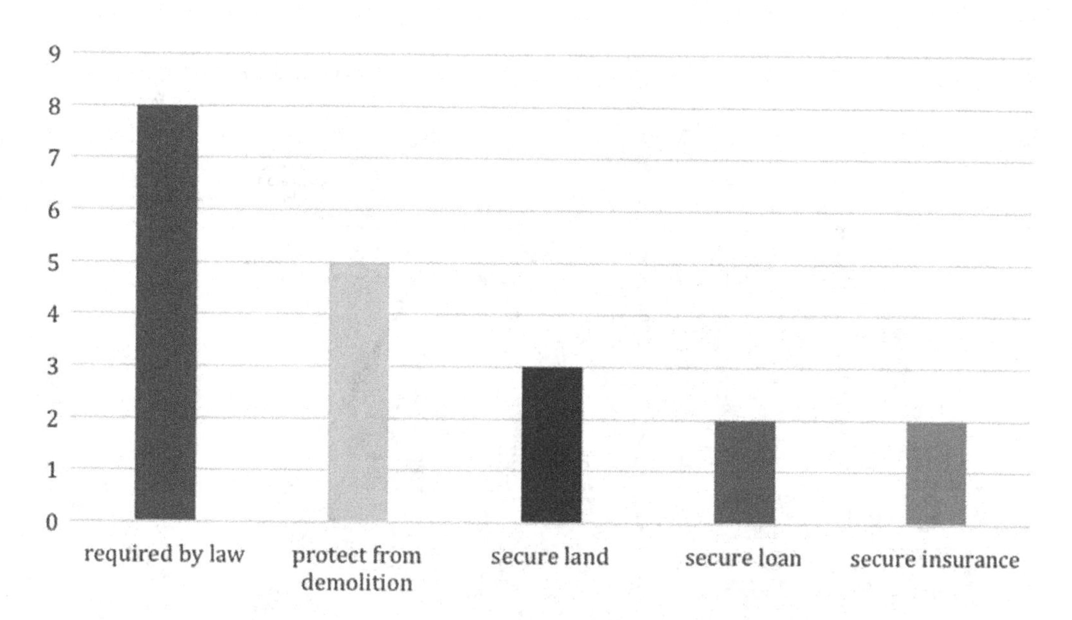

FIGURE 13.7 Reasons for the non-acquisition of building permits during redevelopment. Source: Field survey, 2018.

to it. Some of these respondents mentioned that they did not acquire building permits before they put up the original buildings, hence their reluctance to acquire permits for subsequent expansion of the same buildings. Other respondents also revealed the high cost of the permit acquisition (23%) and the delays and difficulties involved as their reasons for not getting the permits (40%). The findings suggest that most inner-city expansion and intensification in Wa is characterized by unregulated structures to which city authorities turn a blind eye. Previous studies in Ghana indicate that the planning authorities are confronted with the challenges of illegal structural developments giving rise to slum developments as the country continues to urbanize (Botchway et al., 2016).

13.4.5 Spatial Dynamics of Intensification of the Inner Cities: Consequences

In order to fully understand the dynamics of spatial justice in housing expansion and intensification in the inner cities, the study compared images of the selected study areas to see the changes in housing forms between 2005 through 2010 to 2018. To begin with, we compared the approved layout of the Wa inner city area vis-à-vis the contrasted densities. Housing density in this area is the highest in Wa, recording a maximum of 65 units per hectare above the maximum recommended standard of 54 units (Zoning and Planning Standards of Ghana, 2010). However, the part west of the Fongo and Jengbeyiri Streets maintains a medium residential density of 30 units per hectare, as was planned in 1980 (see Figure 13.8). To the east of the Fongo and Jengbeyiri Streets where the densities are above the standards, there are more indigenous groupings than the other half, and this could be the distinctive reason for the high densities. This is because a higher percentage of the respondents have retained their private spaces in these neighborhoods and instead of seeking better opportunities, they are improving the space structures to fit their needs with essential building elements, especially toilets and baths. In Figure 13.9, different shades are used to represent these contrasting densities in the inner Wa area. Even though these developments are not vertical in nature, they bear a resemblance to the saturated stage of urban development as proposed by Abebe (2011).

The local plan of 1980 was superimposed on Google satellite images of 2005, 2010, and 2018 to show the extent of infilling, expansion of units, and the distortions that have occurred over this space of time. In Figure 13.8(a), there were originally 397 individual units and 18 km^2 of clustered units in the area. In 2005, 109 units and 0.1 km^2 of expanded cluster of units were added. Then in

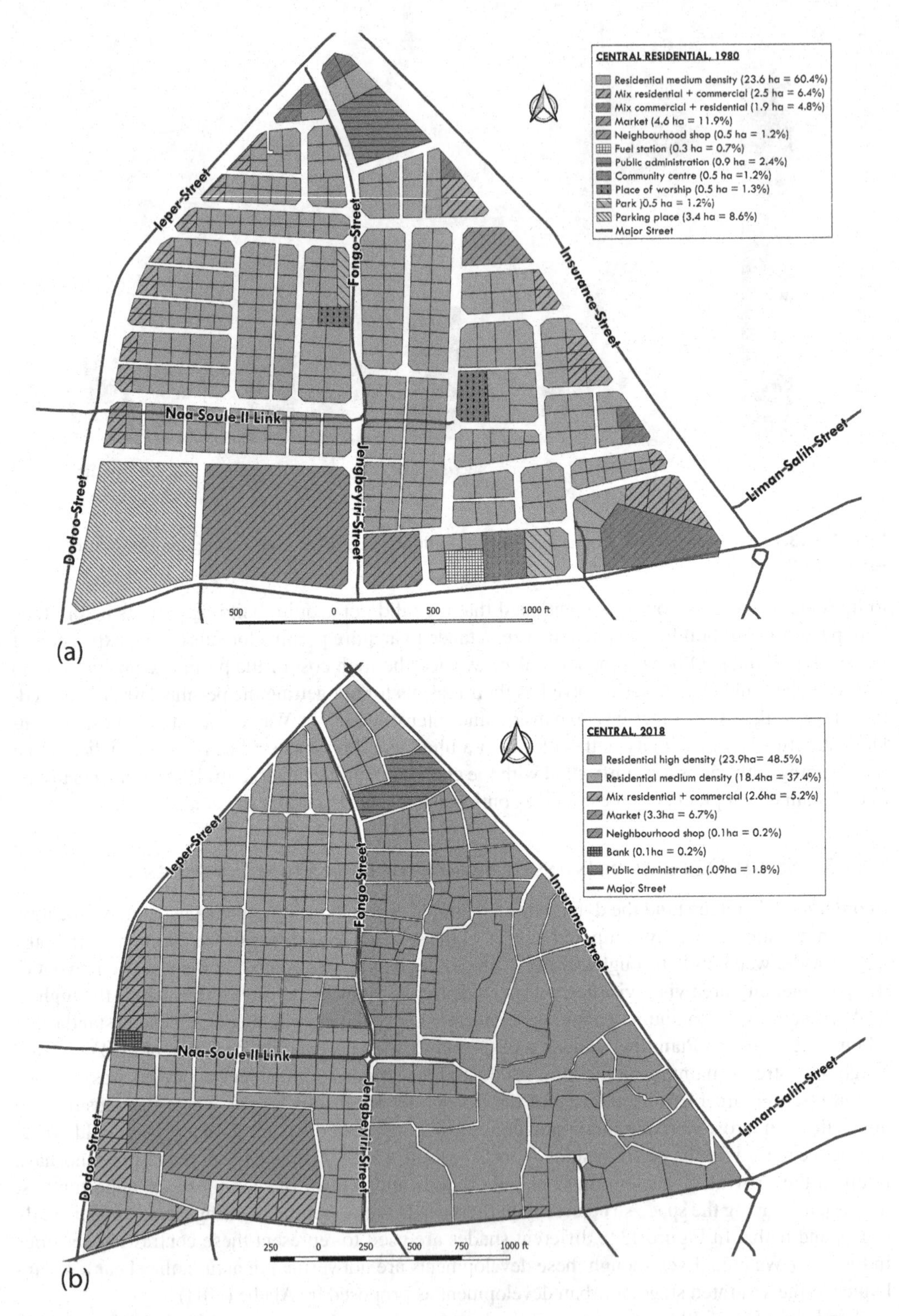

FIGURE 13.8 (a) and (b): Planning of central residential area and contrasting densities. Source: Authors' construct, 2018.

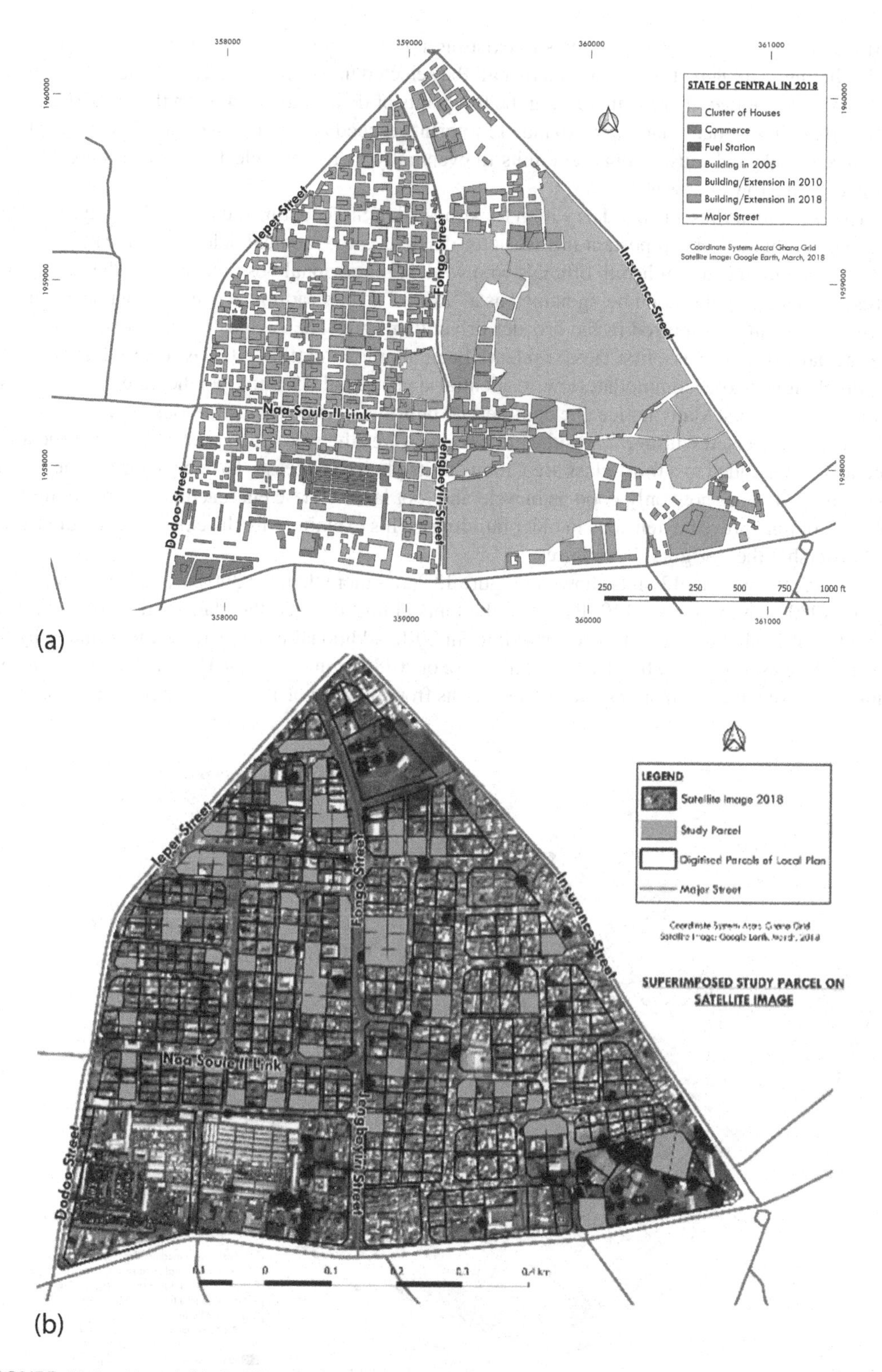

FIGURE 13.9 (a) and (b): Infilling between 2005 and 2018 and expanded footprints with distorted plan in Wa central residential area. Source: Authors' construct, 2018.

2018, additional units and attachments to existing houses numbered 120 with 1.8 km^2 of clustered units. In Figure 13.8(a), it was observed that all the selected units had experienced one form of redevelopment or another, thereby increasing the footprints of their units from when they were first constructed.. It was evident that these expanded footprints exceeded the plots sizes as planned and have narrowed roads and caused separated units to eventually merge into clusters, as in Figure 13.8(b), with a few narrow alleyways.

The Tendamba Crescent and Jerry John Rawlings High Street define the Zongo-Kabanye residential neighborhood. Very important infrastructure including the medical field unit is located here, as well as accommodation of health officials who work at the regional hospital south of the Tendamba Crescent. This neighborhood being named as a "Zongo" is misleading, because the name generally connotes a slum, as captured in the broader urban literature. The Zongo-Kabanye residential area was actually designed as a first-class residential neighborhood in the 1980s, as seen in Figure 13.10. It was designed to accommodate low residential densities of 15 units per hectare to house formal sector workers. Many noticeable commercial buildings, financial institutions, and the Catholic cathedral are located within this residential enclave. Over the years, however, it has been reclassified and downgraded to a third-class area; many of its planned accesses have been encroached upon extensively, and are now only good as bicycle and foot lanes. These encroachments are committed because of unapproved extensions in older buildings. This explains why the existing structures as of 2018 overshot their original boundaries.

In 2005, there were 612 units shown in Figure 13.9(a)—more than the 360 parcels legally provided by the 1980 plan in Figure 13.10. By 2005, the units almost doubled the planned density and it had already lost its first-class status as of that time. In 2010, additional units increased marginally by 51 and increased substantially by 212 units at the close of 2018 (Figure 13.11(a)). Here, too, all the sampled units underwent transformations and the deviations from the plan of 1980, as seen in Figure 13.13(b).

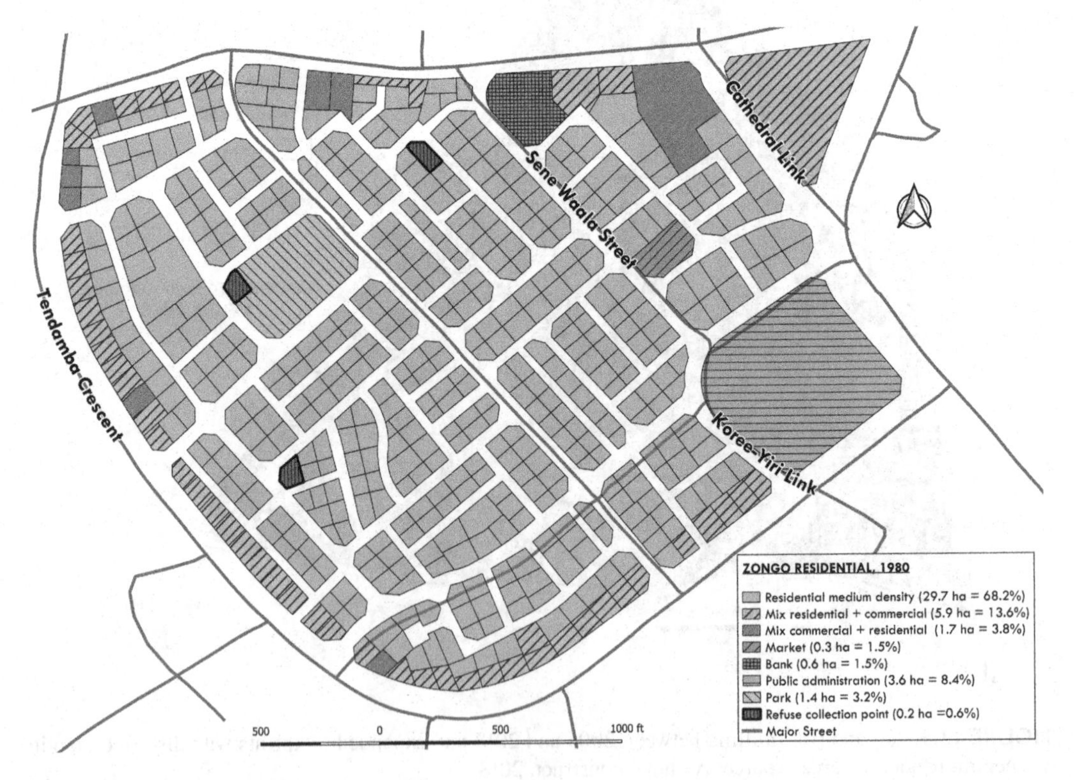

FIGURE 13.10 Original land use plan of the Zongo/Kabanye residential area in 1980. Source: Authors' construct, 2018.

FIGURE 13.11 (a) and (b) Infilling and expansions between 2005 and 2018 and expanded footprints with distorted plan. (Authors' construct, 2018.)

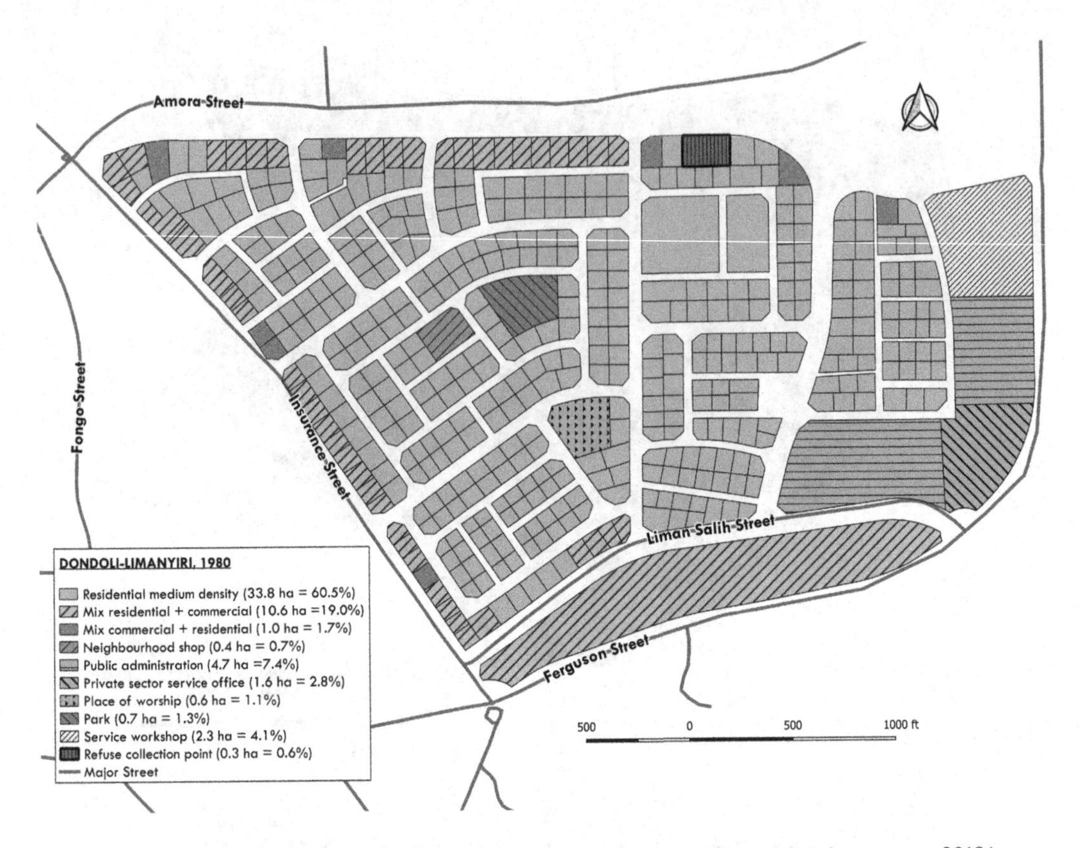

FIGURE 13.12 Original land use planning of Dondoli-Limanyiri in 1980. (Authors' construct, 2018.)

From the Dondoli-Limanyiri, similar trends were observed as in the Wa Central and Zongo-Kabanye residential areas. This is the neighborhood with the lowest number of 280 planned parcels compared to the previous two in 1980 in Figure 13.12. Along the Amora, Ferguson, and Insurance Streets that define this neighborhood, the parcels there are planned for mixed residential and commercial developments because of its proximity to the Wa central market.

By 2005, there were 310 parcels (Figure 13.13) that were legally designated for urban development, 30 more units than the number of parcels. With this increase, the neighborhood, however, maintained its medium density character as of 2005, with a cluster of units to the north of the neighborhood. Five years later in 2010, the number of additional units increased by 123 (colored in yellow) and increasing the density higher than it was planned, and by 2018, an additional 297 units (in Figure 13.13(b)) had been added. Like the others, the extensions distorted the boundaries of parcels planned, as indicated in the orthophoto in Figure 13.13(b).

13.5 CONCLUSION

This chapter sought to understand the nature and extent of inner-city developments in Ghana within the Wa Municipality. Inner city intensification in Ghana goes through a series of informal processes without appropriate development controls. Urban land and house owners expand their housing units as long as there is a bit of space to accommodate the increasing demand for additional shelter without encroaching on another person's private property. These extensions are done on piecemeal terms using family resources. Such extensions are occasioned by the desire to accommodate increasing urban poor populations, and also to improve on convenient habitation by improving private sanitary areas of the houses. These responses of the inner-city dwellers from our study were motivated by diverse personal, social, cultural, and economic factors. It was realized that, over the physical

FIGURE 13.13 Infilling and expansions between 2005 and 2018 with distorted plan. (Authors' construct, 2018.)

lifespan of inner-city properties, house owners modify their properties to incorporate emerging housing needs, sometimes to the neglect of planning regulations or zoning requirements. The consequences are that such neighborhoods are characterized by rising concerns of security, sanitation, accessibility, poor utility delivery, and strained social relationships.

Urban change in the inner cities requires as much attention as developments in the peri-urban region of cities in Africa. From the study, these unauthorized developments take the form of addition of new rooms, construction of new apartments, changes in building materials, construction of fence walls, and extension of existing room spaces. Since these developments have intensified over time, they have moved into all allocated plot spaces including reserved public corridors. Building extensions have either narrowed these spaces or have reduced them to the extent that they become woefully non-conducive to the designated use. Beyond poverty, a plethora of other factors account for the urban intensification, and not the attribution of urban economic growth as has been projected by the previous literature. Autochthonous urban landowners end up extending their ownership of land into public spaces in violation of urban regulations. Informal urban dwellers are oblivious to urban rules and regulations and break them with impunity, but in good faith that they do not require permits to develop their customary lands.

To address some of these trends and consequences, there is the need for concerted efforts by city authorities to prioritize inner cities in their quest to manage urban development. The major challenges of the cities are also found in the inner cities. Since these areas provide significant accommodation needs for the urban poor, there is the need to uphold acceptable standards that are neighborhood-friendly. Strict enforcement of planning and zoning regulations will greatly preserve public spaces, especially access ways. In the longer term, selective demolitions may improve on sustainable livability of inner cities. There will also be the need to properly integrate inner cities into the bigger urban fabric; there is the need to deploy smart land management practices that seeks to deepen spatial efficiency through participatory approaches. These approaches are not only relevant for the Ghanaian context, but for all major cities of Africa where informal urbanization is predominant.

REFERENCES

Abebe, F.K. (2011). *Modelling Informal Settlement Growth in Dar es Salaam, Tanzania*. Enschede, University of Twente Faculty of Geo-Information and Earth Observation ITC.

Baker, J.L. (2008). *Urban Poverty: A Global View*. Washington DC, The World Bank Group.

Botchway, E., Afram, S.O. and Ankrah, J. (2016). International Institute of Science, Technology and Education (IISTE) 11-22 (2014). *Building Permit Acquisition in Ghana: The Situation in Kumasi*, 4(20). Retrieved from http://www.iiste.org (last visited Mar. 01, 2016).

Bradshaw, M. (2000). *World Regional Geography* (2nd ed.). McGraw-Hill Company Ltd.

Cohen, B. (2005). Urbanization in developing countries: Current trends, future projections, and key challenges for sustainability. *Technology in Society*, 28(1–2), 63–80. www.elsevier.com/locate/techsoc.

Connelly, S. and Bradley, K. (2004). Spatial Justice, European Spatial Policy and the Case of Polycentric Development. *Paper for the ECPR Workshop on "European Spatial Politics or Spatial Policy for Europe?"* 13–18 April 2004, Uppsala, Sweden.

Donk, V.M. (2006). Positive urban futures in sub-Saharan African: HIV/AIDS and the need for a broader conceptualisation (ABC). *Environment and Urbanisation*, 18(1), 155–177.

Fainstein, S. (2010). *The Just City* (1st ed.). New York, London, Ithaca, Cornell University Press.

Farvacque-Vitkovic, C., Raghunath, M., Eghoff, C. and Boakye, C. (2008). Development of the cities of Ghana challenges, priorities and tools. Africa Region Working Paper Series Number 110. The World Bank. Retrieved from http://www.worldbank.org/afr/wps/index.htm.

Fodor, E. (1999). *Better Not Bigger: How to Take Control of Urban Growth and Improve Your Community*. Gabriola Island, New Society Publishers.

Ghana Statistical Service. (2012). 2010 Population and housing census: National analytical report, Accra, Ghana.

Ghana Statistical Service. (2014). 2010 population and housing census. District Analytical Report, Wa Municipality, Ghana Statistical Service.

Giliani, S., Gertler, J.P., Undurraga, R., Cooper, R., Martinez, S. and Ross, A. (2017). Shelter from the Storm: Upgrading housing infrastructure in Latin American slums. *Journal of Urban Economics*, 98, 187–213.

Hafeznia, M.R. and Hajat, M.G. (2016). Conceptualisation of spatial justice in political geography. *Geopolitical Quarterly*, 11(4), 32–60.

Harvey, D. (1998). *Social Justice and the City* (1st ed.). Oxford, Basis Blackwell.

Keil, R. and Kipfer, S. (2003). The urban experience and globalization. In: Clement, W and Vosko, L (eds). *Changing Canada: Political Economy as Transformation*. Montreal, McGill-Queen's Press.

Lall, S.V., Henderson, J.V. and Venables, A.J. (2017). *Africa's Cities: Opening Doors to the World*. Washington, DC, World Bank, License: Creative Commons Attribution CC BY 3.0.

Leaf, M. (1995). Inner city redevelopment in China- Implications for the City of Beijing. *Cities*, 12(3), 149–162.

Lerise, F., Lupala, J., Meshack, M. and Kiunsi, R. (2004). *Managing Urbanisation and Risk Accumulation Processes: Cases From Dares Salaam*. Tanzania, University College of Lands and Architectural Studies.

Lévy, J. (2012). Société (working version). In: Lévy, J (ed). *L'Humanité* (Forthcoming) (pp. 123–168.

Lévy, J. (2013). *Réinventer la France. Trente cartes pour une nouvelle géographie*. Paris, Fayard.

Linard, C., Tatem, A.J. and Gilbert, M. (2013). Modelling spatial patterns of urban growth in Africa. *Applied Geography*, 44, 23–32. doi:10.1016/j.apgeog.2013.07.009.

Mahabir, R., Crooks, A., Croitoru, A. and Agouris, P. (2016). The study of slums as social and physical constructs: Challenges and emerging research opportunities. *Regional Studies, Regional Science*, 3(1), 399–419. doi:10.1080/21681376.2016.1229130.

Marx, B., Stoker, T. and Suri, T. (2013). The economics of slums in the Developing World. *Journal of Economic Perspectives*, 27(4), 187–210. doi:10.1257/jep.27.4.187.

Owusu, G. and Afutu-Kotey, R.L. (2010). Poor urban communities and municipal interface in Ghana: A case study of Accra and Sekondi-Takoradi Metropolis. *African Studies*, 12(1), 1–16.

Segrue, T.J. (2014). *The Origins of the Urban Cirses – Race and Inequality in the Postwar Detroit* (Revised ed.). Princeton and Oxford, Princeton University Press.

Soja, E.W. (2008). Taking space personally. In: *The Spatial Turn* (pp. 27–51). Routledge.

Soja, E. (2009). The city and spatial justice. Justice spatiale/Spatial justice, 1(1), 1–5. *Paper Prepared for Presentation at the Conference Spatial Justice*, Nanterre, Paris, 12–14 March 2008.

Soja, E. (2010). *Seeking Spatial Justice*. London, University of Minnesota Press.

Sori, N.D. (2012). Identifying and classifying slum development stages from spatial data. In thesis, Enschede, Faculty of Geo-Information Science and Earth Observation, University of Twente, ITC.

Tacoli, C., McGranahan, G. and Satterthwaite, D. (2015). Urbanisation, rural-urban migration and urban poverty. World Migration Report 2015. London, IIED.

Turner, J.C.F. (1970). *Squatter Settlements in Developing Countries*. London, Macmillan Education Ltd.

UN Habitat. (2003). *The Challenge of Slums. Global Report on Human Settlements 2003*. Nairobi.

UN Habitat. (2004). *Global Campaign for Secure Tenure: A Tool for Advocating the Provision of Adequate Shelter for the Urban Poor'*. Concept Paper (2nd ed.).

UN-Habitat. (2006). *State of the World Cities 2006/7*. Kenya, Nairobi.

UN-Habitat. (2014). *State of Africa Cities: Re-imagining Sustainable Urban Transitions*. Nairobi, UN-Habitat.

UN-Habitat. (2016). *Urbanization and Development: Emerging Futures*. World Cities Report 2016. Nairobi, Kenya.

UN-OHRLLS. (2016). *Criteria for Identification and Graduation of LDCS*. Retrieved June 8, 2016, from http://unohrlls.org/about-ldcs/criteria-for-ldcs/.

Uwayezu, E. and de Vries, W.T. (2018). Indicators for measuring spatial justice and land tenure security for poor and low income urban dwellers. *Land*, 7(3), 84. doi:10.3390/land7030084.

White, M.J., Awusaba-Asare, K., Nixon, W.S., Buckley, B., Granger, S. and Andrzejewski, C. (2007). Urbanization and Environmental Quality: Insights from Ghana on sustainable policies. *Paper Presented to the PRIPODE Workshop on Urban Population*, Nairobi, Kenya, Development and Environment Dynamics in Developing Countries.

Williams, J. (2013). Towards a theory of Spatial Justice. *Paper presented at the Annual Meeting of the Western Political Science Association Los Angeles, CA on "Theorizing Green Urban Communities" Panel Thursday*, March 28, 2013.

Zhang, X., Hu, J., Skitmore, M. and Leung, B.Y.P. (2014). Inner-city urban redevelopment in China metropolises and the emergence of gentrification: Case of Yuexiu, Guangzhou. *Journal of Urban Planning and Development*, 05014001-1-8. doi:10.1061/(ASCE)UP.1943-5444.0000169.

Zheng, H.W., Shen, G.Q., Song, Y., Sun, B. and Hong, J. (2016). Neighborhood sustainability in urban renewal: An assessment framework. *Environment and Planning B: Urban Analytics and City Science*, 44(5), 903–924.

14 The Role of the Social Tenure Domain Model in Strengthening Land Tenure Security in Informal Settlements of Uganda

Lilian Mono Wabineno-Oryema

CONTENTS

14.1 INTRODUCTION

The world population is expected to reach 9.6 billion people by the year 2050, with more than half of this population being in Africa (UNFPA, 2013). In Uganda, the population is rapidly growing at a rate of 3.4% per year which translates into a total human population of 55 million by 2025 (Nakabugo, 2018). Uganda Vision 2040 by the Uganda National Planning Authority (2007) indicates that the population in Uganda may increase to 61 million, with 60% (37 million) of the population being in the urban areas as opposed to 40% (24 million) who will be living in the rural areas by the year 2040. This statistic indicates that the population in Uganda is increasing at a very fast rate, more so in the urban areas, and measures must be taken to curb the negative impact that comes with urbanization. Urbanization, when not planned, regulated, and organized leads to: increased pressure on the available resources (Batte, 2018) which are already scarce; congested urban areas that cannot absorb the growing population in terms of space and employment; which in turn translates into unplanned urban poor settlements which are mainly located along the periphery of urbanized areas with high levels of unemployment and housing deficit.

Manju et al. (2011) observe that urbanization can be a curse when unplanned, whereas when planned, regulated, and organized it is a gift to societies. This condition of urbanization turning into a curse when unplanned is a true scenario for Kampala City and Uganda at large, as slums are evident in most of the well-developed municipalities. According to UN-Habitat (2007b), the emergence of slums in Kampala is due to failure to cater for growth and development in the Kampala Structure Plans. Since planning is no longer one of the reasons for compulsory acquisition in Uganda, it is then hard to plan for such settlements when people have occupied them. In other words, if urbanization is not controlled, the urban areas become a mess, with informal settlements sprouting out everywhere, which makes it hard to plan for such areas. According to Wabineno-Oryema, Mono, and Omondi (2018), once the unplanned settlements are established, they are irrevocable and it is difficult to plan for their regulations and expansion.

Unplanned settlements known as slums act as a living hub for between 49% and 64% of Uganda's urban population (Vision Reporter, 2012). This implies that there is a very big urban population living in an environment characterized by: congestion in terms of population, space, and housing; people with low literacy, low income levels, or/and high unemployment levels; poor sanitation; and shabby housing. According to UN-Habitat (2007b), one of the obvious problems that face people who live in the informal settlements of Kampala and Uganda at large is tenure insecurity. Most people in informal settlements live in temporary shelters whereby they are occupiers and not landlords (Vision Reporter, 2012). This is an indication that most of the slum settlers are liable to eviction at any time, as the landlords can decide to sell their land to developers anytime, even when the occupiers have paid rent (Dimanin, 2012). However, it may not only be an indication restricted to evictions but also acquisition, as well as reclaiming the occupied land for planning purposes, which in the case of Uganda is illegal. The tenure insecurity is that the rights that these people have are to occupy and use, as opposed to some form of title.

In Uganda, there are four different land tenure systems as stipulated in the Land Act of Uganda 1998 which pass as either statutory or customary tenures. The four land tenures are named mailo, leasehold, freehold, and customary. Freehold tenure is one that entails holding land in perpetuity. It is just like the simple freehold system. Customary tenure is one in which the holders hold it by virtual of customs, traditions, and norms they belong to as a community, family, and clan. Leasehold is one where land is held for a particular period of time, and after that duration the land reverts back to the original owner. Leasehold can be granted on customary, freehold, or even mailo land. Mailo tenure is one which is a hybrid of freehold and customary, as it has incidences of both tenures. It is held in perpetuity and has incidences of occupancy rights whereby the owner of the land has to respect the holders of occupancy rights on that given piece of land. Unlike the leasehold, freehold, and customary which are found in every part of Uganda, mailo land is only found in the central part of the country. On each of the land tenures, there are competing and overlapping rights, but the situation gets worse in the informal settlements. In informal settlement, the primary rights of the registered or non-registered proprietor on freehold, customary, leasehold, and mailo land are overlaid by the secondary rights of a tenant, the lawful occupant or the bona fide occupant. Primary rights, also known as ownership rights, are rights that are held and exercised by the owner/registered proprietor of the land, while primary rights which can pass for occupancy rights are rights on land held by other parties other than the owner of the land. These secondary rights are also known as usufruct rights that are exercised by the occupiers/tenants on the land.

According to the Constitution and the Land Act of Uganda, both the bona fide and lawful occupants are tenants by occupancy. A tenant by occupancy should enjoy security of occupancy whether they are on leasehold, freehold, or mailo land. According to section 29 of the Land Act 1998 of Uganda, it defines bona fide and lawful occupants as follows. A lawful occupant is one who either occupies land and pays ground rent, or entered the land with the consent of the registered owner or customary tenant whose tenancy was not disclosed or compensated for by the registered owner at the time of acquiring the leasehold certificate of title. The Act defines a bona fide occupant as either one who entered the piece of land 12 years before the coming into force of the Constitution without interruption from the registered proprietor of the land, or one who has been settled on land by government authority. This then means that any person who does not fall either under the definitions of a bona fide or lawful tenant turns out to be an unlawful tenant whose occupancy status is illegal, as it is not protected by the law. This is the case of most informal settlements, i.e. squatters.

The Government of Uganda acknowledged the need to capture the positive side of urbanization, and thus a Participatory Action Research into issues that affect access to land and increased security of tenure of the urban poor was carried in 2013. PAR involved the researchers and the people being affected by the problem coming together, generating ideas, analyzing the problem, and creating the solutions to the problem that will cause a social change. The PAR by Musinguzi et al.

(2014) recommended that the procedure and cost of formalizing land ownership/rights in informal settlements of the urban poor in Uganda should be simplified and reduced respectively by adopting low-cost but flexible tools such as the Social Tenure Domain Model (STDM), since the available traditional methods are expensive and have failed to meet some of their expected targets. Flexible tools such as STDM act as an incentive for the poor land rights-holders in informal settlements to have their lands and rights registered as such tools are: cheaper, faster, of a participatory nature, and understood by the settlers, as opposed to the traditional methods. The traditional methods that are currently being used have advanced technical standards of adjudication, boundary marking, and field surveys which require accurate surveys of the boundaries, thus making them rigid and costly.

If land regularization in the informal settlements is to be achieved, all the different interests in land have to be streamlined and captured in a land registration system. It is hard to achieve full regularization in informal settlements using the available traditional/conventional land administration methods such as land surveying, since they only consider primary rights that are registrable (mailo, freehold, and leasehold), and they are not affordable to the poor who are the majority in the informal settlements. This causes the problems of lack of planning and tenure insecurities to persist in the informal settlement. As Mitchell et al. (2017) point out, the growth of slums and informal settlements is one of the global challenges that informs and affects the modern land administration system. STDM is one of the modern land administration systems inventions that fills the gap where the conventional cadastral systems have not been adequate (Antonio, Gitau, and Njogu, 2014). This forms part of the solution to lack of planning and tenure insecurities in the informal settlements.

STDM as a form of responsible management system was first piloted in Mbale district in 2011 within two settlements (Bufumbo and Mission). In later years, the use of STDM was extended to cover the whole of Mbale district and other areas such as: Kampala, Wakiso, Masaka, Entebbe, Tororo, and many more, as shown in Figure 14.2. The objective of STDM implementation is to form a basis for proper planning and improved tenure security by: addressing the land information requirement in the informal settlements; building the capacity of the informal settlers to be able to apply and use STDM on their own; and to mainstream the thinking behind the continuum of land rights (GLTN, 2017). The final output of the STDM are a certificate of occupancy and a settlement map which is produced from satellite images. The impact of implementing the STDM in Uganda is improved tenure security, since there is a linkage created between parties (individual, groups, or households) and a specific spatial unit (house, land, or any physical structure on ground), and also STDM forms a basis to initiate dialogue for planning of the settlements by the different stakeholders (Antonio, Gitau, and Njogu, 2014; Antonio, Makau, and Mabala, 2013).

The concept of responsible land management is a framework that helps to ensure that the interventions being proposed to solve land management challenges meet the needs of the local people by measuring the interventions against some indicators known as the 8Rs (Robust, Responsive, Resilient, Respected, Reliable, Retraceable, Recognizable, and Reflexive) (Antonio, Gitau, and Njogu, 2014; Antonio, Makau, and Mabala, 2013). The 8Rs aid in the design and implementation of interventions to address the land management needs of the community where the challenge is faced (Ameyaw et al., 2018). Land management is said to be "The science and practice related to the conceptualisation, design, implementation and evaluation of socio-spatial 'interventions', with the purpose to improve the quality of life and the resilience of livelihoods in a responsible, effective, efficient, consensual and smart manner" (De Vries and Chigbu, 2017). This means that there is involvement of theory experiences, practical experiences, skills, and knowledge in order to address challenges in land management.

STDM in Uganda is an intervention that is being implemented in the urban poor settlements in order to overcome the challenge of lack of information about the settlements. Lack of information on land in informal settlements leads to persistent problems, such as making it hard to plan for the settlement and tenure insecurity. The STDM takes on a participatory approach involving all major stakeholders (government, slum-dwellers, and their leaders) who have a lot of interactions amongst themselves in form of dialogue, sensitization, data collection, data analysis, and capacity-building to manage, maintain, and update the system. The participatory nature and approach of STDM meets the interest and expectations of the slum dwellers as they feel some recognition during regularization of their interests. Based on the design, processes, outcome, and implementation of the STDM, it is evident that this is a responsible and smart land management approach.

14.2 THEORETICAL PERSPECTIVE

Land tenure is the connection people have with the land they are using or staying on. This means that every person, whether a landowner, tenant, or occupier, has a relationship with the land they are staying on or using. However, the relationships must be well-defined, known, and acknowledged by the people or some sort of authority in order to permit good social, economic, and political decision-making for the people and also to avoid conflicts on land or property. There are a variety of relationships that exist between people as individuals or groups with the land. Figure 14.1 shows the continuum of rights.

Figure 14.1 shows the various kind of land rights that exists on land ranging from the formal to the informal. Formal rights are those that are statutory and have documentation, as opposed to the informal. Each range of rights on the continuum has its delineation of responsibilities and degree of security. The formal rights are usually individualized with a paper (land title) that has a set of registered rights by law and demarcated rights of ground by use of conventional methods such as land surveying. The formal rights holder has a full bundle of rights that are exercised on the land they hold (Lemmen et al., 2015), while the informal rights holders have usufruct rights. In Uganda, all the range of rights in Figure 14.1 exist, with formal rights being compared to the freehold tenure holders and the informal rights being an example of clan rights in the customary setting. The degree of security and responsibility is more on the formal side and goes on decreasing as rights move to the informal side in the current situation when conventional methods are used.

There are different rights that exists between the extreme of the continuum of rights and according to Mabikke (2016), citing UN-Habitat (2012), even when the range of rights in the continuum of rights (Figure 14.1) seem to lie in a straight line, in reality these rights are multi-layered and complex, as they do not lie in a straight line and may overlap with one another.

The statistics about land in Uganda shows that 68.8% is customary, 18.6% freehold, 9.2% mailo, and 3.6% is leasehold, with only about 20% of the entire land being registered in the National Land Information System (Wabineno-Oryema, 2016). This implies that only about 20% of landowners in Uganda have land rights that are registered and recognized. The land rights status of the remaining 80% is unknown, and their security of tenure is not guaranteed because customary tenure has different characteristics when compared to statutory tenure, and they do not have any documentation to represent their interests according to the traditional methods of land registration and recordation.

According to Boudreaux and Sacks (2009), citing UN-Habitat (2007a), security of land tenure is the protection the government provides to its people against forceful evictions. One of the ways security of tenure can be guaranteed by government is by issuing of land titles. Whereas the land

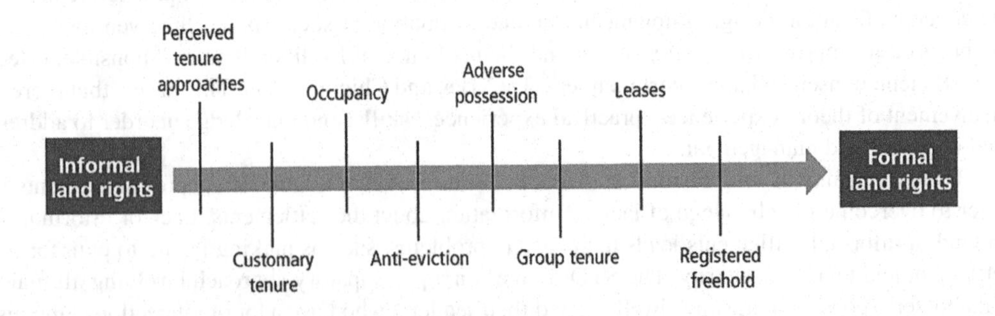

FIGURE 14.1 Continuum of rights (UN-Habitat, 2015).

title gives some form of assurance of people to feel secure on their land, in Uganda this is not necessarily the case. Some people prefer to hold bibanja land, where individuals only have sales agreements and consent from the landlord that shows their occupancy status rather than a land title. Kibanja (plural: bibanja) is land that is held under occupancy rights. The owner of such land is a tenant by occupancy on land which is mainly under mailo tenure. It is believed that a kibanja is more secure than the titled land as it is less liable to evictions. This shows that tenure security goes beyond the issuance of some legal document, as threats may come from different persons such as relatives, government, members of the community, and investors (Yongsi, Blaise, and Wagai, 2017). Security of tenure does not in itself require accurate surveys of the boundaries. However, the important aspect is identification of the land object in relation to the connected legal or social right. Using such advanced technical standards of adjudication, boundary marking, and field surveys are far too costly, too time-consuming and capacity-demanding, and in most cases simply not relevant for providing an initial suitable spatial framework. The focus should therefore be on methods that are fast, cheap, complete, and reliable. The spatial framework can then be upgraded and updated whenever necessary or relevant in relation to land development and management activities. The key focus should be on providing secure tenure for all, and managing the use of land and natural resources for the benefit of local communities and society as a whole. With this argument, it is then possible that there may or may not be a relationship between STDM and tenure security, as is the case for the conventional methods in the Ugandan context.

In areas of informal settlements, relations of people and land are a complex issue, as in most cases there are different overlapping and competing interests in the same piece of land and according to UN-Habitat (2015), citing UN-Habitat (2012), some of the interests may be recognized by law while others are not. The traditional methods of land administration deal with some primary rights such as mailo, freehold, and leasehold but cannot handle other primary rights and secondary rights such as customary, usufruct, tenancy, and occupancy. It even becomes difficult to use the traditional methods of surveying as there are overlaps of rights in most lands. In most African countries, and more so in the slum areas, there are a lot of what are termed informal and grouped rights on the continuum of rights. Most African governments have concentrated on the registration of formal tenures such as leasehold and freehold, forgetting that most of the lands in their countries do not belong to those statutory tenures, but rather they have a mixture of a range of rights.

Conventional/traditional land administration methods have failed to deliver results of tenure security, wide coverage, and affordability for both private and public entities (UN-Habitat, 2012) on the non-statutory tenures in Uganda, because their application is inappropriate in such areas. Conventional methods look at only the owner of the land and no other parties and relations on the land. Because of such failings, innovations of non-conventional methods such as STDM have come up to fill the gap. Such innovations take up all rights on the continuum of rights timeline to enjoy legal and *de facto* recognition, just as with the results of the conventional methods. This ensures enjoyment of the same benefits, such as guarantee of security of tenure, provision of services, and many others from the government and other responsible organizations. The continuum of rights in Figure 14.1 suggests that there are different ways to achieve tenure security and not necessarily through the formal land rights, and thus the development of non-conventional methods.

14.3 EMERGING RESEARCH CONCEPT

Due to urbanization, the proportion of urban poor is growing. However, there is a concern over security of land rights in areas where the urban poor always settle (informal settlements). One of the paths being used to secure land rights in these settlements is to create community cadastres through the use of STDM. Whereas the STDM tool has been implemented successfully in some settlements, it is still in process for other settlements. However, what is not known is how STDM is perceived to strengthen security of land rights by the settlers in the settlements where STDM has been implemented.

Some studies show that there is no relationship between tenure security and STDM, while others credit improved tenure security in informal settlements to STDM. According to Archer (2016), tenure security in the study area (Mbale) was never influenced by the STDM. Tenure security in this particular study was looked at as the perception one has to being able to stay on their land without losing its physical possession. The indicators used to assess tenure security in the study were: the confidence settlers have in holding their land; reduced disputes; motives of informal rights holders to invest in the land they are holding; and finally, the spatial patterns of land use and development. The study by Archer (2016) found out that: the confidence of settlers was attributed to the written agreements the settlers made during land transactions and not STDM; despite the falling levels of disputes, STDM only played a role around in reducing boundary disputes, while the local governance of the settlement played the biggest role in reducing land disputes in the settlements; age and income levels influenced decisions to invest, although the roads were as a result of the STDM.

However, according to GNLT (2018), the introduction of STDM in the informal settlements of Uganda has increased tenure security for the entire 181,604 households that are in the 120 informal settlements. This is in contradiction to what Archer (2016) found, but it concurs with Otieno et al. (2017) who argue that STDM creates tenure security to some extent. In the study done by Otieno et al. (2017), tenure security was measured in terms of: infrastructure development (electricity, water, and sewer), threatened evictions, and perception of tenure (in the form of transactions made, and confidence) by the informal settlers. The study by Otieno et al. (2017) found out that in terms of perception of tenure security, the high level of confidence about staying on the land could be attributed to STDM, although the investment in building was never influenced by STDM. However, STDM could have had a positive impact on land transactions, as they had increased since the implementation of STDM. The study also found that although threats to evictions still exist and government protection by use of STDM was still low, there was a general reduction in threats to evictions, which was attributed to STDM. For issues of investment in infrastructure, Otieno et al. (2017) concluded that although there was an improvement in the water and sewer line services, this was not linked to STDM and that there was no change in the road network and electricity connections, except for the floodlights, as there were still illegal connections. Therefore, the four indicators (confidence; transactions; water and sewers; electricity) imply a positive impact of STDM, with evictions closing in as well, whereas buildings, roads, and government protection are unknown to STDM. It is from the contradiction of whether tenure security is linked to STDM or not that this research concept emerged.

Since STDM has been rolled out in various parts of the country, Figure 14.2 shows the areas where this has occurred. The study has chosen two of the areas per each region where the STDM tool has been applied in Uganda. This makes a total of eight areas, since there are four regions (northern, western, eastern, and central region). Mbale has been chosen as the ninth study area since it is where the pilot study was done, and because of the success story, STDM got rolled out in the other areas of the country.

The first set of districts to be chosen for the study are Kabale and Mbarara in the western region, Arua in the northern region, Jinja and Mbale in the eastern region. This is because these were study areas in Participatory Action Research into issues that affect access to land and increased security of tenure of the urban poor which was carried out by the Department of Geomatics and Land Management, Makerere University on behalf of the Government of Uganda through the Ministry of Lands, Housing and Urban Development. According to the research findings by Musinguzi et al. (2014), there was insecurity of tenure in the studied settlements; this was before the STDM rollout. This finding will form part of the basis for a comparison of before and after implementation of STDM. The other study areas include: Tororo in the eastern region, Gulu in the northern region, then Wakiso and Masaka in the central region.

The main objective of the study is to evaluate the role of the STDM in securing land rights in informal settlements. This will be done through finding out how the people in informal settlements perceive and manage tenure security before and after STDM implementation based on different

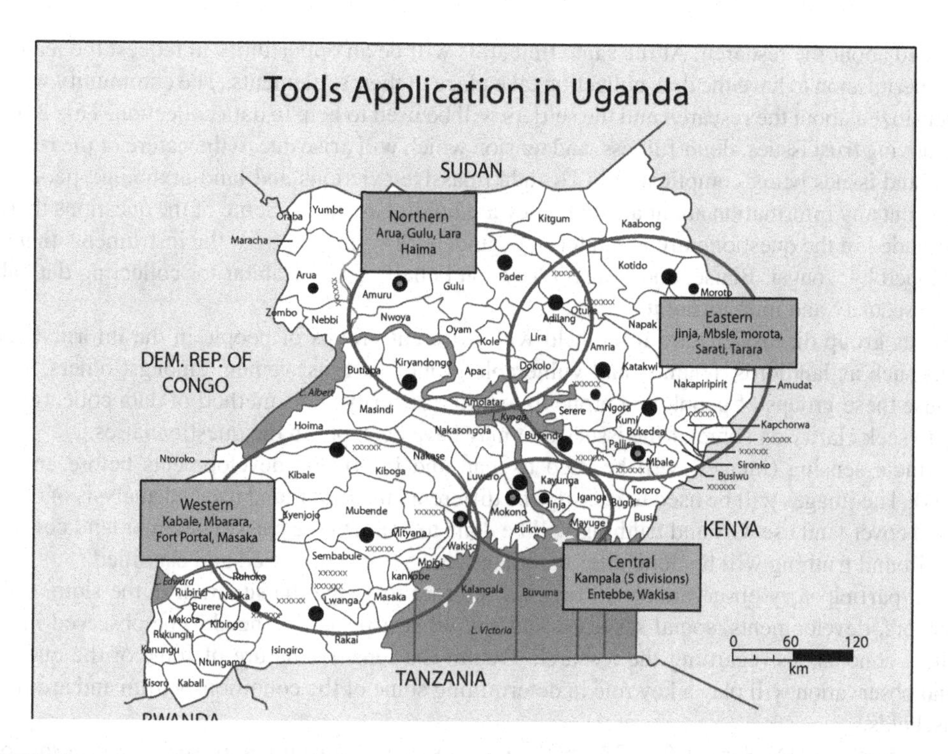

FIGURE 14.2 Areas where STDM has been applied in Uganda (GLTN, 2018).

indicators such as compensation, overlapping, and multiple claims of rights on land, patterns of land development, confidence of holding and transacting on land, forceful eviction, and land grabbing by elites and investors.

The scientific contribution of this research is to contribute to the body of knowledge on STDM and create awareness on the impact of STDM in terms of securing and strengthening land rights in different settlements of Uganda.

14.4 METHODOLOGY

The methods to be used are selected based on the specific objectives of the research. Based on the objectives, there is a mixture of both quantitative and qualitative methods to be used to collect data to accomplish the research, and they are explained below.

Document review is the main source of secondary data that is going to be used to obtain information about tenure security, STDM, and responsible and smart land management amongst others. This data collection method is where the research concept has been generated from.

Face-to-face key informant interviews with people who are knowledgeable about the STDM roll out and implementation will have to be carried out. Individuals for the interviews are to be selected from the different partnering organization of the STDM rollout such as GLTN, Slum/Shack Dwellers International (SDI), ACTogether Uganda, and MoLHUD. Other interviewees will be the local council leaders in the study areas and the leaders of the municipalities where the study areas are located, because they work with the key partners to have STDM rolled out and implemented.

Questionnaires will be distributed amongst the different dwellers of the different settlements within the study area. Questionnaires will be used since the research will need to reach a bigger population, but they will be responding to the same questions about security of tenure. The sample size needed for the study will be calculated later on when the actual data collection is about to start. The leaders of the informal settlements will be contacted before the actual collection of the data and educated about the research. At the same time, this will be an opportunity to request the leaders to grant permission to have the data collection take place in their settlements. The community will first be sensitized about the research and the settlers will be used to help in data collection. This is aimed at removing trust issues, doubtfulness, and tension which will arise due to the nature of the research. With land issues being complicated in Uganda (massive evictions and land grabbing), people fear giving out any information about any land they are holding or using. Some of the questions that will be included in the questionnaire and interview guide will be picked from the instruments that were developed by Yongsi, Blaise, and Wagai (2017) on behalf of UN-Habitat for collecting data about tenure security and monitoring it.

Focus group discussions are to be held with different groups of people in the informal settlements such as landlords, tenants, and vulnerable groups such as women, amongst others. This is because these groups of people are prone to tenure insecurity. This method of data collection will help to seek clarity on some of the issues that may have been put in the questionnaires.

Remote sensing (images) will be used to map and inform on developments before and after STDM. The images will be used to classify the different land cover land use and analysis of changes in land cover land use. Ground truthing will be done to confirm the classification of land cover land use. Ground truthing will be done after the images have been processed and classified.

Non-participatory observation will be used in order to gauge the situation in the slum, such as behaviors, developments, social services, and any other aspects that need to be observed in order to draw conclusions regarding the research. Due to the sensitive nature of some of the questions, visual observation will play a key role in determining some of the conditions within and around the households.

According to UN-Habitat (2007a), methodologies of measuring tenure security in terms of its scope and scale are not largely available, and according to UN-Habitat (2008), various organizations are battling with this problem, including UN-Habitat. There are discussions going on about measurement tools of perception of tenure security based on the Sustainable Development Goals (SDGs). As a result, a questionnaire and interview guide have been developed and piloted in three African countries (Kenya, Nigeria, and Cameroon). The tools have been found to produce adequate information and according to Yongsi, Blaise, and Wagai (2017), the tools can be used to collect data on and monitor tenure security by customizing it based on each country context. This tool permits data standardization in areas of tenure security data collection and monitoring.

14.5 CONCLUSION

Based on the different literature, it is evident that one of the intentions of the STDM is to strengthen tenure security. However, based on the studies by Archer (2016), STDM does not have an impact on tenure security, while some literature states there is a link between STDM and tenure security. This then leads to some additional questions that need to be thought about and tackled in reference to the different findings stated above. The questions include: was it because the research done by Archer was in a pilot area? Is it that the timeframe between the implementation of the STDM and when the research was done by Archer was very short to have an impact assessment done? Since STDM is flexible and can be customized to suit what it is needed for, then was it that when STDM was applied in other areas, it was updated to include other aspects of tenure security which were not included in the pilot area and thus there was no relationship with tenure security? Did GLTN use different indicators from those of Archer to measure tenure security? These are some of the additional questions that need to be answered in order to finally make a conclusion as to whether the STDM as a responsible land management tool has a role in strengthening tenure security in Uganda or not and also to get answers on why there is variation in the study findings. Smart and responsible land management tools have been used to achieve their intentions, and this is to be tested on the STDM.

REFERENCES

Ameyaw, Prince Donkor, Walter Dachaga, Uchendu E. Chigbu, Walter Timo De Vries, and Lewis Abedi. 2018. "Responsible Land Management: The Basis for Evaluating Customary Land Management in Dormaa Ahenkro, in Ghana." In *World Bank Conference on Land and Poverty Conference*, March 19–23.

Antonio, Danilo, John Gitau, and Solomon Njogu. 2014. "Addressing the Information Requirements of the Urban Poor: STDM Pilot in Uganda." http://documents.worldbank.org/curated/en/655271468185045558/pdf/99245-WP-P126966-Box393191B-PUBLIC-tenure-domain.pdf.

Antonio, Danilo, Jack Makau, and Samuel Mabala. 2013. "Piloting the Social Tenure Domain Model in Uganda - Innovative Pro-Poor Land Tools under Implementation." *GIM International*. https://www.gim-international.com/content/article/piloting-the-social-tenure-domain-model-in-uganda.

Archer, Tom Wilberforce. 2016. "Investigating the Impact of the Social Tenure Domain Model (STDM) on Tenure Security- a Case Study of Mission Pilot." University of Twente.

Batte, Edgar R. 2018. "Why Kampala Continues to Grapple with Increasing Urban Poor." *Daily Monitor*, December. https://www.monitor.co.ug/SpecialReports/Why-Kampala-continues-to-grapple-with-increasing-urban-poor/688342-4877494-2bytj0z/index.html.

Boudreaux, Karol, and Daniel Sacks. 2009. "Land Tenure Security and Agricultural Productivity." Mercatus on Policy No 57. https://www.mercatus.org/system/files/Land_Tenure_and_Agriculture.pdf.

Dimanin, Patrick. 2012. "Exploring Livelihoods of the Urban Poor in Kampala, Uganda: An Institutional, Community, and Household Contextual Analysis." http://www.actionagainsthunger.org/sites/default/files/publications/Exploring_livelihoods_of_the_urban_poor_Uganda_12.2012.pdf.

GLTN. 2017. "Social Tenure Domain Model." *GNLT and UN-Habitat*. https://stdm.gltn.net/applications/mbale-uganda/.

———. 2018. "Promoting Tenure Security for the Urban Poor in Uganda through Pro-Poor Tools Implementation in Uganda." In *Presented at the Country Learning Exchange Held during the 7th GLTN Partners Meeting 2018 in Nairobi, Kenya*. Nairobi: Government and Nonprofit. https://www.slideshare.net/LandGLTN/promoting-tenure-security-for-the-urban-poor-in-uganda-through-propoor-tools-implementation-in-uganda?from_action=save.

Lemmen, C., J. Du Plessis, C. Augustinus, P.M. Laarakker, K. de Zeeuuw, P. Saers, and M. Molendijk. 2015. "The Operationalisation of the 'Continuum of Land Rights' at Country Level." In *Proceedings of the Annual World Bank Conference on Land and Poverty 23-27 March 2015*, Washington DC, USA, 1–27, The World Bank.

Mabikke, Samuel B. 2016. "Historical Continuum of Land Rights in Uganda: A Review of Land Tenure Systems and Approaches for Improving Tenure Security." *Journal of Land and Rural Studies* 4(2): 153–71.

Manju, Mohan, K. Subhan Pathan, Narendrareddy Kolli, Kandya Anurang, and Pandey Sucheta. 2011. "Dynamics of Urbanization and Its Impact on Land-Use/Land-Cover: A Case Study of Megacity Delhi." *Journal of Environmental Protection* 2: 1274–83.

Mitchell, David, Siraj Sait, Jean Du Plessis, and Dimo Todorovski. 2017. "Towards a Responsible Land Administration Curriculum." In *FIG Working Week Surveying the World of Tomorrow - From Digitalisation to Augmented Reality*, Helsinki, Finland, May 29–June 2, 2017.

Musinguzi, Moses, Lilian Mono Wabineno-Oryema, Brian Makabayi, and Godfrey Mwesige. 2014. "Participatory Action Research into Issues That Affect Access to Land and Increased Security of Tenure of the Urban Poor." Kampala.

Nakabugo, Zurah. 2018. "Uganda's Population Likely to Hit 55 Million by 2025, Gov't Worried." *The Observer*, March. https://observer.ug/news/headlines/57146-uganda-s-population-likely-to-hit-55-million-in-2025-gov-t-worried.html.

Otieno, Elizabeth O., Monica Lengoiboni, Rohan Bennett, and Samson Ayugi. 2017. "Analysing the Impact of Social Tenure Domain Model (STDM) on Tenure Security in the Informal Settlement.", 64–86.

Uganda National Planning Authority. 2007. "Uganda Vision 2040." Kampala.

UN-Habitat. 2007a. "Ehnansing Urban Safety and Security." https://www.un.org/ruleoflaw/files/urbansafetyandsecurity.pdf.

———. 2007b. "Situation Analysis of Informal Settlements in Kampala." https://unhabitat.org/books/situation-analysis-of-informal-settlements-in-kampala/.

———. 2008. "Enhancing Security of Tenure: Policy Directions."

———. 2012. *Handling Land, Innovative Tools for Land Governance and Secure Tenure*. UN-HABITAT. https://unhabitat.org/books/handling-land-innovative-tools-for-land-governance-and-secure-tenure/.

———. 2015. *Property Theory, Metaphors, and the Continuum of Land Rights.* Edited by Victoria Quinlan. UN-Habitat.

UNFPA. 2013. "World Population to Increase by One Billion by 2025." *United Nation Population Fund.* https://www.unfpa.org/news/world-population-increase-one-billion-2025.

Vision Reporter. 2012. "Kampala Is One Big Slum." *New Vision,* October. https://www.newvision.co.ug/new:vision/news/1309482/kampala-slum.

Vries, Walter Timo De, and Uchendu E. Chigbu. 2017. "Responsible Land Management – Concept and Application in a Territorial Rural Context." In *Fub - Flächenmanagement Und Bodenordnung,* 65–73.

Wabineno-Oryema, Lilian Mono. 2016. "A Real Property Register to Support the Property and Credit Market in Uganda." KTH.

Wabineno-Oryema, Lilian Mono, and Wycliff Ogello Omondi. 2018. "Site Evaluation of Eco-Towns Using GIS and Analytical Hierarchy Process: Case of Greater Kampala Metropolitan Area." *African Journal of Land Policy and Geospatial Sciences,* 1(2): 45–60.

Yongsi, Nguendo, Henock Blaise, and John Ndegwa Wagai. 2017. "Monitoring Tenure Security, Data Collection Questionnaires Modules and Manual." NO. 6 /2017. GLII Working Paper. https://reliefweb.int/sites/reliefweb.int/files/resources/Monitoring Tenure Security CAM-NG-KE Pilot.pdf.

15 Community Response to Land Conflicts under the Transitional Constitution (South Sudan)

Justin Tata

CONTENTS

15.1 INTRODUCTION: BACKGROUND AND DRIVING FORCES

South Sudanese community members, led by church clergy, are calling for responsible and smart land management intervention; a return to a traditional system of land management, where communities take active ownership and management and role (Radio Tamazuzj, 2019; Land Act 2009).

A few months before the January 2011 referendum for secession from Sudan, talks around land reform were already suspended between two camps. On the one hand were the aspiring government elites spearheading campaigns for separation from Sudan, and on the other hand, the community openly demanding resolution of illegal land occupations and threatening to vote for unity if this was not achieved. The voices of the elites echoed the same voices of concerned NGOs at previous functions in Washington D.C. prior to the independence of South Sudan. Just a few weeks after the success of the referendum, the same divided voices resurfaced at the ceremony for handing over the Draft Land Policy 2011 to the Ministry of Justice. Speeches from representatives of the land line NGOs/international agencies and the South Sudan Land Commission (SSLC) echoed the sentiments of those aspiring elites. In contrast, the undersecretary for the Ministry of Justice voiced his reservations consistent with the community's position. The controversy over which systems, government, or community management actually lead to smart land management, set the stage for this chapter on the two conflicting camps.

After attending numerous NGO-sponsored land related meetings, workshops, and conferences with representatives of government and several communities, as well as visiting the National Archive, it became apparent that South Sudan already had a working traditional system of land governance. Those traditional structures have found ways to manage tribal divisions well so they don't lead to extreme conflict (Grazing Agreement, 1958; Brosche, 2012). However, during the war of the Sudan People's Liberation Army/Movement(SPLA/M) against the government of Sudan, traditional land governance systems were suppressed by liberation propaganda, and force was used to influence ownership and land access in Greater Equatoria (Pinaud, 2016).

Faced with the challenges of addressing the land issues such as: illegal occupation and refugees/IDPs, as well as forced marriages and killings to gain access to land, the government initiated land reforms. The success induced by the reforms was the recognition of community ownership of land, while government led the management (Land Act 2009). However, as community land ownership was affirmed, and greater trust in the government began to develop, a few land speculators began to take advantage of the newly forged relationship. Certain groups within the aspiring elites became land speculators; they hid behind the government and began to exploit community land through illicit land deals. When as a result, the Mukaya land scandal erupted (Deng, 2010), and the rampant conversion of public land to private ownership by aspiring elites became known, community trust in the government was lost. Stories of illegal land claims made by refugees/IDPs from the liberation period to the post-independence period began to impact on the government, who were reluctant to resolve the refugee/IDPs/army crisis.

As the government affirmed absolute management of land, issues surrounding land management went from good (Land Act 2009), to bad (Transitional Constitution, 2011), and then to worse (Draft Land Policy 2011). Communities began to avoid government officials suspected of corruption, while continuing to implement collaborative projects with trusted government officials and NGOs. However, the aspiring elites, with their power and influence, began to manipulate the government and fabricate stories insinuating that the communities disrespected the government. They even convinced the government to relocate the capital to Ramciel on the grounds that the Bari community in Juba had refused to hand land to the government. During the land reform period 2005–2011, all previously and newly acquired public land in urban centers (Juba, Yei, Yambio, Bor, etc.) across the country, ended up in the hands of individual aspiring elites.

The aspiring elites were able to influence the key decision-makers assigned to review the Transitional Constitution 2011. With the support of these decision-makers, land ownership under the Transitional Constitution shifted from the community to the people (Article 170(1) TC2011). The government, and no longer the traditional authorities, were in charge of all land management. Under the transitional land reform, via restitution and resettlement programs, rights of (refugees/IDPs) and illegal occupants to land ownership and access in the communities of their choice, no longer by birth, was recognized (Land Policy 2011). The South Sudan Land Commission and the independent government arbitration body can now secure land for the government through powers of eminent domain (Land Policy 2011).

Reviewing the development of land conflict in South Sudan, it becomes clear that it resembles closely the scenarios of land conflicts in neighboring Kenya, and also as far away as Nigeria. In this respect, land grabbing by the ethnic groups of people in power, and taking over of land by the government through constitutional reform, are both apparent in South Sudan. Both scenarios, as foreign as they are, by their adoption have contributed to the initiation and continuity of land conflict; however, the solutions to the land conflict lie within the traditional land governance systems. The community land management system that has an in-built self-regulatory mechanism counterbalancing influences by which aspiring elites are regulated, ensuring that political power to manage and control land is not concentrated in the hands of individuals or politicians. Therefore, the community response to land conflicts under the transitional constitution is to prevent individuals from abusing their community-given privileges while permitting access to land and its resources through community-accepted norms and practices.

15.2 THEORY

The theoretical perspective of rentier state theory defines government policies which derives a substantial portion of national revenues from the local/indigenous resources (Beblawi and Luciani, 1990), a characteristic of a government that does not depend on a productive domestic market for its economy (Beblawi and Luciani, 1990). There is a very small group within the government with direct external links involved in the generation of the rent (Rached, Craissati, and IDRC, 2000), and the government is the principal recipient of the revenue (Smith, 2004).

In regard to land governance in South Sudan, the rentier state theory explains the creation of a new history which places land governance under the absolute control of a minority group within the government (Draft Land Policy, 2011). The rentier state theory fits well with South Sudan's economy that entirely depends on revenues generated from its land resources (Beblawi and Luciani, 1990). A smaller group within the government, the aspiring elites, have direct external links involve in the generation of the rent (Beblawi and Luciani, 1990). The government becomes the principal recipient of the revenue (Transitional Constitution, 2011).

The disadvantage of the state rentier theory is that it replaces a prevailing land administration system under the traditional authorities (World Bank, 2014). The problem with this theory is that it gave rise to a few aspiring elites who walk between the government and the community with impunity (Prendergast, 2016) (Figure 15.1).

The application of the rentier state theory in the conceptual framework is useful in examining the level of land conflicts across the three proposed land governance styles: (1) traditional land authorities with a meritocratic land governance style; (2) condominium and the Sudan Government's permissive land style; and the (3) SPLA/M/Government's problematic style.

Land governance under the traditional land authorities is meritocratic in style because it only accounts for/grants land access and ownership to members of community through birthright, and to some extent through some recognized marriage schemes appropriate to the cultural context of a particular community (De Simone, 2015). The traditional land governance is the same as the customary land administration systems that operate like federal systems of traditional/customary land governance. During the colonial period, headmen at the lowest level reported in the hierarchy to sub-chiefs, and the order went on to the highest, the paramount chiefs. The paramount chiefs were answerable to the district commissioner at the provincial level. The land governance under British colonial and Sudan Governments was regarded as permissive, because it allowed community land

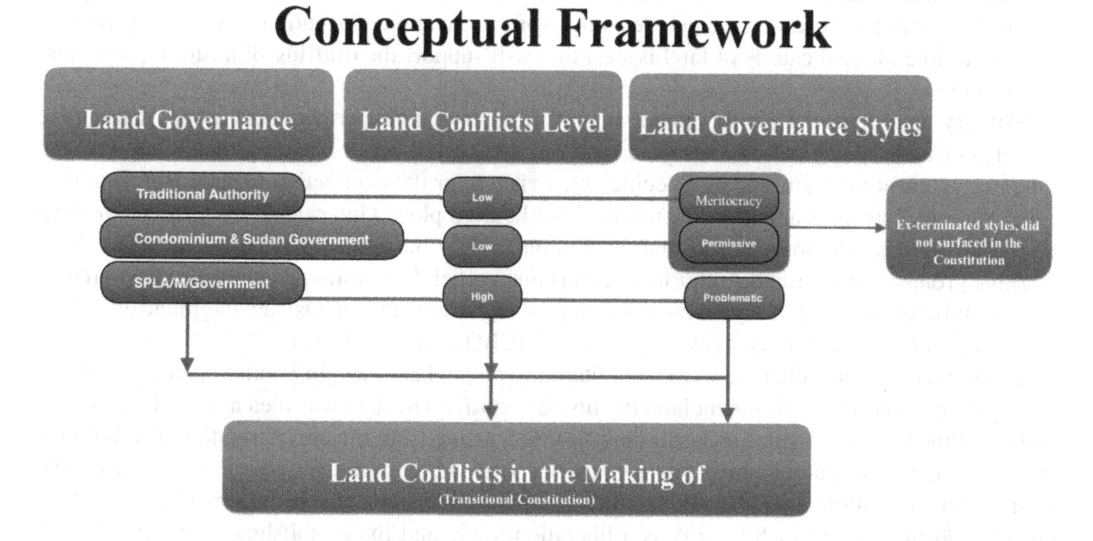

FIGURE 15.1 Conceptual framework.

ownership under management of traditional authorities to co-exist (Owen, 1951). Due to the multiple involvements of authorities in decision-making on land issues, these two styles of land governance registered extremely low levels of land conflicts.

This view of meritocratic and permissive styles can be referred to as the historical or "holistic narrative" (Freeman, 2008) of land conflict in South Sudan. It is an alternative to the prevailing scholarly "selective narrative" approach offered by reports from international agencies and NGOs (World Bank, 2014), as well as governments accounts, which address "alternative issues" instead of the main causes of land conflict in South Sudan. These give scholarly support to the selective narrative, a history produced by politics (Munslow, 2006) which is contrary to the prevailing history (Coraiola, Foster, and Suddaby, 2015, p.211).

However, the land governance under the SPLA/M/Government is regarded as problematic in style because of its heavy reliance on the use of force, political propaganda, and land reform led by a very small group within the government, to maintain absolute control over land access and ownership. Applying the rentier state theory on the SPLA/M/Government problematic style of land governance, the style registered high level of land conflicts.

The creation of a historical land governance style during the transitional period is the real craftwork of the aspiring elites in the recent land reform program. It is the creation of a history to sustain land conflict, and to divide communities along tribal/clan fault lines. The land reform approach is a system designed to benefit only the aspiring elites, not the government nor the community. The theory behind the land reform is to: (1) isolate and continue the suppression of meritocratic traditional systems of land governance, including the permissive colonial systems of land governance, by erasing the history of the country and limiting it to the liberation era, (2) legalize the shift of land ownership from community to people, (3) place government in charge of land management, hence reward the aspiring elites with the accruing benefits, (4) legalize the consequences of the refugee/IDP-related land grabs and illegal occupation by specific community in non-traditional communities across the country, particularly in the greater Equatoria areas, Wau and Malakal.

15.3 METHODOLOGY

Methodology: Problem structuring methodology (PSM) (Ormerod, 2014) is employed to discuss the structure of already messy and multidimensional land conflicts (Rosenhead, 1996). This method allows a rich conceptual framework modeling of land conflict under the transitional constitution. My intention is to identify alternative methods for conflict analysis, as well as tools providing the possibility for future genuine land debates, reinterpretation, and contribution (easy to add, correct, and streamline the real causes of land issues) that will support the drafting of a future permanent constitution.

Primary data collected for this paper is both from my life experience and seven years (2010 to 2017) of anthropological ethnographic field research based on first-hand professional work on land management in South Sudan. Specific areas visited for data collection include Bor, Rumbek, Juba, Yambio, Terekeka, and Yei. The research methods employed interviews, interactions, participant observations, and workshops/funeral ceremonies. Key narrative contributors included: State working groups, County Land Authorities, government officials, traditional authorities, conference/workshop/funeral presenters, and non-governmental organizations (NGOs) and their land coordination forum (LCF), community-based organizations (CBOs), and civil society.

The secondary data collected were from literature on land conflicts in South Sudan (articles and news publications, and government land (archives) records). The data was then assessed against the history of the land governance styles in the country. Starting from the era of traditional authorities' first encounters with Catholic missionaries, which continued through the Colonial government, and then the Sudan government (1847–1980s), I then compared this land use management to that which emerged during the rise of SPLA/M as a liberation force, and the establishment of the SPLA/M government (1980s to date). Tracing land conflicts and related social issues through these structured periods gives a clear image that South Sudan had effective traditional land governance systems before the SPLA/M/government.

This paper is structured as follows: firstly, providing a background to the South Sudan land conflict and related debates/confrontations. Secondly and thirdly, presenting theory and employing problem structured methodology to discuss the theory behind the suppression of traditional authorities' land governance and the rise of land reform. Fourthly, assessing the concept of land governance styles. Outlining the roles played by traditional land management system, as well as issues resulting from legal land reform. Fifthly, discussing events leading to the creation of a history designed to replace traditional systems of land management. Sixthly, providing a conclusion.

15.4 RESULT AND ASSESSMENT

15.4.1 Rentier State Theory in Meritocracy and Permissive Land Governance Style (1854–1980s)

Traditional Authorities: The traditional authorities have been the most dominant and prevalent systems of land governance throughout South Sudan. The chiefs know their land, history, and culture. They know how to accommodate foreign interest. They know how to manage periodic access to land through the corridor arrangement for cattle grazing (Grazing Agreement, 1958). The communities know how to co-exist and collaborate with cattle-keepers and famers. They know how to accommodate and live in peace with clans resettled within their communities during the British Colonial period. They managed resettlement programs related to sleeping sickness eradication and road construction (Owen, 1951), and they provided solutions to the refugee/IDP crisis.

Community Land Transfer Schemes: First records of traditional authority land management were informative, with accurate accounts of local collaboration among the chiefs allotting land to accommodate the interests of the first foreign Catholic missionaries in 1854 (Middleton, Toniolo and Hill, 1975). The Gondokoro Land Acquisition process for the mission house was consultative and inclusive, with a focus on protecting community interests. Moreover, the system was accommodating in nature, balancing the need for foreign investment with the social and economic needs of communities.

Consultative Practice: Findings from interactions with elders and chiefs in Yei and the vicinity revealed that traditional authorities with community participation have maintained the same consultation strategies to allot community land for religious, government, and private projects. They allotted land to two prominent private land investors with ties to communities, the Hagar and Mohamed Jabir. The majority of the private traders were directly connected with markets and trades in agriculture produce. Very few land conflicts were ever recorded in this period (1960s–1980s). The only record they ever had of land conflict was the distant news of cattle raids in Bor and beyond; the closest was illegal encroachment on government game reserves by poachers.

Grazing Corridor Management: While in Terekeka and South Bor, two communities separated by the Nile, chiefs managed periodic grazing corridors for cattle from pastoral communities grazing on designated land in farming communities across Terekeka. The chiefs managed the process by which fee charges, grazing periods, and locations were arranged and agreed upon by both sides. The enforcement of the arrangement was done by both sides as well. This practical control and management of access to community land protected farms from cattle destruction, while still permitting grazing (Grazing Agreement, 1958). The traditional authorities demonstrated that cattle-keepers and farmers could live and co-exist peacefully.

Land Access via Marriages: Apart from the traditional authorities' collaborative arrangement for access to land for cattle grazing, the two communities across the Nile river have had a history of recognized marriage relations. The existing relations between the in-laws would allow them to arrange for cattle grazing among themselves. However, this arrangement would later pave the way for each of these communities to practice cattle looting from within their own community; they would hide then those stolen cattle with their in-laws just across the Nile river, making it hard for the owner to trace their cattle. This practice would gain ground later during the war of liberation and would encourage cattle-keepers to keep stolen cattle away from their own communities across the Greater Equatoria.

***Resettlement Schemes*:** When the Condominium Government began the process of resettlement of sub-community/clan groups within larger communities/clans and the resettlement of communities from the banks of the Nile river and some major road corridors (Owen, 1951), traditional authorities and communities across regions worked collaboratively. In Yei, for example, the Kiliko community were resettled in Senza Asiri on the outskirts of Yei (Justin and Lotje, 2016), and in Western Equatoria, Moru Miza resettled from Bangolo to Jambo (Muro District Native Affairs Report, 1938).

Although the relocation and the resettlement were in essence political, but carried out under the pretext of disease management, traditional authorities in collaboration with the condominium government worked together in administering land for those resettlement programs (Owen, 1951). The processes did not engulf communities in land conflict. There were no physically documented land records under traditional systems, but through oral tradition, both the host and settled communities, to this day, know the origins of the claim to the land, the owner of the land, and those who only have rights to access it (Justin and Dijk, 2017; Radio Tamazuj, 2019).

***Infrastructure Schemes*:** In the missionary era, private infrastructure and services like schools and hospitals were set up and operated by church missions. The lands for such infrastructure and services were negotiated between church leadership and traditional authorities (Wheeler, 1997, 1998; Pirouet, 1999). Land, once allowed for such infrastructure purposes, was in fact delivered for the prescribed purpose and effectively implemented by the receiving end, the community (Governor, Upper Nile Province, 1920). There are no records of any objections and repossessions of such land by the communities in the history of traditional land administration.

***Participatory*:** Some prominent chiefs like Ladu Lolik of Liriya, who was also well-regarded as a road engineer, supervised his subjects on road construction (Owen, 1951). A chief in the former Amadi district and Maridi was also instrumental to the success of the road project. For example, when natives were fleeing from roadsides deep into the rural areas, due to the harsh road labor conditions, the Amadi chief was able to convince the British District Commissioner to review the law (Owen, 1951). The chiefs also recommended their preferred infrastructure to government decision-makers at the provisional headquarters in Juba for assessment and approval (Maridi District Native Affairs Report, 1940). Roads like the Jambo-Rokon-Juba link, which were built in 1960s, were the result of community-led projects (Dingwall, 1950; Muro District Native Affairs Report, 1960). Schools across the regions were also built thanks to collaboration between chiefs and the government.

***Community Collaboration*:** Chiefs also supported the British and Arab governments in maintaining production and supply lines of agricultural goods, ensuring shipment to Juba, and helping pay government taxes (Hassan, 1963). They supported land auctioning initiatives in urban contexts. Land deemed necessary to support the project, for example, trading centers, was auctioned to the general public (Yei District Native Affairs Report, 1951). Similarly, public projects were created and executed for the benefit of all the nation (Beaton, 1948). Traditional authorities also participated in various resettlement programs for victims of natural disasters. For example, during the Bor flood of 1970s, chiefs were instrumental in handling and managing the situation.

***Agricultural Schemes*:** Government collaboration with traditional authorities was exercised in the establishment of much bigger land projects. One example is a United Nations Development Program (UNDP) project: the Project Demonstrated Unit (PDU). This was the second-largest land project in Sudan after the Gizera scheme (Yongo-Bure, 2007). The PDU project was conceived by both Britain and Sudan's governments, World Bank-funded, and managed by the UNDP and 19 NGOs. To this day, whenever elders and chiefs talk about this particular agricultural scheme, the PDU and its subsidiaries across the regions (including Katire and Loka forestry, as well as the

Maridi and Nzara fruits schemes (Maridi Annual Report, 1963; Yongo-Bure, 2007)), they reflect on the spirit of collaboration and participation among communities and the Sudan government.

15.4.2 SPLA/M Problematic Land Governance Style (Impunity Era, 1980s–2005)

***Mass Forced Illegal Marriages*:** With the rise of SPLA/M, the role of traditional authority began to fade away and was eventually replaced by problematic systems of land governance. Use of force to access and own land was introduced. Even social and cultural events like marriage became heavily reliant on the use of force. When SPLA/M was forced out of Ethiopia in the 1990s at the fall of Mangestu's regime (Henze, 2007), the presence of the army took a serious toll on women and girls in the Equatoria region. Women and girls were raped, violated, and forced to marry at gunpoint. The women and young girls who escaped their tormentors were followed to their homes and in most cases brutally murdered, together with their parents and extended families (De Simone, 2015).

15.4.3 Interim Government, 2005–2011

***Influx of Specific Refugee/IDP Groups*:** Land conflict was manufactured by a few aspiring elites who took advantage of the chaos presented by the refugee crisis during the liberation and the interim period. The influx of specific refugees/IDPs and armed individuals from cattle-keeping communities led to illegal occupation of both private and public property in urban centers across the Equatoria provinces. They occupied government/public institutions, residences of government officials, private settlements, homes of orphans, etc. The illegal and forceful occupation went on to claim all lands and houses previous occupied by Arabs from Sudan and Equatorians in exile. All vacant land or occupied buildings adjacent to any army barracks and police station, garrisons and checkpoints, schools and hospitals, churches and mosques instantly became settlements of those cattle-keeping communities. It took authorities five to six years (2005–2011) to finally evict those who occupied the University of Juba, at the price of serious gunfights.

Many faultlines of conflict surrounding land claims were initiated: firstly, forceful taking of the private houses and property by the armed pastoralist groups from defenseless and vulnerable Equatorians, under the false claims that they the pastoralists were "the liberators." Then followed by the land conflict between politicians and the communities; a well-known conflict case is that between those aspiring elites and the Bari community which became known as "the conflict of the capitals." The aspiring elites tried to move the capital city from Juba to Ramciel in an attempt to silence the Bari community. Government resources were then committed to the feasibility study and master plan development for Ramciel city (Radio Tamazuj, 2017).

***Continuous Conflicts*:** When the Land Act 2009 was passed it was believed the new law would ease the tensions about land. However, conflicts continued to rise even to the extent that they seemed to dominate the 2011 referendum about separation, amid fears that many would vote for unity with the Government of Sudan in protest over land conflicts. Several claims were made over land which allowed for temporary settlement of Refugees/IDPs and army. Several boundary disputes arose and much inter-communal fighting over land claims took place. Examples include 2005, when the Azandi community in Yambio, the Moru community in Mundri, and the Kuku community in Kajo Keji, respectively, fought with the cattle-keepers. Large-scale land grabs were first recorded during this period. For example, the Mukaya land scandal involving a few aspiring elites, who teamed up to lease all the land in four counties to a single private investor (Deng, 2013). Public land taken by politicians, the aspiring elites for private purposes has become a tool in the exercise of power and influence. Land conflict related to illegal occupation by cattle-keeping communities extended farther to the southernmost farming communities of Kaya, Kajo Keji, and Magwi in Nimule (Radio Tamazuj, 2017).

In 2016, five years after the passing of the Transitional Constitution, following the outbreak of the fourth civil war in Juba, the youth could no longer swallow the bitterness of the situation. In Yei, some groups took up arms, went into the jungles, and suppressed all routes to Yei. They blocked off all food supplies from Yei, causing panic among the trapped cattle-keeping communities. Those who tried to escape were brutally murdered, and this action immediately provoked the president to publicly claim that his people were being targeted and he would command the army to Yei to stop the mayhem (VOA News, 2017).

15.4.4 Transitional Government (2011 to Date)

When the country moved into the transitional period, the Transitional Constitution became controversial not only over issues of presidential power, but also over allocation of power over land matters. The Transitional Constitution 2011 accorded the SSLC, an independent government body with a mandate to resolve land disputes through adjudications and arbitrations, the new mandates to draft land policy (GOSS, 2011). The Draft Land Policy shifts land ownership from community to "people," land governance from traditional authorities to the "government," and gives the SSLC greater powers of eminent domain to expropriate land for government purposes. Through land reform programs like land restitution and resettlement, the policy was likely to become a clear tool for the legalization of previous land claims over occupied army barracks, refugee/IDP sites, etc. With one hand, the SSLC can secure land through eminent domain, and with the other hand, it can settle land disputes in ways that favor certain groups or individuals.

Conflicts continued to escalate until the signing of the 2018 peace agreement culminating in the assignment of foreign troops and governments over some communities across South Sudan (IGAD, 2018). For example, Yei River State in the south has been assigned to Ugandan troops, and oil fields in Unity State in the north have been assigned to the Sudan government, as per the arrangement and terms of the agreement (IGAD, 2018).

15.5 DISCUSSION

Land conflict in South Sudan has two angles: (1) leadership and ethnicity, and (2) government taking over land through the constitutional reform.

***Leadership and Ethnicity*:** At the peak of the interim period, the land conflict in South Sudan is almost a cut-and-paste copy of the situation in Kenya. During the early days of independence of Kenya, when the British left the country and Jomo Kenyatta assumed the new leadership, men from his Kikuyu tribe unleashed a mass influx of people into lands previously occupied by the British, claiming houses and property and even the highlands that belonged to the Kalenjin tribe (Shilaho, 2017). President Jomo Kenyatta did not address pressing land issues, and neither did the three presidents that followed him: Moi, Kibaki, and Kenyatta Uhuru (Shilaho, 2017). Their regimes instead became involved in illicit acquisitions of public land. They exploited the land for their benefit and the benefit of their supporters (Shilaho, 2017). To these days, land and tribe are very volatile issues in Kenya.

Similarly, in South Sudan, particularly in the Equatoria provinces, occurrences of land grabs are associated with top political leadership and their ethnic groups. The first incident was during the presidency of Abel Alier who presided over the Regional (Southern) High Executive Council from 1972 until 1978, and again in 1980 (Andrew, 1998). Pastoral groups from Abel's Bor community, flooded the region and engaged in nepotism, power struggles, and land grabbing (Andrew, 1998), which led President Nimeri to issue a decree dissolving the regional government and establishing a regional federal government in 1983 (Johnson, 2014). This federal arrangement forced the pastoral communities to relocate back to their traditional land. However, this led to the formation of Anyanya II movement in Upper Nile region, which later was reorganized into SPLA/M (Johnson, 2014).

***Government and Constitutional Reform*:** The existence of land conflict in South Sudan is related to individuals, the aspiring elites, and the government taking over land from the community. As in

Nigeria under Obasanjo military regime (1976–1979), when in 1978, the Land Use Act was enacted, land was arrogated from the community to the government, but failed to achieve the planned objectives, and the land eventually went to the political and economic elite in the country (Brosche, 2012; Nwocha, 2016).

During the drafting of the South Sudan civic/constitutional land reform (Draft Land Policy 2011 and Transitional Constitution 2011), land ownership shifted from community to the "people," and the management of land also shifted from the traditional authorities to the "government." The shifting of land ownership and access from community (under traditional management) to the "people" (under government management) actually took South Sudan in the direction of Nigeria.

Similarly, in South Sudan, during the interim period (2005–2011) almost all public land in Juba, and the PDU Project land in Yei, Yambio, Nzara, etc. were arrogated to private hands. Then two to three years after the passing of the Transitional Constitution, a public site in Hai Guanya, being prime land in Juba city, was quickly surrounded by armed soldiers, and squatters were forcefully evicted. The eviction did not only affect those squatting on public land, but also the owners of the adjacent private properties, who were evicted at gunpoint. One of those evicted from their private property was a senior staff member of the University of Juba. He later died of a heart attack, having suffered the pain of losing his family's land to the men behind the big guns. Hai Guanya land was then surveyed, re-divided into smaller plots, and shared among top Government officials and their chief supporters. Investors then accessed the land through Juba's "build-and-occupy" plan, which was more like a localized "build, operate, and transfer (BOT)"-style of investment arrangement (Yun et al., 2009), where the foreigners provide the finances (Ding, Wyett, and Werker, 2012) and operate the facilities for agreed period of years, while the aspiring elites collect rents.

The recent civic land reform by the current government produced a high level of land conflict (Hirblinger, 2015). The aspiring elites managed to eliminate both the meritocracy and permissive land governance approaches, as they adopted only the problematic approach. Policy-makers in fact induced land conflict through the Transitional Constitution, as the power to subdivide communities and to appoint their leaders now rests with one person, the President (Transitional Constitution, 2011; The Enough Project, 2017). For example, the controversial creation of the 32 states by presidential decree, and the subsequent creation of counties and payams, were seen as solutions to the prevailing land conflict, but instead became the core of tribal land conflict. Across the country, there are now two camps: one camp is for those who want the government to take control of land management, who are supported by some elders in Bhar el Gazal (Radio Tamazuj, 2019), while the second camp is for those who want the traditional authorities to have absolute management of the land, and they are backed by the elders and respected clergies across Greater Equatoria (Radio Tamazuj, 2019).

15.6 CONCLUSIONS

The South Sudanese traditional system of land management is the most widely recognized and practiced land management system in communities across the country (World Bank, 2014). Its practice is meritocratic, as ownership is by birth rites, but again it recognizes marriage relations for automatic land access and usage within the cultural norms and practices. The system is also permissive in style, as historically it supports execution of massive developmental initiatives such as infrastructure and agricultural projects. It is accommodative and flexible with community and refugee resettlement projects. All these make the traditional system very holistic, collaborative, and overall, is more or less free of conflicts.

However, the Government-induced style of land management through the Transitional Constitution and its land policy, the legal land reform, has proven to produce never-ending land conflicts. By its nature, the government land management is power-centered. In practice, it motivates division along tribal lines, and encourages use of force which is easily abused and exploited and turned into sources of conflict by power hungry politicians, the aspiring elites.

The Adventist theologian, Shawn Boonstra, drawing from the argument of medieval theologians, writes that everything has a cause (Boonstra, 2016); in theory, behind every aspect of South Sudan's land reform, there is a cause. Certainly, there is a cause behind all the attempts at creating a history of land for new land governance in South Sudan. But on the other hand, for every cause, there is a solution too. The Arabs say, "understanding a problem is half a solution." The biggest problem the fewer aspiring elites have in South Sudan, and perhaps this might be the same for all African countries, is that the elites, in this age of globalization, love the culture of "cut-and-paste." Could this be because of the issues of African identity? A former minister of lands in Zimbabwe once commented at a land workshop in South Sudan, that Africans love to westernize rather than modernize. If Africans had modernized, it would have been easier for them to enhance their already-established traditions. The Americans say it is better to deal with the devil you know than to deal with the devil you do not know. The South Sudanese are better off dealing with their own problems because they know them, know their causes, and even know the obvious solutions to half their problems. They should want to know better the causes of their own problems, and then learn how to deal with those problems in a manner that will not cause more problems.

However, because of the belief in a selective narrative for creating a history, and a determined hope that a newly created history will prevail, this is simply a blind denial of the reality of life. Throughout the world, unresolved land disputes do not simply go away. No matter what level of westernization a society adopts, the problems will always be there, and yet they seem not to learn at all. The true and viable solutions to the current persistent land conflict in South Sudan lie mostly in the hands of traditional systems of land management and land governance. These traditional systems are already prevalent, with excellent track records of service in land governance that have survived and outlived multiple governmental systems. Since the Land Act 2009 has already recognized traditional authorities, while limiting their roles, influence, and power with regard to land ownership, there is a need to first rewrite the draft land policy to recognize traditional systems more substantially in the Land Act and assign more roles, power, and responsibilities to them.

In the process of drafting the permanent constitution, authority over land ownership should be fully delegated to communities under the management of traditional systems, who can then be monitored and supported by the land institutions of the national government. Policy- and decision-makers should be drawn from a pool of experienced land practitioners and researchers from across the communities in the country who will work under traditional systems guidelines (Scott, 1998). The objective is to avoid constitutional loopholes by which a few aspiring elites with power and influence can interfere with the Transitional Constitution in their search for personal gain, as well as avoiding conflicts resulting from the adoption of foreign land governance systems without a basis in the traditions, history, and reality of the communities of South Sudan.

If the aspiring elites of South Sudan had been led by African wisdom, they would have easily learned from those serious mistakes which now keep Kenya, Nigeria, and many other African nations hostages of their own making. South Sudanese traditional leaders and elders know problem-solving skills from the dawn of ages; as they would say, "the death of the hyena saved the hare his head." If the hare had not been attentive to the recklessness of the hyena in the presence of the lion king, he would have lost his head too. South Sudanese elites should have been like the hare and learned from the mistakes of those African countries, faced with the ever-present challenges of urbanization and globalization, to avoid the obvious pitfalls. They should have seen how the adoption of civic land reform and forced illegal occupation are taking a toll on the economy, politics, and social life of Kenya. Having the wisdom of learning from the mistakes of others gives the ability to avoid problems and their costs.

REFERENCES

Amusan, L. (2018) "Globalisation of Land and Food Security Challenges in Africa," *African Renaissance*, 15(Special Issue 1), pp. 83–99. doi:10.31920/2516-5305/2018/sin1a5.

Beblawi, Hazem Al, and Giacomo Luciani (1990). The Rentier State in the Arab World. In G. Luciani (ed), *The Arab State*, p. 96. London: Routledge.

Bodetti Michael, Austin. (2018). How the South Sudanese Civil War Is Fueling Climate Change. Available at: https://www.offiziere.ch/?p=34948.

Boonstra, Shawn. (2019). *Authentic: Daily Devotional*. Nampa, ID: Pacific Press Publishing Association.

Brosche, J., and Elfversson, E. (2012). Communal Conflict, Civil War, and the State: Complexities, Connections, and the Case of Sudan. *African Journal on Conflict Resolution*, 12(1), pp. 33–60.

Coraiola, D., W. M. Foster, and R. Suddaby (2015). Varieties of history in organization studies. In P. G. McLaren, A. J. Mills, and T. G. Weatherbee (eds), *The Routledge Companion to Management & Organizational History*, pp. 206–221. Abingdon: Routledge.

Deng, F. M. (2010). *Sudan at the Brink: Self-determination and National Unity*. New York: Fordham University Press. 72 pages.

Deng, D. K. (2013). Competing Narratives of Land Reform in South Sudan. In *Handbook of Land and Water Grabs in Africa: Foreign Direct Investment and Food and Water Security* (Editors: Tony Allan, Martin Keulertz, Suvi Sojamo, Jeroen Warner), Routledge: London, UK, pp. 446–455.

De Simone, S. (2015). Building a Fragmented State: Land Governance and Conflict in South Sudan. *Journal of Peacebuilding & Development*, 10(3), pp. 60–73. doi:10.1080/15423166.2015.1085812.

Ding, S., K. Wyett, and E. Werker (2012). South Sudan: The Birth of an Economy. *Innovations: Technology, Governance, Globalization*, 7(1), pp. 73–90.

Dingwall, R. G. (1950, March). Juba. Dry Weather Road Jumbo-Juba. District Commissioner, Equatoria. (EP/58.A.1Jo-S1, File Number, P4). Munuki, Juba: South Sudan National Archive Building.

Freeman, R. (2008). Frameworks for Policy Analysis: Merging Text and Context by Raul Lejano. *Critical Policy Analysis*, 2(2), pp. 161–162.

GoSS (2011). *Laws of South Sudan: The Transitional Constitution of the Republic of South Sudan*. Juba, Southern Sudan: The Government of Southern Sudan..

Governor, Upper Nile Province (1920, March). Letter to the Director of Lands, Sudan Government, Khartoum: Lands re-survey of Mission Stations. Malakal, Upper Nile Province (UNP/38, G2, File Number, P1). Munuki, Juba: South Sudan National Archive Building.

Grazing Agreement (1958, February). Between Tribal and Inter-Tribal Questions, Relations & Disputes. Governor.

Henze, P. B. (2007). *Ethiopia in Mengistu's Final Years*. 1st edn. Addis Ababa, Ethiopia: Shama Books.

Hirlblinger, Andreas (2015). Land, Political Subjectivity and Conflict in Post-CPA Southern Sudan. *Journal of Eastern African Studies*, 9(4), pp. 704–22. http://documents.worldbank.org/curated/en/421901468336313987/pdf/869580WP0P4370nance0in0South0Sudan.pdf. "How the South Sudan Civil War Is Fueling Climate Change." Offizier.ch. January 26, 2019. https://www.offiziere.ch/?p=34948 (Accessed: August 14, 2019).

IGAD (2018). *Agreement on the Resolution of the Conflict in the Republic of Sudan*, Addis Ababa, Ethiopia: The Government of Sudan South Sudan.

Johnson, D. H. (2014). *Federalism in the History of South Sudanese Political Thought*. Londen: Rift Valley Institute (Rift Valley Institute research paper, 1). http://riftvalley.net/publication/federalism-history-south-sudanese-political-thought (Accessed: March 22, 2019).

Justin, P. H., and Van Leeuwen, M. (2016). The politics of displacement-related land conflict in Yei River County, South Sudan. *The Journal of Modern African Studies*, 54(3), 419–442.

Justin, Peter & Lotje de Vries (2017). Governing Unclear Lines: Local Boundaries as a (Re)source of Conflict in South Sudan. *Journal of Borderlands Studies*, pp. 1–16. doi:10.1080/08865655.2017.1294497.

Justin, P., and H. van Dijk (2017). Land Reform and Conflict in South Sudan: Evidence from Yei River County. *Africa Spectrum*, 52(2), pp. 3–28. https://journals.sub.uni-hamburg.de/giga/afsp/article/view/1047/1054 (Accessed: February 1, 2019).

Middleton, D., E. Toniolo, and R. Hill (1975). The Opening of the Nile Basin. *The Geographical Journal*, 141(3), pp. 488–488. doi:10.2307/1796503.

Munslow, B., and O'Dempsey, T. (2008). Chapter 9.5 Complex political emergencies in the war on terror era. In Vandana Desai and Robert B. Potter (eds), *The Companion to Development Studies*, pp. 464–468. Routledge, USA.

Muro District Native Affairs Report (1938) personal communication

Muro District Native Affairs Report (1960) personal communication.

Nwocha, M. E. (2016). Impact of the Nigerian Land Use Act on Economic Development in the Country. *Acta Universitatis Danubius. Administratio*, 8(2), 117–128.

Ormerod, R. J. (2014). Or Competences: The Demands of Problem Structuring Methods. *EURO Journal on Decision Processes*, 2(3–4), pp. 313–340. doi:10.1007/s40070-013-0021-6.

Owen, J. S. (1951, May). Torit. Divisional Engineer, Public Works Department, Equatoria. Road Maintenance (EP/65, File Number, P414). Munuki, Juba: South Sudan National Archive Building.

Pinaud Clémence (2016). Military Kinship, Inc.: Patronage, Inter-Ethnic Marriages and Social Classes in South Sudan. *Review of African Political Economy*, 43(148), pp. 243–259. doi:10.1080/03056244.2016.1181054.

Pirouet, M. L. (1999). Various authors Faith in Sudan Series, Nos. 1-5 (Book Review). *Journal of Religion in Africa/Religion en Afrique*, 29(1), 129.

Prendergast, J. (2016). Roots of famine in Sudan's killing fields. In John Sorensen (ed), *Disaster and Development in the Horn of Africa*, Chapter 6, pp. 112–125. MacMillan press limited, UK.

Rached, E., D. Craissati, and International Development Research Centre (Canada) (2000). *Research for Development in the Middle East and North Africa*. Ottawa, Ont., Canada: IDRC. Available at: https://www.idrc.ca/sites/default/files/openebooks/310-0/index.html. (Accessed: March 21, 2019).

Radio Tamazuzj (2019). Farmers and pastoralists in Yei resolve to cooperate. Published January 29, 2019. https://radiotamazuj.org/en/v1/news/article/farmers-and-pastoralists-in-yei-resolve-to-cooperate (Accessed: March, 3, 2019).

Radio Tamazuzj (2019). Torit Bishop Urges Politicians to Allow Chiefs to Solve Land Disputes. Published January 22, 2019. https://radiotamazuj.org/en/v1/news/article/torit-bishop-urges-politicians-to-allow-chiefs-to-solve-land-disputes (Accessed March, 3, 2019).

Radio Tamazuj report (2019). Shilluk community submits border documents to TBC. Radio Tamazuj report. Published February 22, 2019. https://radiotamazuj.org/en/v1/news/article/shilluk-community-submits-border-documents-to-tbc (Accessed: March, 3, 2019).

Rosenhead, J. (1996). What's the Problem? An Introduction to Problem Structuring Methods. *Interfaces*, 26(6), pp. 117–131. doi:10.1287/inte.26.6.117.

Scott, J. C. (1998). *Seeing Like a State: How Certain Schemes to Improve the Human Condition have Failed*. New Haven: Yale University Press (Yale agrarian studies).

Sgd. B.M. Hassan (1963, May). Maridi. Annual Report Maridi Demonstration Farm for the Years/1962/3. Department of Agriculture, Extention, Equatoria. (EP/62, File Number, P21). Munuki, Juba: South Sudan National Archive Building.

Shilaho, Westen K. (2017). *Political Power and Tribalism in Kenya*. Johannesburg: SA. Publisher, Springer.

Smith, B. (2004). Oil Wealth and Regime Survival in the Developing World, 1960–1999. *American Journal of Political Science*, 48(2), pp. 232–246. doi:10.1111/j.0092 5853.2004.00067.x.

The Enough Project (2017). Ed. Jacinth Planer, Weapons of Mass Corruption: How corruption in South Sudan's Military Undermines the world's newest country. The Political Economy of Africa Wars: No5.

The World Bank (2014). *Land Governance in South Sudan: Policies for Peace and Development*. Washington, DC. ©World Bank. https://openknowledge.worldbank.org/handle/10986/20767 License: CC BY 3.0 IGO.

VOA NEWS. 2017. South Sudan President Tells Cattle Keepers in Equatoria to Go Home. https://www.voanews.com/a/south-sudan-president-tells-cattle-keepers-in-equatoria-to-go-home/4092787.html (Accessed: March 11, 2019).

Wheeler, A. C. (1997). *Land of Promise: Church Growth in Sudan at War*. Nairobi: Paulines Publications Africa (Faith in Sudan, no. 1).

Wheeler, Andrew C. (1998). Christianity in Sudan: The State of Studies, Christian Mission during the Condominium1899-1955, Dictionary of African Christian Biography. https://dacb.org/stories/south-sudan/bullen-herbert/ (Accessed: March 21, 2019).

Yei District Native Affairs Report (1951) personal communication.

Yongo-Bure, B. (2007). *Economic Development of Southern Sudan*. Lanham, MD: University Press of America.

Yun, S., S. Han, H. Kim, and J. Ock (2009). Capital Structure Optimization for Build-Operate-Transfer (bot) Projects Using a Stochastic and Multi-Objective Approach. *Canadian Journal of Civil Engineering*, 36, pp. 777–790.

16 Theory versus Reality of Inter-Agency Collaboration in Land Registration Practices

The Cross-Case of Ghana and Kenya

Uchendu E. Chigbu, Anthony Ntiador, Stellamaris Ogutu and Walter Dachaga

CONTENTS

16.1 INTRODUCTION

Land registration, according to Henssen (1995), is the process of official recording of land rights through deeds or titles. What this means is that there has to be an official record (usually called a register) of land rights or of deeds concerning changes in the legal situation of defined units of land. Modern land registration methods are believed to provide the most reliable path to securing land tenure. This is why it has become an integral part of land management methods, and an issue of global interest. This notwithstanding, there are several barriers to securing land or property rights and ownership through land registration. Many African countries embraced modern land registration either in their colonial or post-colonial periods of nationhood. However, a major problem in these countries is that these land registration (or their land administration) systems rely on silo approaches in recording land rights, due to a lack of inter-agency collaboration in processes leading to property registration (Ntiador, 2009; Williamson et al., 2010). That is why some literature has called for integrated systems and approaches to core land issues such as land management and land administration (Zevenbergen et al., 2013; Biitir, Nara, and Ameyaw, 2017).

In countries like Ghana and Kenya, land management (and specifically, its subsequent registration) processes are characterized by a series of back and forth procedures by different agencies that not only make the processes unnecessarily longer and more expensive, but also open room for corruption or malpractices (Kirk et al., 2015). This situation has been attributed largely to a lack of coordinated inter-agency collaboration in land registration procedures, especially land data-sharing (Agyeman-Yeboah, 2018; Biitir and Nara, 2016). It is one of the reasons both countries encounter challenges in setting up a practice of "one-stop shops." A one-stop shop is a land administration concept and practice that underpins collaboration and integration of business processes, operations, and functions of agencies to make multiple services accessible to clients in one system. Hence, the need for enabling improved land information availability, as well as pro-poor land administration approaches in Ghana and Kenya.

In this regard, this chapter aims to provide insights on how inter-agency collaboration in land administration practices apply in relation to theory and practice in Ghana and Kenya, and provide a path towards improvement. To achieve this aim, the chapter answers three major questions. The questions are: (1) how is a parcel of land registered under the current land registration systems in Ghana and Kenya? (2) How do inter-agency collaborations in land registration practices in Ghana and Kenya align with best practices described in theory? (3) How can inter-agency collaboration challenges in land registration be improved in Ghana and Kenya? In answering these questions, the chapter provides information on the reasons why land information management in Ghana and Kenya still suffers from information redundancy and information management problems that are rooted in poor inter-agency collaboration. Providing these underlying reasons is a starting point for smart and responsible land management interventions that meet local land registration needs.

The rest of the chapter is structured as follows: a literature background to enable grasping what the theory says is provided. This leads to an overview of the methodology, focusing on the assessment of inter-agency collaboration in theory in relation to the realities in the Ghana and Kenya land registration systems. The results of the case studies' assessments are then presented. The concluding section looks at the implications of the results for future practice and suggestions going forward.

16.2 LITERATURE PERSPECTIVE

Despite the well-thought out land tenure (and land administration) reforms that are being implemented in Ghana and Kenya, land registration is still problematic for their citizens, especially the poor. Difficulties in land registration have been blamed on information redundancy and information management problems rooted on inefficient collaborations between the institutionalized agencies whose functions influence land registration processes (Ntiador, 2009; Kirk et al., 2015). Therefore, understanding how these land agencies can collaborate to deliver better land registration services is a concern. Ghana and Kenya aside, emerging land policies in Africa (for instance, in Uganda, Rwanda, South Africa, Malawi, Zambia, Namibia, Zimbabwe, and many others) call for joint efforts towards tackling the land challenges in their various countries.

Various semantics have emerged from many of these calls for land agencies to work together. Some of them include inter-sectoral, inter-agency, inter-professional, and multiagency, to mention a few (Lloyd, Stead, and Kendrick, 2001). These terminologies may imply a range of structures (or arrangements, approaches, practices, strategies, and rationales), but they share one common meaning. That is, "working together" to achieve common land management and administration purposes (Warmington et al., 2004, p.13; Zevenbergen et al., 2016). Rather than focus on semantics, this study adopted the term "inter-agency," because it is one of the few terminologies for collaboration that has gained considerable importance in various aspects of public administration (including land administration). In its simplest meaning, inter-agency collaboration implies a situation where "more than one agency [is] working together in a planned and formal way, rather than simply through informal networking (although the latter may support and develop the former)" (Warmington et al., 2004, p.13). In general, there are possible models of inter-agency collaboration which can apply to Ghana and Kenya. Atkinson et al. (2002) identified them to include decision-making groups, consultation, and training, center-based delivery, coordinated delivery, and operational team delivery models.

The decision-making groups model provides a forum for professionals of different agencies to meet and discuss issues and make decisions at a strategic level (Langemeyer et al., 2016). The consultation and training model engages professionals from one agency to enhance the expertise of those from another agency mostly at the operational level (Ulibarri, 2018). In a center-based delivery model, a range of expertise ia gathered on one site to deliver a more coordinated and comprehensive service (Atkinson et al., 2002). A key feature of this model is exchange of information and ideas, although services may not be delivered jointly. In a coordinated delivery model, a coordinator is appointed to bring together separate services to facilitate a more cohesive response to needs through collaboration between agencies involved in the delivery of services (Atkinson et al., 2002; Ulibarri, 2018). While professionals deliver services at the operational level, the coordinator operates at a more strategic level. Finally, an operational team delivery model involves professionals from different agencies working together on a day-to-day basis, forming a cohesive multiagency team delivering services directly to clients.

The operational components of the land management paradigm are necessary for inter-agency collaboration to work appropriately in Ghana and Kenya. Operational components of the land management paradigm include the range of land administration functions that ensure proper management of rights, restrictions, responsibilities, and risks in relation to property, land, and natural resources (Enemark, 2005). It includes identifying land, defining interests in land, and organizing data about or inventories of land information (Williamson et al., 2010). It also includes devising ways to make land management structures (including procedures and outcomes) work in a responsible way (de Vries and Chigbu, 2017). Effective and efficient land registration systems are necessary for achieving this. It is therefore necessary to understand the importance of inter-agency collaboration among land sector agencies directly involved in land registration in fulling the task of land registration. Hence, a look at best practice, or what features an ideal inter-agency collaboration should have, is necessary.

Within the context of land management (specifically land registration), inter-agency collaboration is necessary because land registration tasks involve interdisciplinary, multidisciplinary, and cross-disciplinary (including inter-sectoral) activities for achieving sustainable policy objectives. This implies that all agencies involved whose activities influence land registration should work together towards achieving efficient outcomes. In this regard, contemporary literature describes what should characterize an ideal inter-agency collaboration. Based on a body of literature reviewed in this study, this chapter identifies some ideal features of an efficient inter-agency collaboration for effective land registration. A synthesis from the literature has been adopted as the indicators for assessing the situation in Ghana and Kenya (see Table 16.1).

Although the indicators presented here are inconclusive (as they have been evolving over time), they generally allude to the idea that inter-agency collaboration in land registration should be done to fulfill objectives that are in accordance with the societal needs for efficient land administration, property rights formation, protection, and tenure security improvement.

16.3 METHODOLOGY

In accordance with the aim of the chapter, the ideal (or theoretical) situation of inter-agency collaboration in land registration systems forms the basis for assessing its practice or application (reality) in Ghana and Kenya.

As the research questions lean towards qualitative case study enquiries, explorative literature and in-depth interviews have been used to probe for answers. A case study approach has been deemed relevant because it allows for the "examination of episodic events with focus on answering 'how' questions" (Chigbu, 2019, p.6). Ghana and Kenya were purposively chosen as case studies because they have had major land reforms over the past two decades with the objectives of ensuring wider land registration coverage. Studying the inter-agency collaboration in these countries will be helpful in understanding the levels of their progress in regard to their operational cultures. It was also necessary to conduct a two-country case study to enable knowledge-sharing and knowledge-transfer between Ghana and Kenya.

TABLE 16.1
Literature Evidence of Features (or Indicators) of Inter-Agency Collaboration for Land Registration

Features of inter-agency collaboration in land registration	Descriptions of these features as indicators for inter-agency collaboration in land registration	Literature evidence
Defined roles	Collaborating agencies should play unique functions. They should not have duplicated, conflicting, and overlapping functions.	Ehwi and Asante, 2016; Rosly, Ahmad, and Tarmidi, 2018
Legitimacy	As service providers, agents in the collaborations should be legitimate in legal, policy, or formal administrative norms.	Tomlinson, 2003; Lundberg, 2017
Co-location	To achieve efficiency, they should operate from one location, building, or in close proximity.	Karikari, Stillwell and Carver, 2005; Forkuo and Asiedu, 2009
Integrated data	Harmonized and standardized land-based information or database system and interoperability of data.	de Vries et al., 2015; Lemmen et al., 2017
Shared data	Shared access to legal, spatial, and ownership data from other agencies.	Biitir, Nara and Ameyaw, 2017; Rosly, Ahmad, and Tarmidi, 2018
Pooled budget	Coordinate finances through common financial resource management systems, pooled funds, and single payment points for services.	de Vries and Ester, 2015
Joint capacity development	Engage in capacity development exercises to share cross-agency experiences.	Chigbu et al., 2018
Shared aims and values	Shared organizational cultures and norms for a common land management goal.	Laarakker, de Vries, and Wouters, 2015
Transparent communication	Ability to deal with ambiguity and misinformation through effective communication strategies	Agunbiade, Rajabifard, and Bennett, 2014
Citizens' participation	Engaging and empowering local citizens and users	Wescott and Fitzsimons, 2016; Chigbu, Alemayehu, and Dachaga, 2019
Flexible to change	Ability to respond to new organizational cultures and societal changes	Williamson et al., 2010; Lee, de Vries, and Chigbu, 2019
Pro-poor services	Promote social inclusion and capability of providing, or aiming to provide, support to all, including the poor	Zevenbergen et al., 2013
Decentralization	Ensure land agencies provide land registration services from the top down to the local level	Ardiansyah, Marthen, and Amalia, 2015; Biitir, Nara, and Ameyaw, 2017
Responsive to women's land rights	Encourage gender equality in the registration process by supporting co-registration in protection of women's (spousal) land rights	Chigbu, Paradza, and Dachaga, 2019; Lundberg, 2018
Co-working	Investing conscious efforts or resources in collaborating in compatible operations	Rosly, Ahmad, and Tarmidi, 2018
Mergers	Cadastre and land registry operations should be are merged or handled by one agency	de Vries, Laaraker, and Wouters, 2015 de Vries, Muparari, and Zevenbergen, 2016

Primary data collection for the case studies was carried out (using face-to-face interviews) during the period August 2018 to April 2019. The interviews were conducted with experts from the land registration institutions in Ghana and Kenya, including practitioners in the field of land administration. These interviews were administered using semi-structured questions designed to discern the realities of inter-agency collaboration in land registration practice in Ghana and Kenya, and how to improve the situation. Desktop research (with focus on literature) allowed for understanding and identifying the key indicators for best practices on the subject.

Qualitative analysis of the "theory versus reality of inter-agency collaboration" practices was conducted by comparing the realties in practice (derived from interviews in Ghana and Kenya) with the good practice indicators (derived from explorative literature as shown in Table 16.1). This way, it was possible to know how the practices in Ghana and Kenya (reality) fare or relate to the best practice posited in the literature (theory). Further interviews allowed for understanding the nature of the situations in practice and the most practical ways towards improving the situation in the two countries.

16.4 RESULTS OR FINDINGS

16.4.1 THE EXHAUSTIVE LAND REGISTRATION PROCESS OF GHANA

In Ghana, land management functions are spread over multiple agencies. They include the four divisions of the country's National Lands Commission (NLC-Ghana). These divisions are the Public and Vested Land Management Division (PVLMD), the Survey and Mapping Division (SMD), the Land Valuation Division (LVD), and the Land Registration Division (LRD). The other agencies are the Office of Administrator of Stool Lands, the Town and Country Planning Department (TCPD), and the District Assemblies and Local Traditional Authorities. For a typical land registration in most parts of Ghana, a client must deal with the four divisions of the NLC-Ghana who execute separate functions in the land registration process, but with indirect encounters with external agencies such as the TCPD and the Office of the Administrator of Stool Lands (OASL). Land registration can involve a stool (skin) land, state land, family lands, or private lands. The registration process can either end with a registered deed or registered land title.

Figure 16.1 illustrates the general processes an indenture involving stool, state, family, or private land is taken through either a deed or title registration. The land registration process (see

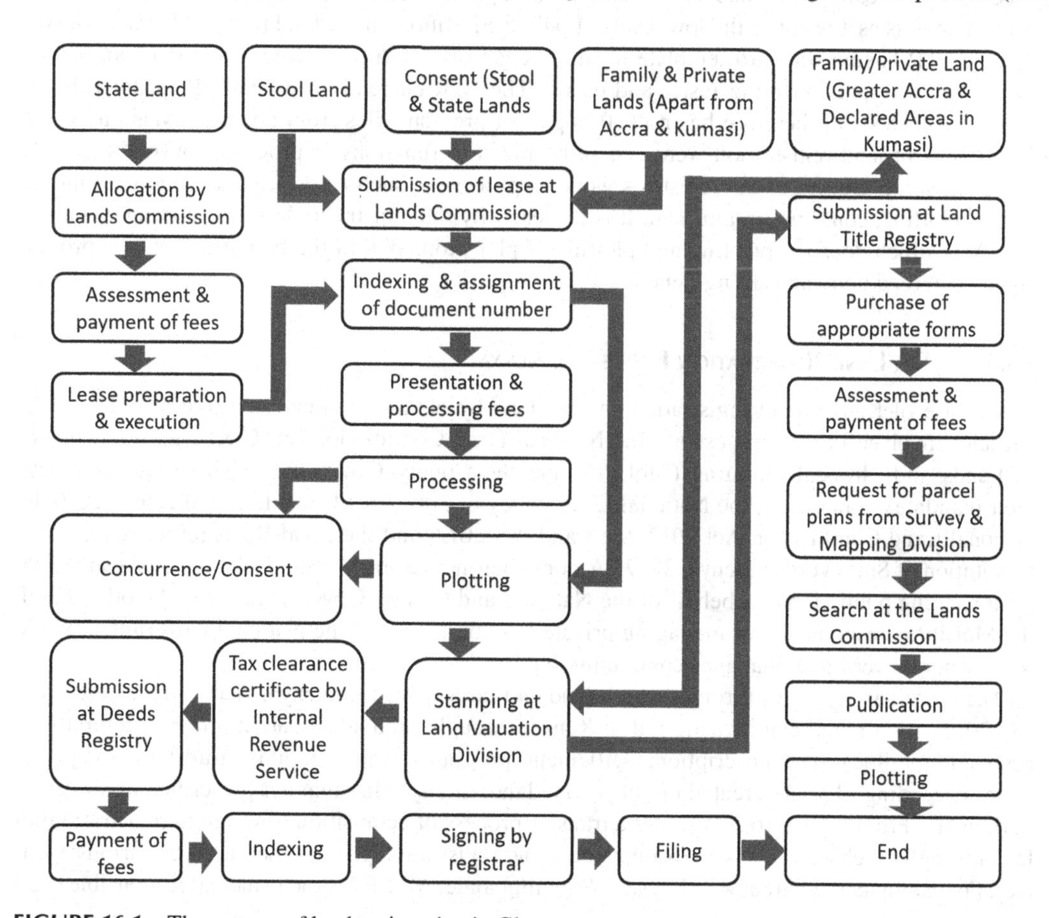

FIGURE 16.1 The process of land registration in Ghana.

Figure 16.1) for a stool land commences with preparation of a cadastral site plan once a land transaction is affected between a stool and a purchaser of land, mostly evidenced by an allocation note. With an approved site plan by the director of surveys at SMD, an indenture is then prepared, executed, and submitted by the applicant to PVLMD for consent and concurrence. The next stage of the process is that the indenture goes to LVD for stamp duty assessment and stamping. The stamped indenture is then submitted to LRD (deeds registry for deed registration or Land Title Registry (LTR) for title registration). Aside these processes, some indentures before their registration require planning comments from the TCPD. After registration of a stool land, responsibility is transferred to OASL to assess and collect annual ground rent on the registered land, which requires ownership information from the other land sector agencies. It can be observed that the registration of land title with stool land is very complex because of the concurrence element required by law. Within the LTR itself, the process is saddled with duplications and filing challenges. Filing systems within the LTR is very poor without clear procedures for the management of data stored from applications received. The office arrangement of receiving and processing applications is confusing within the organization. The first office an applicant encounters is the lodgment unit, and not reception.

At some point in the process, applicants are either asked to send a search request form to the PVLMD or to SMD for parcel/cadastral plan. Upon completion of the services, however, the LTR designated officer goes to the PVLMD and the SMD for the information. The rationale for requesting the applicant to send the requests and not be the one to go for the same information establishes the lack of customer-focused processes in the land title registration regime. The process also breeds extortion and illegal payments from the applicants. Sometimes the public have confused the issuance of "yellow card" from the LTR to mean the acceptance and clearance for a title certificate. This phenomenon is as a result of the low levels of public education on the land process. It was observed that the LTR has no standard template for the preparation of draft certificates, which would have the capacity to make printing faster and easier. The LTR currently uses PowerPoint for printing draft certificates and therefore has no soft copies of previous files from about ten years ago. The bottlenecks of land registration were generally traced to functions or processes beyond the LTR. However, some aspects of the registry's working processes, such as printing and file tracking, as well as plotting, could be streamlined. It is unclear why the registry undertakes plotting when it is the SMD which does the printing and plotting of plans too. Within the two agencies, one process could be agreed upon and strengthened.

16.4.2 The Land Registration Process in Kenya

In the case of Kenya, land registration can be done by obtaining either as a title or a deed. The agencies involved in this process are the National Land Commission (NLC-Kenya), the Ministry of Lands and Physical Planning (MoLPP), and the County Government. These agencies have their mandates drawn from the National Land Policy of 2009, thet Constitution of Kenya 2010, he National Land Commission Act 2012, the Land Act 2012, and the Land Registration Act of 2012 (Institution of Surveyors of Kenya, 2019). Among their numerous functions, the NLC is responsible for managing public land on behalf of the National and County Governments. On the other hand, the MoLPP is responsible for managing private land on behalf of the National Government while the County Government manages community land.

There are various ways through which land can be acquired in Kenya. According to the Land Act 2012, title to land can be acquired in Kenya, through: land adjudication process; compulsory acquisition; allocation; prescription; settlement programs; transmissions; transfers; long-term leases exceeding 21 years created out of private land; or any other manner prescribed in an Act of Parliament. From Figure 16.2, it shows a transfer process of acquisition (through to registration) of land through purchase. This is an example of a land registration process that involves various agencies. For example, the Director of Physical Planning under MoLPP is the principal responsible head

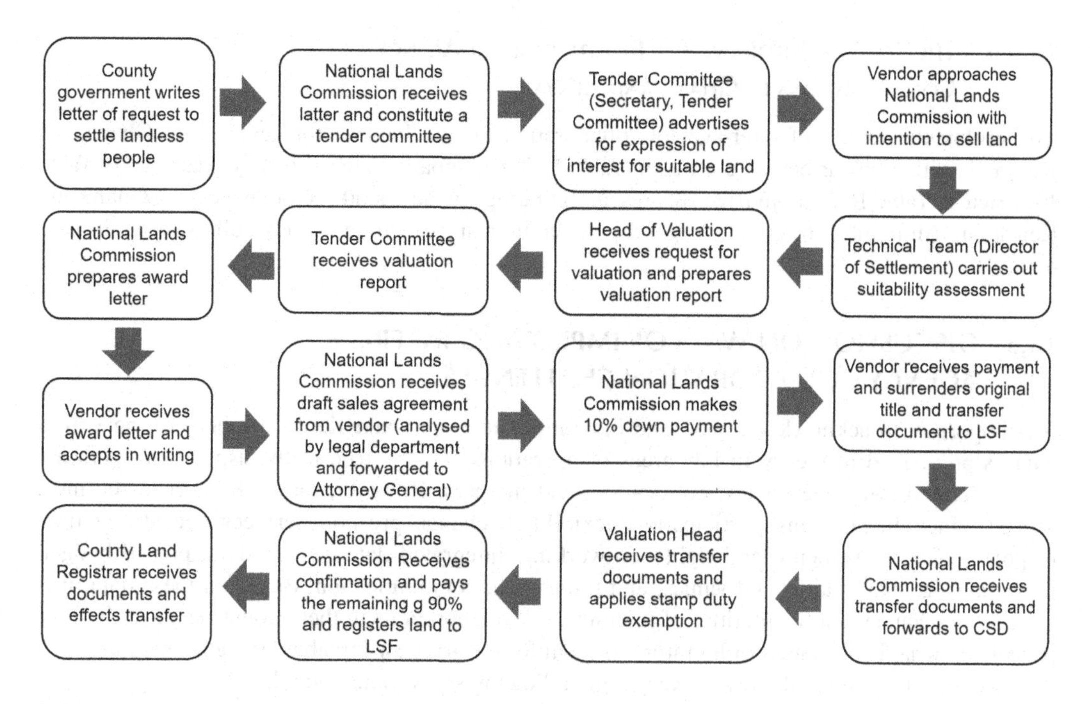

FIGURE 16.2 The process of land registration in Ghana.

for the effective and efficient implementation and maintenance of the procedure, while the county government is the initiator of the process through their elected leaders who identify a piece of land and request NLC to settle the landless there.

16.4.3 Land Registration in Ghana and Kenya is Based on a Mix of Inter-Agency Collaboration Models

An important question that arises is what model of collaboration drives the land registration process in Ghana and Kenya? In both Kenya and Ghana, the different agencies involved in the land registration exercise tend to have adopted a mix of three (out of four) of Atkinson et al.'s (2002) typology of inter-agency collaboration models, comprising decision-making groups, center-based delivery, coordinated delivery, and operational team delivery models. This is the situation because the various agencies have a strong focus on exercising operational independence in their various roles in the land registration process. However, they have not embraced the consultation and training model (a model that engages professionals from one agency to enhance the expertise of those from another agency at the operational level). The land registration operations in Ghana and Kenya are based on an unconscious mix of inter-agency collaboration models that exclude the consultation and training model.

The application or practice of inter-agency collaboration models in both countries have been identified as unconscious, because they are not based on any organizational policy of these agencies in both Ghana and Kenya. It has consequences on how the realities relate to the theory of inter-agency collaborations these countries. This manifests mostly in the lack of data standardization and integration, lack of data-sharing, duplications of roles, overlaps, and conflicts in the functions and business processes concerning land registration and related functions. In both countries, even though the level of cordiality among the agencies is good and encourages collaborative activities, some functions and activities result in conflicts and overlaps among and within the land sector agencies.

16.4.4 The State of Theory versus Reality in Inter-Agency Collaborations in Ghana and Kenya

Concerning the reality of inter-agency collaboration among land sector agencies in Ghana and Kenya, the situation has been presented in Table 16.2 in comparison to the theory presented in Table 16.1. Hence, Table 16.2 summarizes the research findings on how land sector agencies in Ghana and Kenya fare (in reality) in relation to good practice indicators of inter-agency collaboration in land registration.

16.5 DISCUSSION ON WAYS OF IMPROVING INTER-AGENCY COLLABORATION CHALLENGES

It is important to acknowledge that land institutions in Ghana and Kenya are performing within what is prescribed in theory in few areas of operations. This is true in the aspects of operating within defined roles, legitimacy, co-location, and mergers. However, they fall short in the most aspects—including citizens' participation, flexibility to change, pro-poor services, decentralization, responsiveness to women's land rights, co-working, integrated data, shared data, pooled budget, joint training, shared aims and values, and transparent communication. Based on these data, it is logical to conclude that the realities of inter-agency collaborations in these countries differ in most parts from what is presented in literature as ideal for ensuring sustainable land administration. To improve the situation in Ghana and Kenya, the following steps should be taken:

- Embrace a common land management vision of national importance: the first criterion for improving situations in Ghana and Kenya (as well as in any other African country) for all land agencies to base their operational goals on a common land management vision of national importance, e.g. secure land use and tenure security based on the principle of collaborative operations. Their ability to focus on an agreed land management vision could become a link among these agencies that will make them work together without putting in duplicative efforts.
- Ensure a systematic approach to land registration: systematic land (title) registration is a necessity in many African countries in order to provide an efficient, low-cost, less time-consuming, and secure approach to land registration across the country, and overcome the difficulties that citizens face when formalizing their land rights. Only efficient collaboration between land agencies would make this possible.
- Improve data integration, management, and sharing in land registration operations: making the current land registration process effective requires a systematic approach of integrating all available data held by the various agencies into a common system for ease of access and pro-customer service delivery. This situation will not be possible under the present situation in which all the land agencies are acting independently of each other in terms of data-sharing. Integrating the various data used and generated by the agencies would require inter-agency collaboration as a matter of policy.
- Institutionalization of good governance measures in land registration: effective land registration is not achievable unless it is hinged on good governance practices. Inter-agency collaborations can facilitate practical knowledge exchange or experience-sharing between land registration agencies in the aspects of decision-making and leadership processes on how to implement (or not implement) land information recording, use, and management, thereby helping to institutionalize good governance measures in land registration.

These steps are necessary to ensure the sustainability of an effective land registration system. They are important because the sustainability of a land administration system is hinged on strong internal administrative controls that ensure that checks and balances are incorporated to prevent loss or

TABLE 16.2
Comparison of Reality of Inter-Agency Collaboration with Best Practice Indicators Posited in Theory

Features of inter-agency collaboration theory	Reality in Ghana	Reality in Kenya
Defined roles	Agencies have defined roles through legislation. However, administrative processes of registering land (e.g. in site inspections and data acquisition) overlap and are repetitive	Mandates are clearly stipulated in laws but there is no clear distinction on the categories of land to be handled by different agencies. This has led to duplicating, conflicting, and overlapping roles
Legitimacy	All agencies are legitimized by the Ghana National Land Policy and Lands Commission Act 2008	All agencies have the backing of the Constitution and specific Acts
Co-location	Land registration agencies operate in one building, and where they are not in one building, are located within the same premises or within close proximity to each other	National level agencies (such as NLC-Kenya and MoLPP) are located within the same premises (and in some Counties within the same building). However, the County government offices are usually in a different location
Integrated data	Agencies (such as PVLMD, SMD, LVD, LRD, TCP, OASL) operate with different databases. Lack of standardization of data and processes makes data interoperability either difficult or impossible	There is no integrated database operated or shared by the different agencies. However, these agencies are making efforts to create independent digital databases that could lead to improved collaborations or an integrated system
Shared data	There is no sharing of data or information between agencies. They engage in separate data acquisitions and storage (different databases)	There is no sharing of data within the different agencies, resulting in duplicated efforts in data acquisition and management.
Pooled budget	Agencies have separate financial budgets and financial coordination, with multiple payment points for land registration services	The management of the agencies are independent of each other and this is the same with the budget allocation, expenditures, and coordination.
Joint training	There is no organized joint training. However, staff who are members of land professional organizations do get engage in joint workshops, write-shops, mentoring programs, and continuous professional development. This sometimes makes shared cross-agency experiences possible	There is no organized joint training. However, staff who are members of land professional organizations do get engage in joint workshops, write-shops, mentoring programs, and continuous professional development. This sometimes makes shared cross-agency experiences possible
Shared aims and values	Agencies continue to focus on their core functions and not the common goal for collaboration	Agencies have different organizational cultures. There is no common goal and this limits the room for collaboration
Transparent communication	There is effective internal communication mechanisms and ineffective public information dissemination. There is a reliance on second-hand agents (usually referred to as the "goro boys") for public information dissemination. This usually leads to misinformation of in the public domain	Since the land registration process is manual, communication between the applicant and the agency is only through face-to-face visits. This is time-consuming and discourages citizens from active participation

(Continued)

TABLE 16.2 (CONTINUED)
Comparison of Reality of Inter-Agency Collaboration with Best Practice Indicators Posited in Theory

Features of inter-agency collaboration theory	Reality in Ghana	Reality in Kenya
Citizens' Participation	There is knowledge gap between local citizens and land agencies on land registration processes. Citizens hardly participate in the formal registration systems	Most citizens do not know what processes and procedures to undertake to carry out a land registration. This makes them vulnerable to middle-men and corrupt agents
Flexible to change	There are entrenched professional silos, norms, and cultures among the surveyors, registrars, valuers, land officers, and administrative staff	The system lacks flexibility to change. There is resistance to change every time there is an attempt to change the norm
Pro-poor services	The system encourages social exclusion, especially for vulnerable groups due to high-end surveying requirements and the high processing cost of land registration	The land registration system does not promote inclusion of all social status. The high conveyancing and surveying fees charged for transactions makes it unaffordable for the low-income earners in society
Decentralization	Land administration is decentralized using customary land secretariats	Opportunity for decentralization is through the County government, but they have not been empowered to participate in land registration.
Responsive to women's land rights	Registration processes do not restrict registration of women land rights but laxities in spousal conditions for land registration do not promote formalization of women' land rights	The laws and policies recognize all genders to have equal right to own and register land. However, there is no mechanism in place to encourage women to own and register their land
Co-working	There is no co-working. It is threatened by unhealthy competition between agencies for funding, and among staff for clients for financial gains	There is no co-working. There are efforts to create co-working awareness by land professional bodies, but this has not been successful
Mergers	Land registration agencies have merged, but cadastre and land registry are run by separate agencies with disintegrated data for deed-parcel identification	Cadastre and land registry are managed by same agency (MoLPP) but their data are not integrated. The different departments handling them are geographically separated

destruction of land records. It also includes conducting professional land registration activities and ensuring the participation of citizens in the land registration process. As many agencies are usually involved in the land registration process, a policy is necessary to determine the procedures for these agencies to work together collaboratively towards achieving these objectives.

Such a policy should clearly define procedures for setting up the rules to guide inter-agency collaboration in all aspects of land registrations operations, including data integration and sharing, and the business processes and functions of various units, departments, and divisions. The policy should identify and specify the roles, obligations, rights, and procedures of each of the agencies. The clarity and cost associated in integration, data acquisition, and management should be factored into the policy.

16.6 CONCLUSION

A land registration system, like any other system, is a monolithic structure consisting of several aspects or subsystems. Therefore, the development of any new registration system or their improvement thereof requires the critical evaluation of all aspects of the subsystems. In that regard, this chapter has provided answers to the "how" questions about land registration processes, in the context of inter-agency collaborations in Ghana and Kenya. The chapter has shown that what is prescribed in theory is not always the reality in countries' land administration systems. The chapter has limited its focus to discussion on land registration because the registration of land forms the crucial aspect of securing tenure security through the land administration systems in Ghana and Kenya. Other aspects of land administration in Ghana and Kenya on which it has not put focus include land use issues, land taxation, land valuation, and spatial planning.

It is important to note that for any of the suggestions provided in this chapter to work towards improving the inter-agency collaboration challenges identified in Ghana and Kenya, procedures that will ensure that the processes for formalizing and subsequently transferring property rights are as simple and efficient as possible must be put in place. This in itself requires renewed land policies in both countries that are grounded. It is only through such an effort that Ghana and Kenya will be able to construct land administration institutions that will effectively collaborate in their efforts to address the constantly evolving land administration challenges the two countries. In this way the structures, processes, and outcomes of land registration will be streamlined to meet local needs, which is central to responsible land management.

REFERENCES

Agunbiade, M.E., A. Rajabifard, and R. Bennett. 2014. Inter-agency land administration in Australia: What scope for integrating policies, processes and data infrastructures for housing production? *Journal of Spatial Science* 59(1): 121–136.

Agyeman-Yeboah, S. 2018. Land rights documentation in Ghana: Experiences from low-income key workers. *European Journal of Research in Social Sciences* 6(1): 13–26.

Ardiansyah, F., A.A. Marthen, and N. Amalia. 2015. *Forest and Land-Use Governance in a Decentralized Indonesia: A Legal and Policy Review*. Bogor Barat: CIFOR.

Atkinson, M., A. Wilkin, A. Stott, P. Doherty, and K. Kinder. 2002. *Multi-Agency Working: A Detailed Study*. Slough Berkshire: National Foundation for Educational Research.

Biitir, S.B., and B.B. Nara. 2016. The role of Customary Land Secretariats in promoting good local land governance in Ghana. *Land Use Policy* 50: 528–536.

Biitir, S.B., B.B. Nara, and S. Ameyaw. 2017. Integrating decentralised land administration systems with traditional land governance institutions in Ghana: Policy and praxis. *Land Use Policy* 68: 402–414.

Chigbu, U.E. 2019. Visually hypothesising in scientific paper writing: Confirming and refuting qualitative research hypotheses using diagrams. *Publications* 7(1): 22.

Chigbu, U.E., Z. Alemayehu, and W. Dachaga. 2019. Uncovering land tenure insecurities: Tips for tenure responsive land-use planning in Ethiopia. *Development in Practice* 29(3): 371–383.

Chigbu, U.E., W.T. de Vries, P.D. Duran, A. Schopf, and T. Bendzko. 2018. Advancing collaborative research in responsible and smart land management in and for Africa: The ADLAND model. In: *World Bank Conference on Land and Poverty,* Washington, DC, March 19–23.

Chigbu, U.E., G. Paradza, and W. Dachaga. 2019. Differentiations in women's land tenure experiences: Implications for women's land access and tenure security in Sub-Saharan Africa. *Land* 8(2): 22.

de Vries, W.T., and U.E. Chigbu. 2017. Responsible land management-Concept and application in a territorial rural context. *fub. Flächenmanagement und Bodenordnung* 79: 65–73.

de Vries, W.T., and H. Ester. 2015. Inter-organizational transactions cost management with public key registers: Findings from the Netherlands. *International Journal of Public Administration in the Digital Age (IJPADA)* 2(2): 22–32.

de Vries, W.T., P.M. Laarakker, and H.J. Wouters. 2015. Living apart together: A comparative evaluation of mergers of cadastral agencies and public land registers in Europe. *Transforming Government: People, Process and Policy* 9(4): 545–562.

de Vries, W.T., T.N. Muparari, and J.A. Zevenbergen. 2016. Merger in land data handling, blending of cultures. *Journal of Spatial Science* 61(1): 191–208.

Ehwi, R.J., and L.A. Asante. 2015. Ex-post analysis of land title registration in Ghana since 2008 merger: Accra lands commission in perspective. *Sage Open* 6(2): 2158244016643351.

Enemark, S. 2005. Understanding the land management paradigm. In: *Proceedings of the FIG Commission 7 Symposium on Innovative Technologies for Land Administration*, Madison, Wisconsin, USA, June 24.

Forkuo, E.E., and S.B. Asiedu. 2009. Developing a one stop shop model for Integrated land information management. *Journal of Science and Technology (Ghana)* 29(3): 114–125.

Henssen, J.L.G. 1995. Cadastre and its legal aspects. In: *First Joint European Conference and Exhibition on Geographical Information*, Maastricht/the Netherlands.

Karikari, I., J. Stillwell, and S. Carver. 2005. The application of GIS in the lands sector of a developing country: Challenges facing land administrators in Ghana. *International Journal of Geographical Information Science* 19(3): 343–362.

Kirk, M., S. Mabikke, D. Antonio, U.E. Chigbu, and J. Espinoza, eds. 2015. *Land Tenure Security in Selected Countries: Global Report*. Nairobi: United Nations Human Settlements Programme / Global Land Tool Network.

Laarakker, P.M., W.T. de Vries, and R. Wouters. 2015. "Land registration and cadastre: One or two agencies?" In: *Proceedings of the 16th Annual World Bank Conference on Land and Poverty 2015: Linking Land Tenure and Use for Shared Prosperity*, Washington, DC: The World Bank.

Langemeyer, J., E. Gómez-Baggethun, D. Haase, S. Scheuer, and T. Elmqvist. 2016. Bridging the gap between ecosystem service assessments and land-use planning through Multi-Criteria Decision Analysis (MCDA). *Environmental Science and Policy* 62: 45–56.

Lee, C., W.T. de Vries, and U.E. Chigbu. 2019. Land Governance Re-Arrangements: The one-country one-System (OCOS) versus one-country two-System (OCTS) Approach. *Administrative Science* 9(1): 21.

Lemmen, C., P.J.M. van Oosterom, M.O.H.S.E.N. Kalantari, E.M. Unger, C.H. Teo, and K.E.E.S. de Zeeuw. 2017. Further standardization in land administration. In: *Proceedings of the 2017 World Bank Conference on Land and Poverty: Responsible Land Governance–Towards an Evidence-Based Approach*, Washington, DC: The World Bank, June 25–26.

Lloyd, G., J. Stead, and A. Kendrick. 2001. *Hanging on in There: A Study of Inter-Agency Work to Prevent School Exclusion in Three Local Authorities*. National Children Bureau, London, UK.

Lundberg, A.K.A. 2017. *Handling Legitimacy Challenges in Conservation Management: Case Studies of Collaborative Governance in Norway*. Norwegian University of Life Sciences.

Lundberg, A. K. A. (2018). Gender equality in conservation management: Reproducing or transforming gender differences through local participation? *Society & Natural Resources*, 31(11): 1266–1282.

Ntiador, A. 2009. Development of an effective land (Title) registration system through inter-agency data integration as a basis for land management. Master's thesis. Technical University of Munich.

Rosly, M.A., A. Ahmad, and Z. Tarmidi. 2018. Collaboration for enabling coastal geospatial data sharing: A review. *IOP Conference Series: Earth and Environmental Science* 169(1): 012018.

Tomlinson, K. 2003. *Effective Interagency Working: A Review of the Literature and Examples from Practice*. Slough: National Foundation for Educational Research.

Ulibarri, N. 2018. Collaborative model development increases trust in and use of scientific information in environmental decision-making. *Environmental Science and Policy* 82: 136–142.

Warmington, P., H. Daniels, A. Edwards, S. Brown, J. Leadbetter, D. Martin, and D. Middleton. 2004. Interagency Collaboration: A review of the literature. *Bath: Learning in and for Interagency Working Project.*

Wescott, G., and J. Fitzsimons, eds. 2016. *Big, Bold and Blue: Lessons from Australia's Marine Protected Areas*. Clayton South: CSIRO Publishing.

Williamson, I., S. Enemark, J. Wallace, and A. 2010. *Rajabifard. Land Administration for Sustainable Development*. Redlands, CA: ESRI Press Academic.

Zevenbergen, J., C. Augustinus, D. Antonio, and R. Bennett. 2013. Pro-poor land administration: Principles for recording the land rights of the underrepresented. *Land Use Policy* 31: 595–604.

Zevenbergen, J., W. T. de Vries, and R. M. Bennett, eds. 2016. *Advances in Responsible Land Administration*. CRC Press. 305p.

Section V

Land Management Methods and Tools

17 The Use of Land Value Capture Instruments for Financing Urban Infrastructure in Zimbabwe

Charles Chavunduka

CONTENTS

17.1 INTRODUCTION

Developing countries like Zimbabwe have been facing the challenge of rapid urbanization. The rapid urbanization rate has far outstripped low levels of economic growth. Rapid urbanization has resulted in a situation where sub-Saharan African countries have been experiencing enormous pressure from the demand for urban infrastructure and services (Ukaid, 2015; Berrisford, Cirolia, and Palmer, 2018; UN-HABITAT, 2012). The problems manifest in huge urban infrastructure deficits. This accounts, partially at least, for the slowed economic growth throughout the region (UN-HABITAT, 2014). Most African cities have not been able to fund infrastructure and services for people living in urban areas (Kihato, 2012; Berrisford, Cirolia, and Palmer, 2018; UN-HABITAT, 2012, Brown-Luthango, 2011; Turok, 2016; White and Wahba, 2019). High levels of poverty and inequality on the one hand, and constrained fiscal decentralization on the other, have limited the resources available to African cities to fulfill expanding functions and to address urban growth. Resources to invest in capital development have been particularly lacking (Berrisford, Cirolia, and Palmer, 2018).

Traditional methods of urban infrastructure finance, such as capital transfers from national governments, capital finance provided by parastatals, the private sector, and contributions from donors have been constrained by the downturn in national economies. Thus, in most African cities, urban development has been accompanied by minimal, if any, investment in infrastructure, a situation that has not been sustainable from economic, social, and environmental perspectives. The dire urban finance situation in most African cities highlights the importance of finding new and innovative methods of capital finance, with land value capture dealt with in this chapter requiring specific attention.

This chapter examines how land value capture has been used to finance urban infrastructure and services in Zimbabwe. It also looks into how the more effective use of land value capture instruments could contribute to responsible and smart urban land management in the country. Land management is "the science and practice related to the conceptualization, design, implementation and evaluation of socio-spatial 'interventions', with the purpose to improve the quality of life and the resilience of livelihoods in a responsible, effective, efficient, consensual and smart manner" (de Vries and Chigbu, 2017, p.66). Land value capture is a concept for collective methods and approaches used to collect and obtain revenues from land owners and private developers as a result of the increment in their land and property values due to public investment in infrastructure (El-Nagdy, El-Borombaly, and Khodeir, 2018; Suzuki et al., 2015). It has been an important mechanism for financing urban infrastructure in European, Latin American, and Asian countries; and if effectively used in Zimbabwe, can potentially become a sustainable, robust, adaptable, and reliable source of revenue for urban infrastructure finance, thereby contributing to responsible land management.

17.2 THEORETICAL FRAMEWORK

Land value refers to how much a parcel of land is worth. Since the 16th century in Portugal, a kind of income which constantly tends to increase without any exertion or sacrifice on the part of the landowners has been observed. Where an increase in land value is as a result of public investment or decision, the general consensus has been that the increment should be shared between the private land owner and government (Ingram and Hong, 2012; United Nations, 1976; UN-HABITAT, 2012; El-Nagdy, El-Borombaly, and Khodeir, 2018; Berrisford, Cirolia, and Palmer, 2018). The general argument has been that private landowners are not entitled to retain increments resulting from such public initiative because they are "unearned" (Ingram and Hong, 2012; Kitchen, 2013; Mathur, 2013; Walters, 2012; Walters, 2016; Munshifwa, 2017; Fainstein, 2012; Liu et al., 2018). An unearned increment is any rise in land values, whether due to public decisions or the general economy, where the rise is not due to the landowners' own initiatives and efforts (Alterman, 2012).

The land value capture concept arose out of the idea that landowners should share some of the increased value of their land with society (Alterman, 2012; Hendricks et al., 2017). Land value increase can be brought about a variety of factors, such as public investment in infrastructure, land use regulation, demographic changes, and economic development. The general application of the land value capture concept has been in cases where public investment in roads, water supply, sanitation, or local amenities such as streetlights results in increased real estate values. Thus, out of all the factors, the greatest attention has been paid to the effect of public infrastructure investment on land value (UN-HABITAT, 2012; Suzuki et al., 2015; Ingram and Hong, 2012; El-Nagdy, El-Borombaly, and Khodeir, 2019; UCLG/Committee on Local Finance for Development, 2012).

A comprehensive definition of land value capture is provided by Suzuki et al. (2015). Land value capture is defined as a public financing method by which governments a) trigger a rise in land values through regulatory decisions, e.g. change in land use and/or infrastructure investments, such as rapid transit; b) institute a process to share this increase in land value by capturing part or all of the increment; and c) use of land value capture proceeds to finance infrastructure investments such as the subsidization of public transportation or other improvements required to offset impacts related to rapid urbanization, e.g. improving water and sanitation.

Figure 17.1 shows the land value capture process. It provides a useful conceptual framework for land value capture. The process begins with the creation of land value. As can be seen from the diagram, this can involve public action in the form of construction of a highway, regulatory decisions

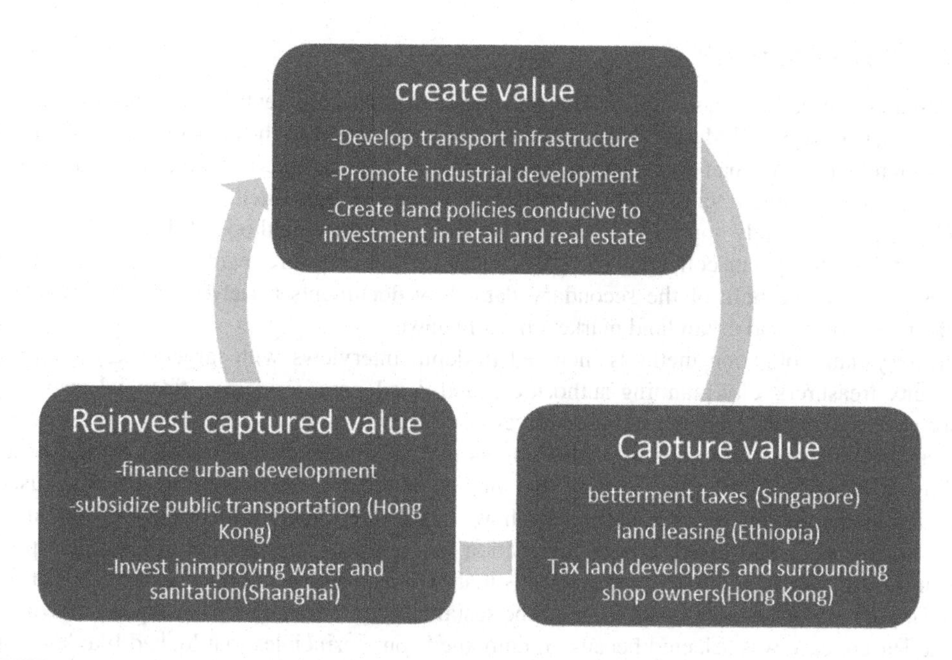

FIGURE 17.1 The land value capture process. Source: Siba and Sow, 2017.

(e.g. change in land use), and payment of subsidies to developers. Gains in land value resulting from public action are derived from the collectivity and the benefits need to be enjoyed collectively.

The created value is captured through a range of instruments, with the most commonly used being the betterment tax or levy (Sietchiping, 2011). Betterment tax is based on imputed increases in property value due to public investment or changes in property rights. Value capture can involve land acquisition and re-sale based on increase in property value due to changes in property rights. It can also involve indirect capture mechanisms, such as developer exactions, that shift the costs of public services onto developers.

The captured value is given back to society through investment in urban infrastructure and services. Land value capture proceeds can be used to improve the road network and water supply and subsidize the public transportation system, as well as implement public policies to promote equity (e.g. provision of affordable housing to alleviate shortages). The reinvestment of captured value effectively redistributes income in society while creating a potentially sustainable cycle of public financing.

17.3 UNDERLYING ASSUMPTIONS FOR LAND VALUE CAPTURE

A review of the use of land value capture instruments for financing urban infrastructure in Zimbabwe can be guided by underlying assumptions about the concept. These are set out as follows:

1. Citizens have the capacity to absorb an additional tax.
2. Local authorities have effective mechanisms for capturing the value of urban property in a way that is sustainable, redistributive, and developmental.
3. Central government bureaucratic and financial capacity for supporting the process.
4. An established financial sector.
5. A functional land administration system, including property registry.
6. Efficient, accurate, and timely land valuation.
7. Existence of enabling policies and legislation.
8. Decentralized and capable local governments.
9. Political support for the process.

17.4 METHODOLOGY

The research is based on a case study of Harare and a scan of developments in two other Zimbabwean cities, namely Gweru and Mutare. Data collection sought to establish the principles guiding the creation of land value in urban areas, identification of land value capture instruments—that is, if any were in existence—and tracing of how the captured value was reinvested.

Within the case study approach, both primary and secondary methods of data collection were employed. Secondary data collection was based on an intense literature search. Consultancy reports/studies provided the bulk of the secondary data. Key documents reviewed were on land-based financing in Harare and urban land markets in Zimbabwe.

Primary data collection methods included in-depth interviews with targeted key informants being city treasurers, city planning authorities, land developers, academia, financial institutions, senior government officials, and representatives of civil society. Key informants were asked questions about the types and trends in public actions that have been creating value in cities so as to establish principles of land value capture that may be applicable, the range of instruments used in capturing land value and their effectiveness, how the captured land value has been used, and the policy objectives so addressed. Direct observation was made of two housing development project sites in Harare, that is, Budiriro and Damafalls that exhibited aspects of the use of developer exactions for purposes of gaining insights about the sustainability of the land value capture instrument.

The Harare case was selected because it embodied some principles that looked like land value capture. It became necessary to establish whether observed revenue-raising mechanisms were a form of land value capture and their potential contribution to sustainable urban infrastructure finance. In other words, the research sought to draw insights about land value capture from the Harare case. A scan of two other Zimbabwean cities, that is, Gweru and Mutare, was done to assess the transferability of insights gained from the Harare case.

Secondary and primary data were analyzed through content analysis. This was achieved through coding for concepts and themes. Meanings were inferred from the data that helped to either confirm or refute instances of land value capture. Where observations did not fit land value capture, it had to be revised in line with the findings.

A limitation of the study was that for reasons of a financial nature, it was not possible to develop more case studies. Nonetheless, the research design has been to situate the Harare case as an instance in a set of parallel instances (Burawoy et al., 1991). Therefore, despite the limitation, the study managed to generate credible information that has immense value to both researchers and policy-makers.

17.5 RESULTS

17.5.1 Creating Land Value

In the first decade of independence (1980–1989), at a time when the national economy was relatively stable, public action created value in Zimbabwe's cities. In Harare, for instance, around 1982, the government acquired privately owned farmland on the western boundary of the city at rural land values. Land acquisition was done in accordance with the willing buyer-willing seller policy, and fair and just compensation was paid to the farm owners. Acquired farmland became state land that government donated to the Harare City Council for urban development. The land was incorporated into the Harare municipal boundary and in the process changed its use from rural to urban. Through this public action, that is, change in land use, there was a gain in value sufficient in size to cover the infrastructure costs of developing the new Kuwadzana residential area and produce a surplus for investment in low income housing by council.

In the early 1980s, the Zimbabwean economy was relatively stable following the end of the civil war, the ushering in of reconciliation, and the nascent democracy. The Zimbabwe Conference on

Reconstruction and Development (ZIMCORD) attracted donor funding to the country and a robust public sector investment program (PSIP) was put in place. In the first decade of independence, cities depended on central government grants and soft loans for infrastructure investment. Much of the land value that was created was a result of state investment and regulatory decisions. Soft loans were set lower than the market rate and cities could borrow money from prescribed assets of pension funds. All this worked well because on their part, cities paid back their loans without default. Cities have a mandate to provide roads, water, sewage disposal, public lighting, and other infrastructure funded through charges to residents—property rates. Funds that are raised through rating of property are not meant for new investment, but during the first decade of independence the money collected from property rating reached amounts above cost recovery level and council could deposit the excess in a capital development fund.

In the 1990s Zimbabwe implemented the economic structural adjustment program (ESAP). ESAP was launched in 1991 and the passage of years witnessed the demise of the PSIP and other forms of central government funding to cities. Since the late 1990s, the central government has lost its capacity to invest in urban infrastructure. This was accompanied by the demise of central government grants and soft loans to local authorities. At the local authority level there has been no capacity to mobilize funds for infrastructure investment. The last batches of central government loans and grants for urban infrastructure were received in the 1990s. Land reforms that began in 2000 triggered a further economic downward spiral, and because of the unstable economic environment, cities experienced high default rates in the repayment of their loans. Upon realization of the drying up of loans, the government made a Constitutional provision of 5% of the national annual budget for funding infrastructure and services in local authorities, provincial, and metropolitan councils (Government of Zimbabwe, 2013). Seven years on, these lower tiers of government are still to receive the 5% Constitutional provision.

A recent study found that in Harare the proportion of revenue spent on capital works was 1% (Berrisford, Cirolia, and Palmer, 2018). In Harare, access to finance was constrained by the national economic downturn. With both internal and external sources of funding constrained, the Harare City Council has been going into partnership with financial institutions to mobilize funds for infrastructure investment. The partnership has seen the Harare City Council contributing land at its intrinsic value while the financial institution provides funding for infrastructure development.

17.5.2 Land Value Capture Instruments

Three forms of value capture were identified in the study: "in kind" contributions, endowment fees, and property rates/taxes. By far the most common instrument used for capturing value has been "in kind" contributions. Harare negotiates with developers to make an "in kind" contribution at the time they apply for planning permission. It is the responsibility of the city to fund off-site infrastructure, but for lack of financial capacity it has become the norm for developers to be asked to fund such infrastructure as the water main, pumping main, reservoirs, and trunk roads. The city gains from the "in kind" contribution of off-site infrastructure by developers. In partnership with the Central Africa Building Society (CABS), the city asked the financial institution to install water reservoirs and water pumps and upgrade the major distributor road in the Budiriro housing project. In another Harare housing project (Damafalls), the city asked the land developer to install similar infrastructure. The same was observed at the Woodlands housing project in Gweru. "In kind" contributions are akin to developer exactions. An exaction is a condition which is placed on a development, usually following negotiation between the authority and developer (Casner, Leach, and French, 2004). It can take various forms such as the payment of money, the installation of infrastructure, and other such contributions in kind.

The second form of land value capture has been endowment fees. Harare, like other local authorities in Zimbabwe, requires developers to pay a prescribed percentage of the property's value upon the granting of a subdivision of land permit. The created subdivisions are expected to be of higher value; hence the applicant is required to pay 10% of the parent property value to the council. This can be paid in cash, a contribution of land equivalent value, or a combination of the two. Property can only be transferred after full payment of endowment fees. It is a one-off payment to the local authority that should ring-fence the money in a capital account for public benefit such as land acquisition. To ring-fence is common parlance for securing the money in a capital account specifically for purchase of land, estate development, savings, and investment as provided by the Urban Councils Act (Chapter 29:15). A local authority can require both "in kind" contribution and an endowment, as in the case of the Budiriro housing project where the Harare City Council asked the CABS Building Society to install off-site infrastructure and set aside land for schools, clinics, and shopping centers.

The third value capture instrument identified was property rates/taxes. This is in line with Hui, Ho, and Ho's (2004) definition of land value capture as the collection by the government of all kinds of taxes derived from the land. As part and parcel of the preparation of Harare's Avondale Local Development Plan, in 2014 the city re-zoned Avondale residential area to multiple land use. Valuation of properties is done every ten years for rating purposes. In the case of Avondale, a new valuation roll is expected to reflect much higher property values. But the impact of the economic crisis seems to be affecting city authorities' perception, whereby they have been increasingly seeing property rating as less of land value capture, but a mechanism for funding public amenities like street-lighting, swimming pools, recreational halls and parks, and other such services.

17.5.3 Use of Captured Value

Procedurally, endowment funds are supposed to be deposited into the city's capital account and used to provide infrastructure in future areas, infrastructure upgrading, and land acquisition. In practice, following years of economic crisis, Harare has not been ring-fencing the money collected from land subdivision applicants. This funding goes into the city's operating account, rather than the capital account, and is spent on salaries or general operations. The trend has been the same in the other two cities studied, that is, Gweru and Mutare. In all the three cases, the capital account became defunct. For reasons of poor rates collection, there has been no inflow into the capital account from this source. Payments of endowments in land have been of greater benefit to society through the provision of amenities.

17.6 DISCUSSION

17.6.1 Developer Exactions, Endowment Fees, and Property Rates

The results of this study confirm the theoretical framework that attributed land value creation to public investment and regulatory decisions. This had to do with the need for the community to benefit from a share of landowners' unearned value increments. The results also validated the idea of property rating as a potential land value capture mechanism (Hui, Ho, and Ho, 2004). Although the term land value capture has not been used in Zimbabwe, some of its principles are inherent in "in kind" contributions, endowment fees, and property rating.

As in the rest of sub-Saharan Africa, the trend in Zimbabwe has been towards the use of developer exactions for urban infrastructure financing. Exactions require developers to actually invest in off-site infrastructure which is transferred to the local authority at the end of the project. There is broad consensus that developer exactions are close to land value capture since they trade some public action (permission to develop land) for provision of off-site infrastructure. Others, for example, Alterman (2012), would regard them as an indirect mechanism of land value capture. What seems to be the case in many instances though, is that the magnitude of the value capture is limited to the value of the infrastructure. These developer exactions have come to be used as a cost recovery mechanism by local authorities.

Endowment fees were long provided for in British town planning law. These were borrowed, as Zimbabwean planning law has its basis in earlier versions of British legislation. The granting of a planning or subdivision permit increased land value and generated a need for public facilities. The developer was obliged to contribute a share of the increment to society.

In spite of respondent differences, property rating emerged as one of the identified land value capture instruments. Monthly property rates are charged in Zimbabwean cities and are typically used first to cover non-revenue-generating services and infrastructure. However, it is also possible for them to be used as a source of funding for infrastructure. Indeed, in developed countries, when the operating account is in surplus, rates have been used as an infrastructure financing measure. For example, in examining the practicality of using property taxes to capture land value in the US, the annual property tax was found to be an effective value capture instrument (Ingram and Hong, 2012).

The scope of land value capture in Zimbabwe has been defined by contextual factors. These closely relate to the underlying assumptions outlined earlier in this chapter. Unlike in developed countries, urbanization in Africa has not correlated with poverty reduction. In the majority of African cities, there are no structures capable of laying down, formalizing, nor implementing land and urban policies which would ensure that the public benefit from land value capture. Hence, cities face institutional challenges in applying the concept. Using land value capture as a capital mobilization instrument necessitates a number of requirements being in place, including enabling policies and a legal framework. This entails legal reform to provide other mechanisms of land value capture, in addition to endowment fees and property taxes.

To be successful, the land value capture process needs central government support. In Zimbabwe, public administration tends to be weak and lacks the financial capacity to provide funding for urban infrastructure investment. Likewise, local governments have lost some of their traditional revenue-raising powers such as electricity generation and vehicle licensing to central government and have been incapacitated in the context of general political and economic crisis. Administrative weaknesses include the lack of a property cadastre over large parts of cities and poor collection of endowment fees due to inefficiency and corruption (City of Harare, 2010). Another problem relates to absentee landlords, whereby a significant proportion of the estimated 3.5 million Zimbabwean diaspora own properties back home, but have been defaulting on rates payments (ZimStats, 2018).

17.6.1.1 The Emerging Model: Developer Exactions

The emerging model for potential land value capture in Zimbabwe would necessarily be based on developer exactions. As a model for land value capture, it holds considerable promise. Its strength is in forging partnerships between local authorities, land developers, and financial institutions. A weakness of the development exaction procedure is that it has not been guided by any legal or regulatory framework. The absence of a legal framework has created room for differential application, favoritism, and corruption. Local authorities have, at times, not enforced the conditions of land subdivision, consolidation, and development permits, and they have not been able to maintain the infrastructure handed over to them by their development partner. These weaknesses have been exacerbated by institutional and technical constraints to assess value increment.

17.6.1.2 Endowment Fees

Endowment fees are paid by developers to the council. In Harare, the contribution of endowment fees to council coffers is minimal. Between 500,000 and 1 million USD are collected annually from endowment fees (Berrisford, Cirolia, and Palmer, 2018). The low amount of fees collected can be attributed to three factors: a) the unavailability of finance for long-term infrastructure investment as a result of the adverse macro-economic environment; b) Harare does not ring-fence the funds it collects from developers. Instead of the capital account, the money is deposited in the operations account and not reinvested in infrastructure development; c) due to politics, most peri-urban development has been undertaken by cooperatives that do not need to make the same contributions as urban land developers.

17.6.1.3 Property Taxes

Rates can only be seen as an infrastructure financing measure when the operating account is in surplus, which is rarely the case in African cities. Property rates are generally used to cover non-revenue-generating services such as street-lighting and recreational parks. In Addis Ababa, property tax is a miniscule proportion of city revenue by any standards (Goodfellow, 2017). In Zimbabwe, property taxes have not been contributing to capital development, leading some respondents to not consider them as a land value capture mechanism. The economic crisis, backlog in land registration, and outdated valuation rolls are some of the factors that explain low levels of rates collection. About 23% of land parcels in Harare are not registered and occupants have not been paying rates, prejudicing the city council in potential revenue (The Herald, 2019). Property valuation is supposed to take place every ten years, but cities have not been able to comply with the requirement.

17.7 CONCLUSION

The financing of urban infrastructure in developing countries has become a challenge under the impact of rapid urbanization and rising levels of urban poverty. This chapter has examined how the traditional model, whereby local governments depended on central government funding of urban infrastructure, is no longer tenable in the context of poor economic performance in most developing countries. For many reasons, including unstable macro-economic environments, the failure of decentralization, and lack of institutional capacity, local governments in these countries have not been able to mobilize funding for infrastructure investment. The need to deal with a fiscal gap has been clearly evident in cities. This chapter examined how a western concept, land value capture, that has been effectively used to finance urban infrastructure abroad, could benefit Zimbabwe.

The evidence gathered in the study showed that although the concept of land value capture has not been used in Zimbabwe, its principles have been applied in the form of endowment fees and property rates. Revenue collected through these value capture mechanisms has, however, been found to be insignificant and not enabled cities to cover infrastructure maintenance costs, let alone embark on capital development programmes. In the absence of central government funding, cities have increasingly relied on developer "in kind" contributions (developer exactions) for infrastructure finance. Developer exactions are an indirect form of land value capture. In the case of Zimbabwe, there has been no policy and legal framework to guide the use of "in kind" contributions for urban infrastructure finance. Local authorities have depended on negotiating with partners when participating in joint ventures that have at times lacked transparency and accountability. In the joint ventures, local authorities have been contributing land as equity while financial institutions funded infrastructure. Although the "in kind" developer contributions model offers promise, much work needs to be done to increase public participation and address sustainability aspects of infrastructure development and management.

Land value capture as implemented in developed countries has been difficult to replicate in sub-Saharan Africa because of many factors. For a start, it has not received enough political support in Zambia, Rwanda, Ethiopia, and South Africa (Munshifwa, 2017; Goodfellow, 2017; Brown-Luthango, 2011). In African countries, rapid urbanization has been accompanied by immiseration. Economic decline in most countries has eroded central government capacity to invest in urban infrastructure. At the local level, municipalities have not been able to raise funds for infrastructure investment. The procedure for value capture requires the authority to determine the value increment, but a large proportion of city properties are usually not on the property cadastre. Most local authorities do not have employees with the skills to determine value increments—a measurement or value assessment problem. Capturing land values in African cities remains a challenge (Berrisford, Cirolia, and Palmer, 2018).

This chapter provides innovation by examining a model, "in kind" contributions, that has potential for development into public private partnerships (PPP) which offer promise as an alternative approach to urban infrastructure financing (Salaj et al., 2018). The model can be designed for affordability, efficiency, and sustainability, relies on co-financing, and is responsive to community needs. For greater societal impact, the model needs to address the weak policy environment and technical and political conditions. It also needs to be supported by friendly and accessible data platforms on land and property.

The theoretical framework of the land value capture process (UN-HABITAT, 2012) guided the analytical procedure that was followed in the chapter. From an earlier perspective, land value capture referred to the collection of all kinds of taxes by the government derived from land and its attached structures (Hui, Ho, and Ho, 2004; Nguyen et al., 2018; Qun et al., 2015). From a later perspective, land value capture was a method of financing urbanization through the recovery of profit generated by the value of property as a result of public infrastructure investment. These varied perspectives enabled a nuanced and context-based understanding of the land value capture concept and provided a robust analytical framework for the study. The case study methodology proved most useful for studying a policy issue such as land value capture. Given that land value capture is a poorly understood concept in Zimbabwe, the case study methodology was most appropriate for addressing the research question (Leedy and Ormrod, 2001). In Zimbabwe, government and civil society have used case studies as an important means for policy dialogue and formulation. The chapter aimed not only to find out how land value capture has been used in Zimbabwe, but to understand it within a particular context.

This research derives a model about how best an evolving approach for funding urban infrastructure can be designed given the Zimbabwe context. It presents a decentralized and adaptable solution to the challenge of infrastructure financing in rapidly urbanizing poor countries. It contributes innovative and alternative ideas to the funding of infrastructure in urban areas, thereby challenging conventional forms of urban finance. It contributes to responsible land management, because the supported developer exactions are voluntary and provide scope for civil society and citizen participation in building sustainable urban communities. The encouragement of shared responsibilities among cities, civil society, and citizens contributes to sustainable urban infrastructure management. The research contributes a pragmatic and contextual solution to urban infrastructure finance.

REFERENCES

Alterman, R., 2012. Land use regulations and property values: The "Windfalls Capture" Idea Revisited. In: *The Oxford Handbook on Urban Economics and Planning*, eds. N. Brooks, K. Donangby and G. Knapp, 755–786. Oxford: Oxford University Press.

Berrisford, S., Cirolia, L. R. and Palmer, I., 2018. Land-based financing in sub-Saharan African cities, *Environment and Urbanization* 30(1): 35–52.

Brown-Luthango, M., 2011. Capturing land value increment to finance infra-structure investment – Possibilities for South Africa, *Urban Forum* 22(1): 37–52.

Burawoy, M., Burton, A., Ferguson, A. A. and Fox, K. J., 1991. *Ethnography Unbound: Power and Resistance in the Modern Metropolis*. Berkeley, CA: University of California Press.

Casner, A. J., Leach, W. B. and French, S. F., 2004. *Cases and Text on Property*. New York: Aspen Publishers.

City of Harare, 2010. Special investigations committee's report on City of Harare's land sales, leases and exchanges from the period October 2004 to December 2009, *Presented by the Chairperson of the Special Committee on 23 March 2010*, Harare.

de Vries, W. and Chigbu, U. E., 2017. Responsible land management – Concept and application in a rural territorial context, *Flächenmanagement und Bodenordnung* 79(2): 65–73.

El-Nagdy, M., El-Borombaly, H. and Khodeir, L., 2018. Threats and root causes of using publicly-owned lands as assets for urban infrastructure financing, *Alexandria Engineering Journal* 57(4): 3907–3919.

Fainstein, S. S., 2012. Land value capture and justice. In: *Value Capture and Land Policies*, eds. G. K. Ingram and Y. Hong, 21–40. Cambridge, MA: Lincoln Institute of Land Policy.

Goodfellow, T., 2017. Taxing property in a neo-developmental state: The politics of urban land value capture in Rwanda and Ethiopia, *African Affairs* 116(465): 549–572.

Government of Zimbabwe, 2013. *Constitution of Zimbabwe Amendment (No. 20) Act*. Harare: Government Printers.

Hendricks, A., Kalbro, T., Llorente, M., Vilmin, T. and Weinkamp, A., 2017. Public value capture of increasing property values – What are "Unearned Increments?" A comparative study of France, Germany and Sweden. In: *Land Ownership and Land Use Development – The Integration of Past, Present and Future in Spatial Planning and Land Management Policies*, eds. R. Dixon-Gough, R. Mansberger, J. Paulsson, J. Hernik and T. Kalbro, 257–282. Zürich: Erwin Hepperle, (vdf-Hochschulverlag).

Hui, E. C., Ho, V. S. and Ho, D. K., 2004. Land value capture mechanisms in Hong Kong and Singapore: A comparative analysis, *Journal of Property Investment and Finance* 22(1): 76–100.

Ingram, G. K. and Hong, Y.. eds., 2012. *Value Capture and Land Policies*. Cambridge, MA: Lincoln Institute of Land Policy.

Kihato, M., 2012. *Infrastructure and Housing Finance: Exploring the Issues in Africa*. Parkview, South Africa: Centre for Affordable Housing Finance in Africa. A Division of the Finmark Trust.

Kitchen, H., 2013. Property tax: A situational analysis and overview. In: *A Primer on Property Tax Administration and Policy*, eds. W. J. McCluskey, G. C. Cornia and L. C. Walters, 1–40. West Sussex: Blackwell Publishers.

Leedy, P. D. and Ormrod, J. E., 2001. *Practical Research: Planning and Design*. Upper Saddle River, NJ: Prentice Hall.

Liu, Y., Fan, P., Yue, W. and Song, Y., 2018. Impacts of land finance on urban sprawl in China: The case of Chongqing, *Land Use Policy* 72: 420–432.

Mathur, S., 2013. Land value capture to fund public transportation infrastructure: Examination of joint development projects' revenue yield and stability, *Transport Policy* 30: 327–335.

Munshifwa, E. K., 2017. Land value capture in support of city infrastructure in Zambia: Challenges and doorstep conditions for adoption, *Paper presented at the 2017 World Bank Annual Conference on Land and Poverty*. Washington, DC: The World Bank, March 20–24, 2017.

Nguyen, T. B., van der Krabben, E., Musil, C. and Le, D. A., 2018. "Land for infrastructure" in Ho Chi Minh City: Land-based financing of transportation improvement, *International Planning Studies* 23(3): 310–326.

Qun, W., Yongle, L.. and Siqi, Y., 2015. The incentives of China's urban land finance, *Land Use Policy* 42: 432–442.

Salaj, A. T., Roumboutsos, A., Verlic, P. and Grum, B., 2018. "Land value capture strategies in PPP – What can FM learn from it?" *Facilities* 36(1/2): 24–36.

Siba, E. and Sow, M., 2017. Financing African cities: What is the role of land value capture? *Africa in Focus*, Thursday, December 14, 2017.

Sietchiping, R. ed., 2011. *Innovative Land and Property Taxation*. Nairobi: United Nations Human Settlements Programme (UN-HABITAT).

Suzuki, H., Murakami, J., Hong, Y. and Tamayose, B., 2015. *Financing Transit-Oriented Development with Land Values; Adapting Land Value Capture in Developing Countries*. Washington, DC: The World Bank.

The Herald, 2019. 80 000 properties not on city database. 10 August 2019.

Turok, I., 2016. Getting urbanization to work in Africa: The role of the urban land-infrastructure-finance nexus, *Area Development and Policy* 1(1): 30–47.

UCLG/Committee on Local Finance for Development, 2012. Land value capture: A method to finance urban investments in Africa? *Special Session of the United Cities and Local Governments, Africities*, Dakar, Senegal.

Ukaid, 2015. Urban infrastructure in Sub-Saharan Africa-harnessing land values, housing and transport: Literature review on land-based finance for urban infrastructure. Report number 1.4. African Centre for Cities.

UN-Habitat, 2012. *Handling Land: Innovative Tools for Land Governance and Secure Tenure*. Nairobi: United Nations Human Settlements Programme (UN-Habitat).

UN-Habitat, 2014. *State of African Cities 2014: Reinventing the Urban Transition*. Nairobi: UN-Habitat.

Walters, L. C., 2012. Land value capture in Policy and Practice [Online] Available from: www.landandpoverty.com/agenda/pdfs/paper.walters_full_paper.pdf. Accessed: 01/06/2019.

Walters, L. C., 2016. *Leveraging Land: Land-Based Finance for Local Governments – A Reader*. Nairobi: United Nations Settlements Programme (UN.Habitat).

White, R. and Wahba, S., 2019. Addressing constraints to private financing of urban (climate) infrastructure in developing countries, *International Journal of Urban Sustainable Development*. doi:10.1080/19463138.2018.1559970.

Zimbabwe National Statistics Agency (ZIMSTATS), 2018. *Migration in Zimbabwe: A Country Profile 2010 – 2016*. Harare, Zimbabwe: International Organization for Migration.

18 Is the Social Tenure Domain Model a Suitable Land Reform Tool for Zambia's Customary Land

A Case Study Approach

Fatima Mandhu

CONTENTS

18.1 INTRODUCTION

Security of tenure and access to land for all, including minority groups such as persons with disabilities, the youth, the aged, and women, are strategic prerequisites for the responsible development of agriculture. It is the role of every government to promote the provision of an adequate supply of land in the context of responsible land use policies. Land reform in Africa is based on two fundamental principles: enhancing the social capital where customary tenure exists, and enhancing individual responsibility in individual tenurial systems. Zambia operates a dual land tenure system based on the two principles where statutory land is under the administration of the government and customary land under more than 270 traditional leaders called chiefs. The government in Zambia is promoting land reforms aimed at enhancing both types of land tenure systems alongside each other, in a social context applying the Social Tenure Domain Model (STDM).

Through a case study approach, this chapter evaluates whether or not STDM will provide a solution to the problem of tenure insecurity and promote responsible land management in Zambia. The case study will analyze the decisions in the two cases on the conflict between statutory land rights held under a certificate of title against land rights of the community held under customary tenure (*Mpongwe Development Corporation Ltd v Kamanda*, 2007; *Mpongwe Development Company Ltd v Kamanda*, 2010). It is acknowledged that the Supreme Court is the final court of appeal, and its decisions are binding on the parties to the matter. This chapter will not question the judgment of the Supreme Court, but will analyze how STDM as a land reform tool can be used to assist the subsistence farmers holding land under customary tenure to enforce their rights to land in cases of conflicts like the one under the *Mpongwe* cases.

The objectives of this chapter are, firstly, to discuss the dual land tenure system currently being applied in Zambia and the gap created as a result of non-registration of customary land tenure. Secondly, to consider the position of STDM as a flexible tenure approach, by discussing the five pilot projects in Zambia. Finally, the chapter will consider the application of the STDM in resolving land tenure disputes between large-scale land investment and customary land use using the decisions from the *Mpongwe* cases. In addition, the chapter will, to a limited extent, evaluate the STDM approach in terms of the responsible development of agricultural land. Thereafter a conclusion will be drawn and recommendations given.

18.2 USE OF STDM IN LAND ADMINISTRATION AND MANAGEMENT

This chapter will adopt the understanding of the STDM approach on three levels as presented by Lemmen; STDM as a concept, STDM as a model, and STDM as an information tool (Lemmen, 2010, 2012). The first two levels are discussed below, while level three will be explained under the discussion of the five pilot projects under Section 18.5.

There is a gap in the conventional land administration systems in Zambia such that customary tenure cannot be easily identified and conceptualized. The need for an alternative approach in land administration has been discussed over a period of time. The concept of STDM is being applied to bridge this gap by providing a standard for representing "people–land" relationships independent of the level of formality, legality, and technical accuracy. STDM can represent all types of "people" and all types of "people–land" relationships, and can represent such linkages or relationships by various types and/or combination of location-based elements or "spatial units."

Du Plessis in his writings further argues that:

> the main value added by the continuum concept is that it offers a shift away from a preoccupation with titling and individually held private property, is a simple call for change in complex contexts, a way of describing and representing new forms of practice, and a foundation for inclusion and building on existing practice. In addition, and perhaps most important of all, the continuum is an aid to both identifying and advocating for where more fundamental reforms are needed to the land policy, law and administration systems (Du Plessis et al., 2016, pp.33–34).

STDM is an initiative of GLTN, as facilitated by UN-Habitat, to support pro-poor land administration, and to offer an alternative against the conventional tenure approach which does not cater for customary tenure.

18.2.1 Understanding STDM as a Concept

In order to conceptualize STDM, it is important to understand what the continuum of land rights is and the process it provides for the gradual shift from informal land rights to formal land rights. It has been argued that STDM in its current development can be used generally in all contexts and situations covering both tenures. It can also serve as an alternative to the current conventional Land Administration System (LAS). However, STDM's development for now is meant specifically for developing countries where there is very little cadastral coverage in urban areas with slums, or in rural customary areas which are not surveyed. The focus of STDM is on all relationships between people and land, independent from the level of formalization, technical accuracy, or legality.

"People–land" relationships can be expressed in terms of persons (or parties) having social tenure relationships to spatial units. The relationships can be expressed in different ways; for example, in the pilot project in Uganda, the main spatial unit used is the house or structure occupied or owned by slum-dwellers. In the pilot implementation of STDM in Uganda, when undertaking participatory enumeration in informal settlement, a household is taken to represent a single party. The flexibility of STDM as a concept allows for its adaptation to a particular situation.

18.2.2 STDM as a Model for the Pilot Projects in Zambia

The STDM being piloted in Zambia is viewed by NGOs implementing the pilot projects as a more flexible, fit-for-purpose, and inclusive approach based on recognition of a diversity of rights, within a context of pro-poor and gender responsive land management and administration. The project on land reform initiatives in Zambia, Phase 1, supported by UN-Habitat (United Nations Human Settlements Programme (UN-Habitat), 2017), was implemented from January 2018 to December 2018. Phase 2 will commence soon as a build-up of the achievements recorded in the first phase is implemented.

There were four interrelated components that had been identified as areas of concentration under Phase 1. Firstly, awareness building and partnership strengthening, which under Phase 1 focused on introducing and supporting the use of fit-for-purpose land tools and approaches in land administration and management. The second phase will build up on the gains achieved under the first area of concentration and focus on advocating for and strengthening the implementation of the fit-for-purpose land tools and approaches among key stakeholders, and the establishment of a multi-stakeholder platform on land. The second component of Phase 1 concentrated on the need to finalize the draft land policy in Zambia so that it can provide comprehensive guidance on land management and administration issues. The second phase will provide the technical support to the Ministry of Lands, Natural Resources and Environmental Protection (MLNREP) to finalize and launch the National Land Policy. The third component during Phase 1 supported scaling-up customary land certification interventions. Eleven villages in the chiefdom of His Royal Highness, Senior Chief Chamuka IV of Central Zambia were supported with the issuance of customary land certification. Further, 70 local women, men, and youths were trained to carry out the participatory enumeration process and mapping using the STDM. In total, 530 customary land certificates covering 5% of the Chief's land were issued. The second phase will focus on strengthening scale-up efforts by training more people, establishing and equipping local offices in the Chiefdoms, and building the capacity of the local people to manage and share relevant data with key stakeholders.

The final part of Phase 1 was to provide support to issue occupancy licenses in selected informal settlements involving Ward 10 of Kanyama informal settlement in Lusaka (UN-Habitat, 2017). Technical and financial support was given to the Lusaka City Council (LCC) to improve the process of issuing the Occupancy Licenses under the new and more progressive legislative framework provided under the Urban and Regional Planning Act (Urban and Regional Planning Act, 2015). In total 18,400 households were mapped and enumerated and a database was established that could be accessed by Lusaka City Council (UN-Habitat, 2017).

This chapter will concentrate only on customary land certification and the conflicts that STDM will resolve between statutory land rights and customary land rights holders. Five main pilot projects in Zambia will be considered, on the implementation of the STDM. The emphasis will be on evaluating whether STDM is a suitable land reform tool for Zambia's agricultural land using a case study approach. The main question is whether STDM will provide a solution to the problem of tenure insecurity caused by conflicts between customary and statutory rights to promote responsible development of agricultural land in the rural areas.

18.3 LAND REGISTRATION AND TITLING SYSTEM IN ZAMBIA

In Zambia, over 70% of the land covering rural and peri-urban land is outside the formal Lands and Deeds Registry records (Tembo, Minango, and Sommerville, n.d.). The records are maintained by MLNREP, the government unit responsible for dealing with land administration and management. Land outside the government records is not registered or recorded in any way and therefore title is not issued to a landholder by the government (Mandhu, 2015). Land that is not recorded is situated in the rural areas. Lack of a recording and titling system for customary land has posed several challenges on issues of tenure security relating to conflicts between customary landholders and statutory title. The lack of registration and formal titling of most customary land areas attests to the limitations of the current land administration system (LAS) which only covers statutory land regulated by legislation and institutional frameworks situated mostly in urban areas. The current registration system and the land administration system has been developed to provide for statutory tenure only and has not been extended to customary areas (Mandhu, 2016).

For agricultural land in particular, two major theories have been used to describe the tenure. The family farm theory of tenure and the farm business theory of tenure (Schiclele, 1952). Subsistence farming is what is described as the family farm theory of tenure where the farmers own and operate small family farms as independent entrepreneurs. These farmers rely on traditional forms of farming on land that is generally passed on from one generation to another, regulated by customs and traditions of the particular tribes. "The source of rules governing customary tenure is custom and usage of the community supplemented by statute in instances where the state has legislated on the issue" (Hansungule, 2001, p.24). The right to own land under customary tenure arises from the fact that an individual is a legitimate resident of a particular area and a member of a community where the particular tribe resides (Hansungule, 2001). In Zambia the only statutory intervention regarding customary tenure is under the Lands Act 1995, sections 7 and 8 which recognize customary tenure and provide for conversion of customary tenure to statutory tenure. The Act does not state the nature of customary rights or how the rights can be acquired, exercised, or recorded and registered.

On the other hand, the business of farming is conducted on a commercial basis and free market forces are allowed to determine the tenure status and farm size. Zambia has a long history of large-scale farms that have co-existed alongside smallholder communities (Chu, 2013). In recent years, the country has experienced a sudden increase in the demand for land to be used for large-scale agricultural purposes. The Land Matrix estimates that 26 deals covering an area of 389,774 ha have been concluded since 2000 (Lay, Nolte, and Sipangule, 2018).

At the national level, policy and institutional changes through the land reform program have not been easy to formulate and even more difficult to implement since they have not been supported or recognized as a priority by the Zambian government. The adoption of STDM is designed to bridge the technical gap by allowing for recording of all possible types of tenures without making changes to the legislative or institutional frameworks (Lemmen, 2010). The choice of STDM over other forms of land tenure reform is supported, since it does not require changes to the policies and legislative as well as institutional frameworks at the initial stage. The other forms of tenure reform are very time-consuming and require lengthy procedures and government resources which in the case of Zambia are not available. Land reform programs involving tenure reform are shelved before they can be finalized at policy level due to the prerequisite of legislative changes. Evidential proof in Zambia is the draft land policy initiated in 2006 which as at 2020 remains to be finalized and launched.

In Zambia, conflicts between statutory and customary land are common; the *Mpongwe* cases are examples of such conflicts. The first case involves a South African Company, Dar Farms Ltd, a company that owns about 60,000 ha of land in Chieftainess Lesa's Chiefdom. The initial land allocation to Dar Farms was during the first republic and the then-President Kenneth Kaunda did not consult the chiefs in the area. As Dar Farms continued to expand its commercial business into the boundaries of land held by Chieftainess Lesa without her consent, the contention was that this expansion was illegally done. Several villagers occupying the land under customary tenure had been, and continue to be, displaced. This conflict between Dar Farms Ltd, a commercial farm, and the community carrying on subsistence farming is what led to the court action. In this case the Supreme Court ruled against the community.

The second case (*Mpongwe Development Corporation Ltd v Kamanda*, 2007), involved ETC Bio Energy, a South African company which was the former Mpongwe Development Company Ltd (MDC), dealing in biofuel plantation. Once again, the 46,000 ha of land was under the Mimbolo farming area, a part of Chieftainess Lesa's chiefdom. The facts in this case are different, in that the President in 1995 directed MDC to give part of its land to the community so that they could continue their farming activities. The company responded positively, but conflict arose regarding the boundaries, which led to the community being shifted to a new piece of land from the one initially offered to the community. At a meeting held in 2004 where all the stakeholders were present, a resolution was passed that the land in question belonged to the chief and that the community should stay and put up permanent structures. Two weeks later, MDC stopped the community, stating that the company was the owner of the land in question. The community took the matter to the High Court and in 2006 the court gave a ruling in favor of the community. On appeal by MDC, the community lost which resulted in the community being forcefully removed from their land. The Supreme Court ruled that the trial judge could not order cancellation of the title since there was no evidence to show that the land was acquired fraudulently, and the company was the beneficial owner of the land.

On the ground, both these cases of displacements have left over 5,600 members of the community without land, shelter, water, and a source of livelihood as the conflict between the companies and the community continues. The conflict relates to security of tenure under statutory land and customary land. In addition, disruption of community life and more importantly, the disruption of the relationship between land and the local people. This chapter will analyze these cases to determine if STDM is an appropriate land reform tool that can be used to resolve conflicts relating to agricultural land in Zambia.

18.4 ZAMBIA'S DUAL TENURE SYSTEM: THE RELATIONSHIP BETWEEN STATUTORY AND CUSTOMARY TENURE

In Zambia, land is divided into two tenure categories designated as leasehold tenure and customary tenure. Customary tenure has been misunderstood as a tenure and has been dismissed as being inferior to statutory tenure and an inefficient means of owing land (Hansungule, 2001). It has been argued that under customary land tenure, the rights are use or usufruct in contrast to ownership as defined under English land law (Hansungule, 2001). The legislative framework in Zambia provides for the conversion of land from customary tenure to statutory tenure, but leaves a gap on whether the rights acquired under customary tenure are simply extinguished as soon as the parcel of land is converted, or continue to exist. A possible answer could be that the rights are extinguished since once the parcel of land is converted to leasehold, it can never be reconverted back to customary tenure. STDM with its core recognition of both formal and informal tenure arrangements can provide an alternative to conversion of land from customary to leasehold tenure to secure tenure rights.

Although the leasehold tenure system provides a state-guaranteed "right" of ownership and security of tenure owing to how it enables holders to pass on the property freely to their designated heirs, the existence of the two tenure systems still presents unequal rights to land among citizens, as clearly elaborated in the draft land policy as follows:

> The maintenance of two tenure systems, customary and leasehold tenure, present unequal rights to land among citizens, an indigenous form of land holding which is generally communal in character and another land holding that is regulated by statutes. In addition, this classification of the land tenure has to a certain extent created a vacuum of how to deal with land reserved or to be reserved in the public interest (MLNR, 2017, p.16).

The question is whether STDM as a flexible tenure approach can provide a solution for these unequal rights to land among citizens. Some of the general challenges with respect to leasehold tenure are characterized by widespread lack of knowledge of land alienation procedures; over-centralization of the institutional structures responsible for dealing with land administration and management; and the lack of willingness by some chiefs to consent to the process of conversion. Additionally, the practical difficulties involved in conversion of customary land to leasehold and the scarcity of information on land availability and under-development, including non-utilization of land, has led to land disputes for land held under customary tenure.

When compared with leasehold tenure, different types of challenges arise with respect to customary tenure. Some of the more specific challenges include lack of clear guidelines on the role and functions of traditional authorities, leaders, and local authorities in the administration of customary land. There is a lack of clear assignment of land rights and responsibilities, especially with regard to gender and social status, since customs and practices vary from one area to another. Comparison of the two tenures has created a false sense of superiority of the statutory tenure over the customary tenure. Furthermore, the rights of landowners under customary tenure are undefined due to overlapping and sometimes contradictory controls as well as user rights to a parcel of land. Therefore, there is a desire by customary land occupants to legally document their rights to land in order to enjoy similar legal protections as those provided to leasehold land rights holders, which protections STDM proposes to provide. The only available option of conversion of customary landholding to leasehold tenure is not seen as a desirable option by most land rights holders under customary tenure.

Analyzing the current position under the dual system in Zambia would reveal that statutory leasehold rights have precedence over customary land rights. However, the law does recognize the existence of customary land rights under section 7 of the Lands Act 1995. The gap is that there has been no attempt to define recognition under the Act, and further there have been no attempts made to provide appropriate documentation for adequate recognition and protection of individual, community, and communal land rights under customary land. While private, individual land rights are well-acknowledged within customary tenure, they are not adequately recognized in law (MLNR, 2017). The problem that requires addressing is what happens when there is a conflict between the unregistered rights of a person holding land under customary tenure and the rights held by a registered proprietor under a certificate of title or leasehold tenure over the same piece of land.

18.5 TRAINING UNDER THE FIVE PILOT PROJECTS IMPLEMENTING THE STDM APPROACH IN ZAMBIA

There are about five main pilot projects in Zambia, utilizing four institutions doing spatial mapping using three different platforms for the certification process under the STDM approach. The first pilot activity to implement the STDM approach is identifiable with a workshop that was conducted from July 21 to 24, 2014, in Chibombo district, central Zambia. The target group comprised 21 women from the Katuba Women's Association and representatives from neighboring communities, and eight men representing the local council, federation leaders, and the Katuba community as participants. The training involved a review and refinements of the enumeration questionnaire to suit local conditions, training on data entry, and mapping using handheld GPS. The workshop generated a lot of interest from the participants, who felt the land tool would be instrumental in expounding land challenges faced by them.

18.5.1 The Case of Mungule Chiefdom in Chibombo District as the First Pilot Activity

The first pilot activity covered Mungule, a Chiefdom in Chibombo District a rural community situated 30 km North of Lusaka under the traditional administration of Chieftainess Mungule. The second STDM training and planning workshop was held in Chibombo District from September 28 to October 2, 2014 to pursue the data collection process. Chieftainess Mungule's area was chosen as the pilot site and it was proposed that 50 villages should be profiled, enumerated, and mapped out of a total of 231 villages. Chieftainess Mungule's chiefdom was chosen because of the rampant land disputes existing in the area. The focal point of the data collection was the social tenure relationships of women over land, housing, and natural resources within the customary areas (Social Tenure Domain Model, n.d.). The project attracted 24 grassroots participants from Chieftainess Mungule's chiefdom who received training in STDM, of which 19 were women. The workshop was facilitated by the UN-Habitat/GLTN, the Ministry of Lands, and the Huairou Commission, in collaboration with local partners including Chieftainess Mungule, Mungule Ward Council, village headmen and women, Child Fund Zambia (Mungule Site Office), local community members, People's Process on Housing and Poverty in Zambia/Zambia Homeless and Poor People's Federation (PPHPZ/ZHPPF), and the Katuba Women's Associations (Social Tenure Domain Model, n.d.). The workshop assessed the outcome of the test enumeration that was conducted in late September 2014, where a total of 50 households were enumerated in the test run with exciting results that led to the finalization of the data collection tools. A further 15-day enumeration and mapping exercise later ensued in the second week of October 2014, after which the data was fed into the SDTM tool. The nature of enumeration and mapping process is that it seeks to document household landholding and conflicts in relation to social tenure arrangements among the community members (Social Tenure Domain Model, n.d.).

18.5.2 Objectives of the Training under the First Pilot Project

The short-term objectives of the project included profiling, mapping, and enumerating 50 villages and, to complete data entry and analysis of seven villages which had already been covered, while the medium- to long-term objectives focused on the government recognizing STDM as one of the efficient and effective pro-poor land management tools. A further objective was to take STDM to scale across the country by galvanizing support from government and other stakeholders, in addition to enhancing networking on women's land rights through the creation of backward and forward linkages with strategic partners, and improving land rights, particularly for women and other vulnerable groups in society. It is clear from the above pilot and the trainings that formed the basis of the STDM approach that participation was to commence from the grassroots level. In other words, the top down imposition of the system was being avoided. The acceptance of the approach from the bottom meant that the system would gain acceptance from its inception.

18.5.3 Achievements of the First Pilot Project

After conducting the training, several achievements had been noted. Amongst them, the main ones were that STDM had been well understood and widely accepted in Mangule Chiefdom and had received massive support from the traditional leadership and the area councilor and community members. The training was a forum for the women to voice their concerns over land matters, which is not the generally accepted custom in most rural settings. In terms of figures, initially 50 villages was the target group, but surprisingly, at the end of the training, at least 122 village headmen and women wanted and supported the implementation of STDM in their areas. The total number of households covered was 434, targeting a population of 1,337 (UN-Habitat/GLTN and Huairou Commission: GLTN-Tool, n.d.).

An additional achievement from the training was the interest generated about STDM by the office of the Surveyor General who had expressed a keen desire for the officers at the government department to learn and understand the STDM approach. The facilitators of the training workshop promised to engage with the government on how this training can be expanded in terms of the target groups to include the personnel from the government departments as well.

18.5.4 Challenges Faced by the First Pilot Project

The biggest obstacle faced during the piloting of the first project was that of getting local chiefs on board to support the process. Initially, the project triggered mixed reactions among the chiefs, including local leadership. Subsequently, however, the Katuba Women's Association and the People's Process on Housing and Poverty in Zambia (PPHPZ) succeeded in moderating discussions with the local leadership with the effect that the majority of them accepted the process. With respect to the inhabitants of the village under the first project, it took the facilitators two months to have the wider community understand and accept STDM. The initial reaction at both levels, including that of the leaders, being the Chiefs, as well as that of the residents of the villages was that the project was to be handled with caution before being accepted. Another practical challenge faced during the project implementation was that the facilitators needed to cover long distances due to vastness of the spacing between some of the villages and farmlands. One other challenge that was faced by the first project was a general lack of good quality satellite images. These challenges were addressed as the subsequent pilot projects were being launched.

18.5.5 The Case of Chamuka, the Second Pilot Project

Chamuka chiefdom is located in Chisamba District, Central Zambia, and sits between two rapidly growing urban areas: Kabwe in the north and Lusaka in the south. It comprises 207 villages spanning a spatial extent of approximately 300,000 ha. In this area, 11 villages have been enumerated and mapped using the STDM approach (UN-Habitat/GLTN and Huairou Commission: GLTN- Tool, n.d.). The project's success is owed to a vibrant and effective partnership between the GLTN/UN-Habitat and partners in Zambia who include the Government of Zambia, PPHPZ and its grassroots alliance partner, ZHPPF, affiliated to Shack Dwellers International. Also, His Royal Highness' leadership and support to the initiative has been pivotal to the success of this project from the beginning. What is important to note about the second project is that this undertaking was an off-shoot of a similar project in Mungule chiefdom and Chibombo District in the periphery of Lusaka which commenced in 2015, spearheaded by Huairou Commission and local partners in Zambia. In other words, this would be the rollover effect from one district to another where the project's initial acceptance had already been secured.

Chief Chamuka lauds the process for having contributed to a reduction of land dispute cases, while at the same time monitoring the allocation of land by headmen (indunas) and providing spatial and socio-economic data for future infrastructural and services planning. Most of Zambia's rural landmass such as the one in Chamuka is managed by traditional authorities through an informal and undocumented land administration system, while the statutory system is largely absent in rural areas. Customary land tenure is perceived to be least secure type of tenure because it is largely undocumented, which makes inhabitants of such land susceptible to forced displacements, and frequent land disputes among villagers, head persons, and even chiefs pertaining to boundaries.

The STDM initiative sets a good precedent for communities having their *de facto* claim to land recognized, and with it securing their future and that of their children in terms of tenure security of their landholdings. The projects' process has also helped reduce internal conflicts on land ownership boundaries through the availability of land maps validated by the communities. Lessons and experiences from this project are also feeding into the ongoing land policy process in Zambia to inform on approaches that can be used for customary land administration and on improving coverage of land records. The third, fourth, and fifth project were implemented using the same processes and procedures, and resulted in the same findings as the first and second projects described above in detail.

In concluding on the five projects, the training and the implementation of the STDM through the pilot projects had similar experiences, and the next question that requires to be addressed is upscaling of the STDM approach at a country level in Zambia. The major issue for consideration is that STDM uses several institutional frameworks and at the moment there are four institutions doing spatial mapping using different platforms for the certification process. There is lack of information and data on the most appropriate platform, institutional framework, and the link between the different pilot projects to up-scale STDM at a national level in Zambia.

18.6 APPLICATION OF STDM IN RESOLVING THE LAND DISPUTE IN THE TWO MPONGWE CASES

The facts in the two Mpongwe cases have been elaborated. The main issue in both the cases is the concept of beneficial ownership under statutory tenure. In the 2010 case (*Mpongwe Development Company Ltd v Kamanda*, 2010), the judgment was against the community, and in arriving at that decision, the ruling was:

> A permit is not different from a licence to occupy land. It can only be valid if it is issued by the owner of the land. In this case the owner of the land, the appellant, did not issue the permits. And the permits cannot override the interests of the appellant as a holder of a certificate of title.

The appellant in this case was the large-scale farmer (MPD), and as beneficial owner, it had to give permission to the members of the community to continue using their land. The court did not consider whether the community had any rights under customary tenure.

In the 2007 case (*Mpongwe Development Corporation Ltd v Kamanda*, 2007), the Supreme Court found that the trial judge could not order cancellation of title where there was no evidence that the land was acquired fraudulently or where a party has not specifically pleaded fraud. In arriving at this decision, the Court relied on the provisions of the Lands and Deeds Registry Act, 1994, section 33.

It is clear from the rulings given by the Supreme Court that the large-scale farmers in both the cases had obtained a certificate of Title by converting customary land to statutory land (Lands Act 1995, section 8). As such, this was conclusive evidence that they were the beneficial owners of the land. The communities that had been residing on this land could only continue to carry out subsistence farming if they were given permission to occupy and use the land by the beneficial owner, in this case the large-scale farmer (MDC). The customary tenure rights of the community were not taken into account.

These cases had been decided before the introduction of the STDM approach in Zambia. Had the communities been issued with the flexible tenure certificates advocated by STDM, the question regarding the large-scale farmer's certificate of title overruling the customary tenure rights of the community and permitting the displacement of the community would have been addressed. However, the current standing of these cases is that the non-governmental organizations dealing with land rights have been tasked to assist both the communities affected by the decisions in the above cases. The two ways they could assist the communities are to either retrieve their land, or to find alternate land for them to settle. Additionally, the cases have attracted a lot of international interest, even from Oxfam. However, the state seems not to be interested in meeting its obligation of protecting its people and their right to land and food security (Mujenja and Wonani 2012). STDM and certification of customary tenure as a flexible, pro-poor tool can assist these communities in strengthening their fight for a right to land and food security in the future (Herre, 2013).

18.7 CONCLUSION

A certificate provides greater clarity over boundaries and inheritance rights, increasing security against land disputes or wrongful eviction for land rights holders. STDM, as part of the continuum of the land rights approach, provides a solution to the problem of tenure insecurity and promotes responsible development of agricultural land in rural areas in Zambia. The five pilot projects implementing STDM have been assessed and the findings are positive in that this approach is suitable for promoting tenure security and reducing land disputes relating to statutory and customary landholding. However, the up-scaling of STDM approaches to a national level will require further investigations on both the institutional and policy frameworks that can support its implementation. Further, the two cases decided by the Supreme Court of Zambia show that STDM and certification of customary tenure rights will strengthen the land rights of the communities that have been displaced. In other words, the communities will be better placed to negotiate their rights to use and occupy land without being arbitrarily displaced.

REFERENCES

Chu, J. M. 2013. Creating a Zambian Breadbasket: 'Land grabs' and foreign investments in agriculture in Mkushi District, Zambia. LDPI Working Paper 33. Rotterdam: The Land Deal Politics Initiative.

Du Plessis, J., Augustinus, C., Barry, M., Lemmen, C., Royston, L. 2016. The Continuum of Land Rights Approach to Tenure Security: Consolidating Advances in Theory and Practice. *World Bank Conference on Land and Poverty.*

Hansungule, M. 2001. 1995 Lands Act: An Obstacle or Instrument of Development? ZambiaLand Alliance Report.

Herre, Roman. 2013. Fast Track Agribusiness expansion, Land Grabs and the role of European Private and Public Financing in Zambia. *A Right to Food Perspective Hands off the Land Alliance.* Retrieved from https://www.tni.org/files/download/13_12_fian_zambia_en.pdf.

Lay, J., Nolte, K., Sipangule, K. 2018. Large-scale farms and smallholders: Evidence from Zambia (No. 310). GIGA Working Papers.

Lemmen, C. 2010. *The Social Tenure Domain Model A Pro-Poor Land Tool.* Copenhagen: International Federation of Surveyors.

Lemmen, C. 2012. A Domain Model for Land Administration. PhD thesis, Netherlands: published, NGC. Retrieved from www.ncg.knaw.nl.

Lemmen, Christiaan, van Oosteromb, Peter, Bennett, Rohan. 2010. The Land Administration Domain Model. Retrieved from www.elsevier.com/locate/landusepol.

Mandhu, Fatima. 2015 A Hybrid System of Land Titles and Deeds Registration as a New Model for Zambia: A Case Study of the Lands and Deeds Registry Lusaka. PhD thesis. University of Africa.

Mandhu, Fatima. 2016. Land Titles or Deeds Registration as It Relates to the New Model of Land Registration System for Zambia. *Zambia Law Journal* 45, 57.

Mpongwe Development Company Limited v Francis Kamanda and Others SCZ Judgment No.14 of 2010.

Mpongwe Development Corporation Limited v Francis Kamanda and 51 Others Appeal No.137 of 2007.

Mujenja, Fison, Wonani, Charlotte. 2012. *Long-term Outcomes of Agricultural Investments: Lessons from Zambia.* London, UK: International Institute for Environment and Development.

Schiclele, Rainer. 1952. Theories Concerning Land Tenure. *Journal of Farm Economics*, 34(5) 734–744.

Urban and Regional Planning Act No. 3 of 2015.

19 Quantification of Tree Crowns in Urban Areas Using Very High Resolution Image

Germain Muvunyi

CONTENTS

19.1 INTRODUCTION

Trees constitute the most important component of green infrastructure in urban areas, providing numerous environmental, economic, and social benefits. Many studies have corroborated the benefits of planting trees in urban areas (McPherson et al. 1994; Tyrväinen et al. 2014; McHale et al. 2007; Ball 2012). Therefore, decision makers and planners, including land managers and administrators, are aware of the balance that should exist between urban infrastructure and the green space. This explain the relevance of monitoring the state of natural resources in cities. There is a need for a quick and reliable inventory of urban forest resources because it has been always difficult to update the urban green ecosystem using traditional field survey methods. Efficient urban forest management demands detailed, timely, repeatable, and spatially explicit information. In recent years, the launch of very high spatial resolution satellites (IKONOS, QuickBird, WorldView, among others), as well as the development of new classification algorithms, have initiated a new era in forest management using remote sensing technology (Plantier et al. 2006). In 2009, WorldView-2, a new satellite-borne sensor, was launched by Digital Globe. The very high spatial resolution (0.5 m in panchromatic bands and 2.0 m in multispectral bands) and four new spectral bands (Coastal, Yellow, Red Edge, and Near Infrared 2) in addition to the four standard bands (Blue, Green, Red, and Near Infrared 1) of this satellite confirm the expectation that this sensor has a high potential for tree mapping. The data provider postulates that all four new bands are strongly related to vegetation properties (Koukal et al. 2012). With the increasing availability of high spatial resolution data and the computational power to process it, remote sensing research in forestry has focused more and more on detecting and measuring individual trees as opposed to obtaining stand level statistics (Cabello-Lebric 2015). This technology provides opportunities for investigating and quantifying the structure and floristic of forests at both the stand and individual tree level (Bunting and Lucas 2006).

Although the use of high resolution satellite data is the best solution for tree quantification in urban areas, some factors, such as the increase of intra-crown spectral variance found in very high resolution (VHR) imagery and the low spectral separability between tree crowns and other vegetated surfaces in the understory (Gougeon and Leckie 2006; Hirschmugl et al. 2007; Pouliot et al. 2002) can comprise its efficiency. Furthermore, the coexistence of trees with urban infrastructures results in spatial arrangements that can complicate the interpretation of an image. In this chapter, an attempt has been made to quantify trees in urban areas using very high resolution WorldView-2 imagery using both a Normalized Difference Vegetation Index (NDVI) threshold and image segmentation techniques. A quick and accurate way to obtain inventory data on individual trees and clustered trees to replace the traditional surveying method that is less accurate and time-consuming is demonstrated. The resulting database can be used in urban planning to recognize the importance of green spaces in urban clusters. Land administrators should integrate remote sensing and geographic information system techniques in urban green space monitoring sectors, which will lead to better decision making as well as time-saving.

19.2 METHODOLOGY

19.2.1 Study Area Description

The study was conducted in Kigali city, Kicukiro District, Nyarugunga sector, Kamashashi village, which is located between 1°46'19"S to 2°05'14"S and 29°46'40"E to 30°28'36"E. (Figure 19.1). Kigali city consists of a complex of urban infrastructures including buildings, roads, sidewalks, canals, and so on, mixed with trees distributed in different patterns (block plantations, evenly spaced, or in arbitrary spatial pattern) and other green areas.

19.2.2 Data Collection and Processing

19.2.2.1 Satellite Data

To perform this study, WorldView-2 satellite image was used. The WorldView-2 satellite provided high spatial resolution data in eight spectral bands, with 0.46 m and 0.5 m spatial resolutions for panchromatic band and 1.84 m to 2 m for multispectral bands at nadir and off-nadir, respectively. Images were recorded in the spectral range of 450 to 1040 nm. Further details about the sensor can be found on the WorldView-2 website (http://worldview2.digitalglobe.com/). The high spatial resolution data provided by WorldView-2 satellite enables the viewer to discriminate and map fine details like shallow reefs, individual trees, and other diverse information such as the quality of urban infrastructures, cadastral information, the health of plants, and to manage various environmental resources. Through different methodologies, WorldView-2 data provides a rapid, standardized, and objective assessment of the biophysical impact, in terms of vegetation cover and restoration interventions (Mariana et al. 2014). Photo interpretation of a time series of aerial photography is generally used to qualitatively evaluate the long-term effectiveness of restoration interventions in terms of persistency, including for recognizable structures such as terraces, grubbing patterns, revegetated areas, and so on (Rango et al. 2002).

19.2.2.2 Image Processing

Image fusion as a process of merging several images acquired with different spatial and spectral resolution at the same time together to form a single image to enhance the information extraction

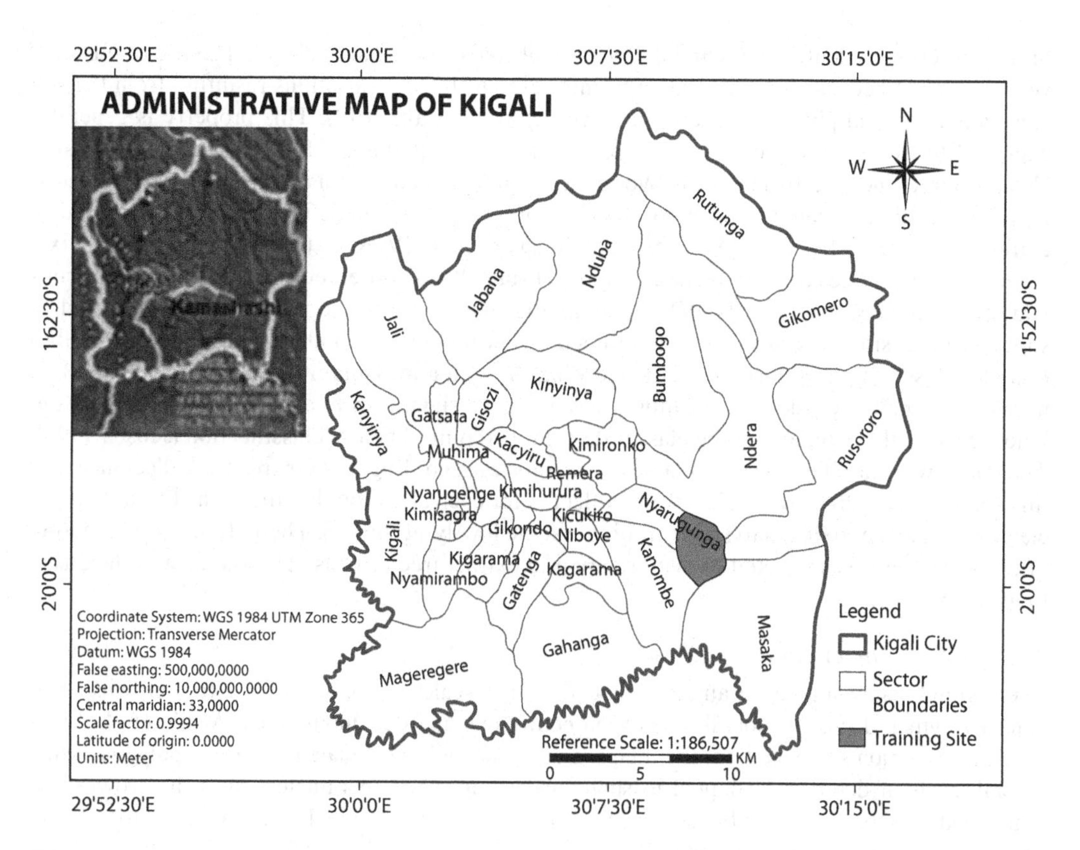

FIGURE 19.1 Study area location.

(Sarup and Singhai, 2010) was performed to enhance the quality of the WorldView-2 image. Panchromatic and multispectral images of the WorldView-2 satellite were fused using the principal component (PCA) technique. In this process, the first principal component of low resolution data is replaced by high resolution data (Shamshad et al. 2004). The PCA is the easiest and most useful of an eigen vector–based multivariate analysis, as it helps to reveal the internal structure of data in an unbiased way (Priya and Chudasama, 2015). The PCA technique helps to retain the spectral characteristics of the multispectral imagery to enhance the spatial resolution of the fused image.

19.2.2.3 Shadow Enhancement

Based on the direction of the sun, the shape and position of shadows can serve as additional information in tree shape detection (Hung et al. 2006). The green cover of the study area is composed of a mixture of mature trees, small trees, and bushes. Therefore, shadow was used to distinguish the mature trees from low-lying vegetation (shrubs). The first principal component analysis was performed to enhance the shadow and the output (decorrelated image) was classified using an image-clustering algorithm. The maximum iterations were limited to 10 with two classes (i.e., shadow and nonshadowed areas). Due to less spectral variability in the shadow and water body, NDVI was calculated using the original image of WorldView-2 to extract water bodies (pixel values > −1 and < 0), and this was used to mask out water body areas from the classified image.

19.2.2.4 Tree Crown Extraction

19.2.2.4.1 NDVI Approaches

The tree crown identification process is affected enormously by spectral separability of tree crown pixels with respect to their background. Therefore, it is better to remove all other features that hinder the correct identification and delineation of tree crowns using NDVI. The use of NDVI is very important because it is insensitive to intra-crown shadow variations resulting from the sun elevation angle and physical structure of trees (Ardila, et al., 2012). This property is generally exploited to extract tree crowns with circular and compact shape. NDVI was computed using Near Infrared and Red bands in the WorldView-2 image, and then a threshold (0.5 < threshold > 0.22) that differentiated trees from other features was generated. The threshold was used to extract crown areas from the pan-sharpened image of WorldView-2; the output image has pixel values of mature trees as well as the low-lying shrubs. It was observed that the low-lying shrubs had approximately the same NDVI values as the mature trees and crowns were subjected to supervised classification (maximum likelihood). Maximum likelihood assumes that the distribution of a class sample is normal. While working through a maximum likelihood algorithm, it is necessary to have well-defined training areas and pure signatures to acquire an expected result. Knowledge of the data, and of the classes desired, is required before classification (Jensen 1996). The image was classified in seven classes or categories according to their spectral reflectance. For this study, at least 20 training sites were collected for each class for classification. The number of classes was further reduced to six to eliminate some low-lying shrubs. Thereafter, the final output of classification was subjected to vectorization to extract tree crowns. The vector was then used to quantify trees.

19.2.2.4.2 Segmentation

The optimal segmentation parameters depend on the scale and nature of the tree crowns to be detected, which differ considerably between coniferous and deciduous trees (Ardila et al. 2012). By multiresolution segmentation, the image was segmented to evaluate tree crown polygons using several levels of detail and adapted to shape and compactness parameters until the crowns and segmented polygons become almost congruent in order to get individual trees and tree clusters with similar shapes and spectral properties.

19.2.2.4.3 Vectorization

Vector and raster are two types of spatial data structures used in a geographic information system (GIS). With the development of GIS and remote sensing (RS) technologies, it is easy to rapidly convert raster to vector data and establish topological relations among vectorized polygons is becoming a bottleneck in data integration between GIS and RS (Chen and Zhao, 2005). Therefore, based on the previous process, a vectorization method was proposed to classify, segment, and texture RS raster data quickly in order to automatically establish topological relations.

19.2.2.5 Accuracy Assessment

A field survey was conducted using a handheld GPS to collect ground truth points, which were used in accuracy assessment and validation. Accuracy assessment is a general term used to compare the classification to geographical data providing the assumptions are true, so as to determine the accuracy of the classification process (Madhura and Suganthi 2015) (Figure 19.2).

19.3 RESULTS PRESENTATION AND DISCUSSION

19.3.1 Digital Image Classification

The land use/land cover map provides a comprehensive data set in terms of the overall landscape across the study area. The Kamashashi cell is covered by 10 land use categories (classes): agriculture land occupies 3%, bare soil 1%, built-up land 17%, crop land 43%, grassland or shrub 6%, wetland 10%, open ground 4%, trees area o 11% and water bodies 2% of the total area (Figure 19.3).

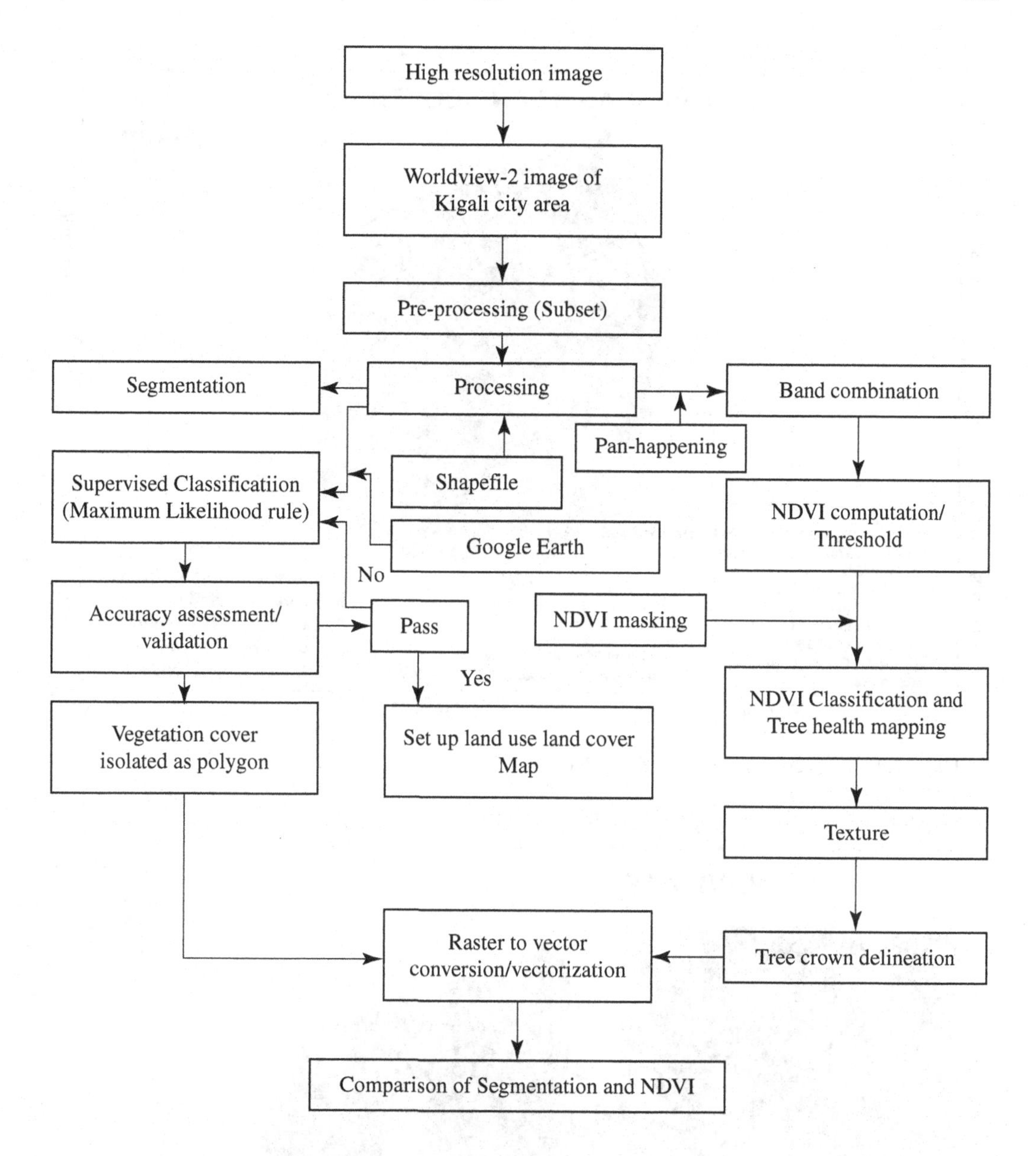

FIGURE 19.2 Methodological flow chart.

19.3.2 Vegetation Mapping and Tree Crown Delineation

Based on NDVI thresholds, vegetated area and nonvegetated area were separated. According to Meera et al. (2016), NDVI values vary between −1 and 1, where negative values are nonvegetation and postivie values represent vegetation. The NDVI for vegetation generally ranges from 0.3 to 0.8, with larger values representing "greener" surfaces (Lachowski 1996). A threshold of 0.41 was used to mask out nonvegetated areas from the NDVI image leaving trees and grassland/shrub (Figure 19.4). The spectral reflectance of vegetation is completely different from the reflectance properties of the background material (i.e., water, soil, and settlements). This unique character of the vegetation spectrum makes it possible to separate vegetation from background material with

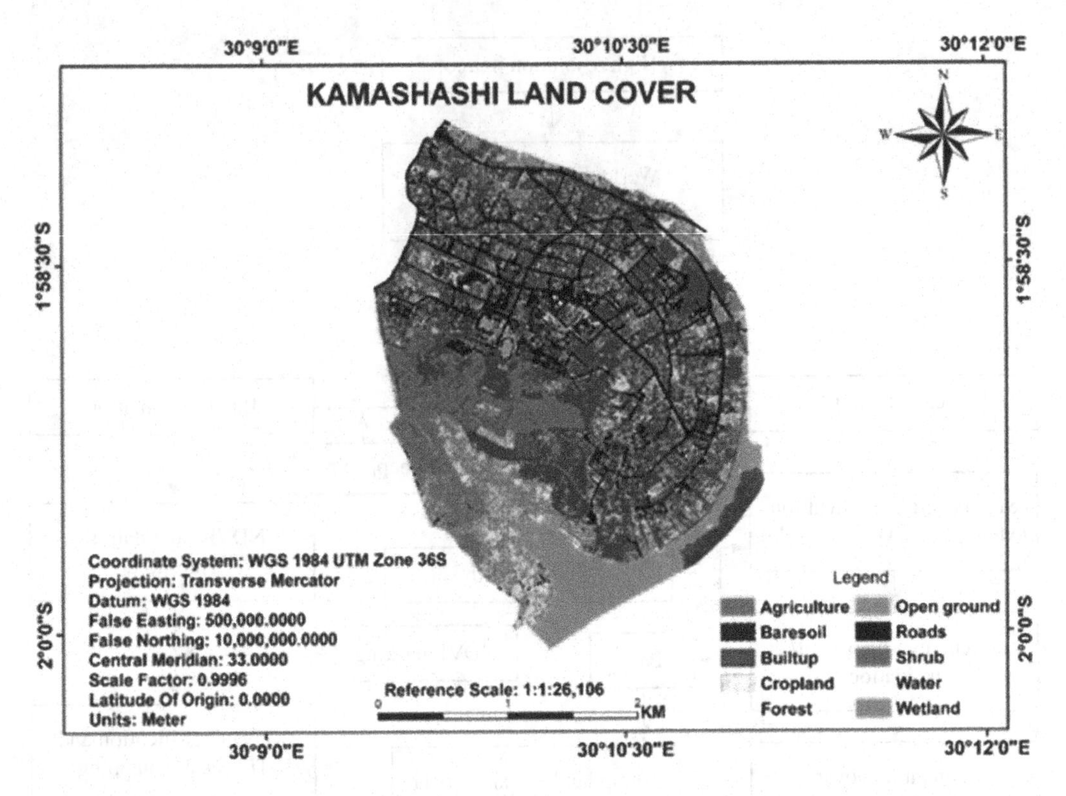

FIGURE 19.3 Land cover map of the Kamashashi cell.

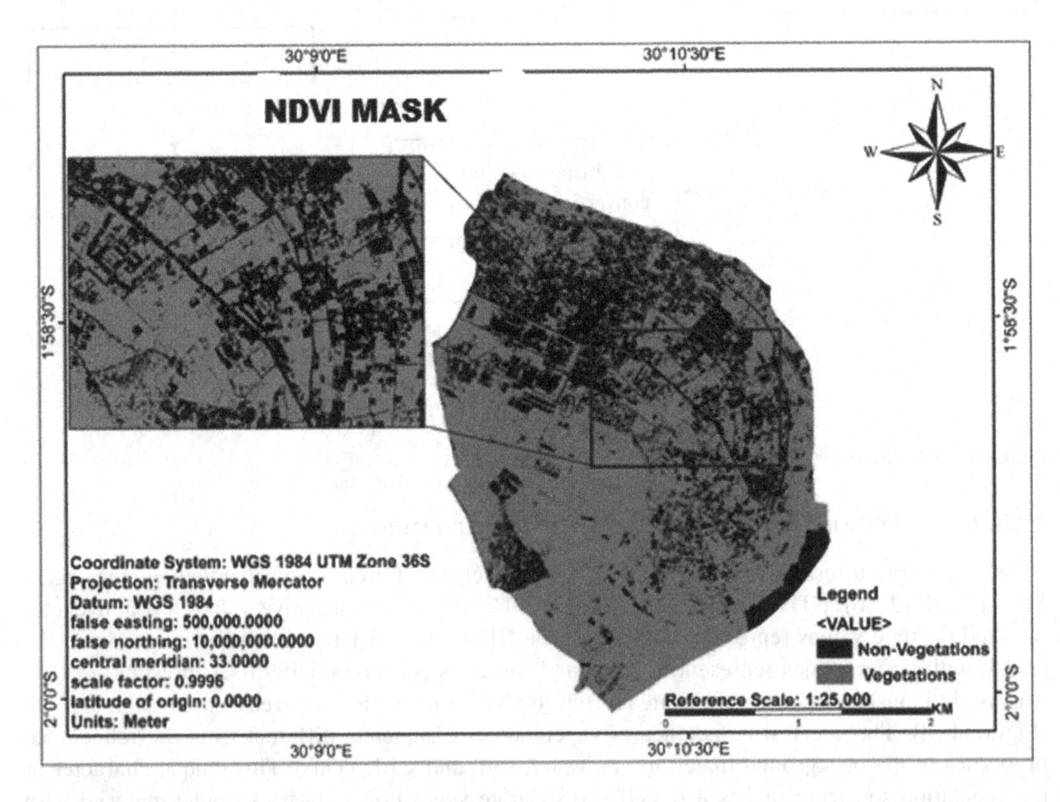

FIGURE 19.4 NDVI masking for vegetation and nonvegetation areas.

remotely sensed multispectral data that at least includes NIR and red region reflectance. NDVI is a good index for distinguishing vegetation and nonvegetation covers (Holme et al. 2008). In the Kamashaski cell, the nonvegetated area covers 215.73 hectares, while the vegetated area covers 583.27 hectares.

To separate tree cover and low vegetation, a threshold of (0.7 < threshold > 0.41) was achieved. This was based on image texture analysis, which is considered to be a measure of the spatial variation of image tone or intensity (Haralick 1979). This is useful to identify grassland areas since the spectral variance of grassland objects is smaller than that of tree crowns in very high resolution imagery. Finally, tree canopies (Figure 19.6) were also masked out from the NDVI image. A total of 864,002 trees were mapped and counted through tree canopies in the Kamashashi cell (Figure 19.5).

19.3.3 Image Segmentation Process

Image segmentation was introduced to extract neighborhood information and, preserving homogeneity throughout the satellite image, segmentation algorithms were used to subdivide the entire image at the pixel level (Baatz and Schäpe 2000; Benz et al. 2004). Segmentation algorithms ideally generate image objects that match the target objects; the shape value was 0.2 and the compactness was 0.8. The threshold of the NIR band was used for masking tree crown with values of 0.41 and 0.68. The features on the resulting image were then classified as trees

19.3.3.1 Urban Tree Canopies Classification and Counting

According to Jennings et al. (1999), the degree of tree density is expected in percentages. From different NDVI thresholds, this study classified trees canopies as follows: Very High Forest, Canopy Density, High Forest Canopy Density, Moderate Forest Canopy Density, Low Forest Canopy Density, and Very Low Forest Canopy Density (Figure 19.6). The Table 19.1 indicates different

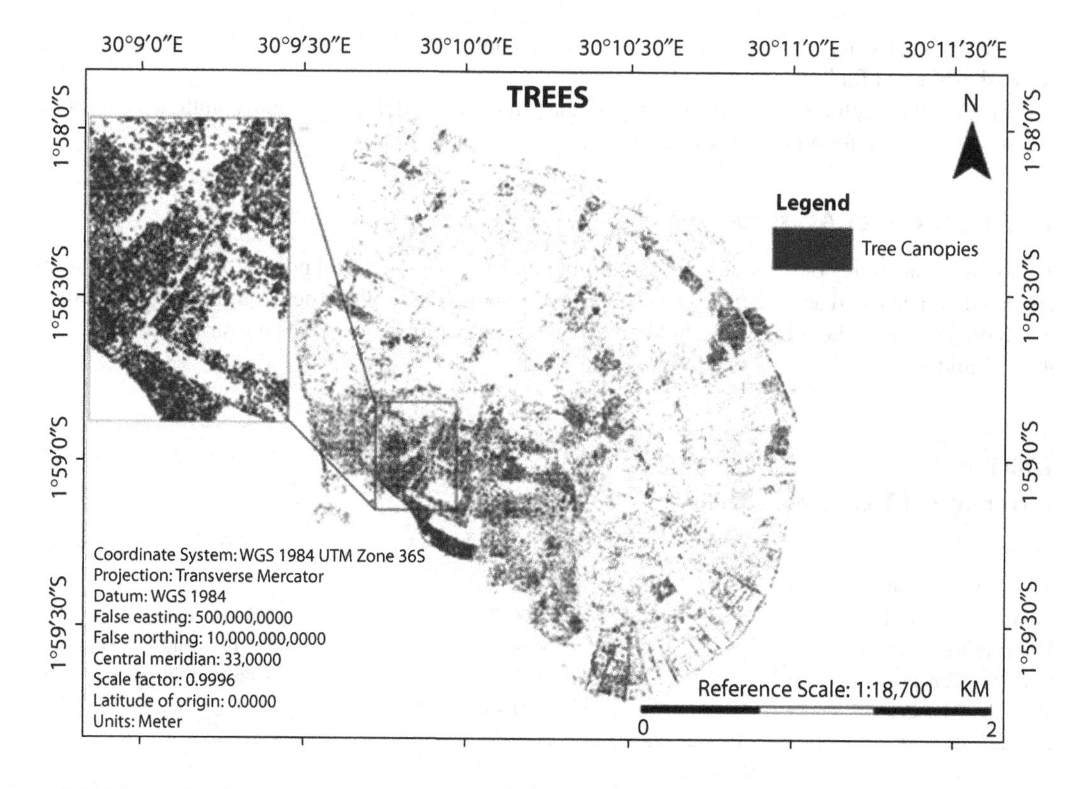

FIGURE 19.5 Tree canopies.

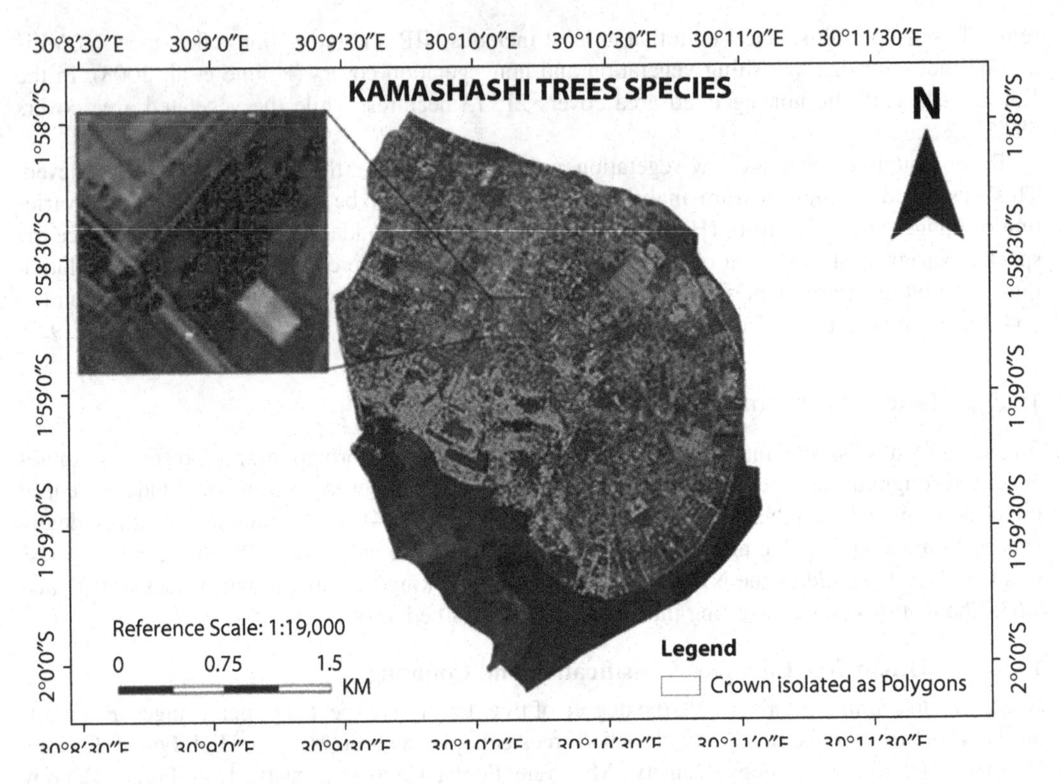

FIGURE 19.6 Tree canopies isolated as polygons.

statistics for each category. The tree canopy cover occupies 49.40 hectares out of 799 hectares in the Kamashashi cell (Table 19.1).

Through the vectorization process, individual trees were delineated and counted. A total of 806,484 trees were found in the Kamashashi cell.

19.3.4 Accuracy Assessment

To assess the accuracy of the study, a total number of 500 ground control points (tree locations) were collected and grouped according to the size of tree crowns. These were then compared to the results obtained from both the NDVI threshold and image segmentation results. The overall accuracy was 99.70% and kappa statistics was 0.9963 (Table 19.2).

TABLE 19.1
Percentage of Tree Canopy Density

Classes	Canopy Density (%)	NDVI Threshold
Very High Trees Canopy Density	70–100	0.70 ≥ trshld ≥ 1
High Trees Canopy Density	60–70	0.60 ≥ trshld ≥ 0.70
Moderate Trees Canopy Density	50–60	0.50 ≥ trshld ≥ 0.60
Low Trees Canopy Density	46–50	0.46 ≥ trshld ≥ 0.50
Very Low Trees Canopy Density	41–46	0.41 ≥ trshld ≥ 0.46

TABLE 19.2
Accuracy Assessment Table

Items	Class Names	Producer Accuracy (%)	User Accuracy (%)
1	Group 1	100.00	100.00
2	Group 2	100.00	100.00
3	Group 3	100.00	100.00
4	Group 4	100.00	99.52
5	Group 5	100.00	100.00
6	Group 6	100.00	100.00
7	Group 7	100.00	100.00
8	Group 8	100.00	100.00
9	Group 9	100.00	100.00
10	Group 10	100.00	99.11

Tree quantification was successfully achieved using pan-sharpened WorldView-2 data since the experiments provided the results with high accuracy (99.70% overall accuracy) in identifications of tree crown using digital images processing techniques. Both NDVI and segmentation-based methods provided reliable results for individual trees mapping. Counting from complex mixed infrastructures in urban areas using very high resolution satellite images despite a certain number of factors that can comprise the satellite image analysis in urban areas, such as shadow of tall buildings, clouds, and calibration and radiometric errors. Compared to existing traditional methods that are time-consuming, less accurate, and labor-intensive, the present method constitutes a quick and accurate way to obtain inventory data on individual trees and clustered trees as well. Such a database can be used optimally in urban planning, recognizing the importance of green spaces in the Kigali Master Plan implementation. The data set can also include and incorporate other factors to predict or model land resources management processes.

REFERENCES

AlShamsi, Meera R. (2016, October). Vegetation extraction from high-resolution satellite imagery using the Normalized Difference Vegetation Index (NDVI). In *Image and Signal Processing for Remote Sensing XXII* (Vol. 10004, p. 100041K). International Society for Optics and Photonics.

Ardila, J. P., Bijker, W., Tolpekin, V. A., & Stein, A. (2012). Context-sensitive extraction of tree crown objects in urban areas using VHR satellite images. *International Journal of Applied Earth Observation and Geo-Information (JAG), 15*(1), 57–69.

Baatz, M., & Schäpe, A. (2000). Multiresolution segmentation – an optimizationapproach for high quality multi-scale image segmentation. In Strobl, J., Blaschke, T., Griesebner, G. (Eds.), *Angewandte Geographische Informationsver-arbeitung XII*. Heidelberg: Wichmann-Verlag, pp. 12–23.

Ball, L. (2012). Models, forests, and trees of York English: Was/were variation as a case study for statistical practice. *Language Variation and Change, 24*(2), 135–178.

Belgiu, M., Drăguţ, L., & Strobl, J. (2014). Quantitative evaluation of variations in rule-based classifications of land cover in urban neighbourhoods using WorldView-2 imagery. *ISPRS Journal of Photogrammetry and Remote Sensing, 87*, 205–215.

Benz, U., Hofmann, P., Willhauck, G., Lingenfelder, I., & Heynen, M. (2004). Multi-resolution, object-oriented fuzzy analysis of remote sensing data for GIS-readyinformation. *ISPRS Journal of Photogrammetry and Remote Sensing, 58*, 239–258.

Bunting, P., & Lucas, R. (2006). The delineation of tree crowns in Australian mixed species forests using hyperspectral Compact Airborne Spectrographic Imager (CASI) data. *Remote Sensing of Environment, 101*(2), 230–248.

Cabello-Leblic, A. (2015). Tree crown delineation: AusCover good practice guidelines: A technical handbook supporting calibration and validation activities of remotely sensed data products. Version 1.2. *TERN AusCover, 1666*(1–2), 197–207.

Chen, Y. H. & Zhu, W. Q. (2005). Estimating net primary productivity of terrestrial vegetation based on GIS and RS: A case study in Inner Mongolia, China. *Journal of Remote Sensing-Beijing, 9*(3), 300–312.

Gougeon, F. A. & Leckie, D. G. (2006). The individual tree crown approach applied to IKONOS images of a coniferous plantation area. *Photogrammetric Engineering and Remote Sensing, 72*(11), 1287–1297.

Haralick, R. M. (1979). Statistical and structural approaches to texture. *Proceedings of the IEEE, 67*(5), 786–804.

Hirschmugl, M., Ofner, M., Raggam, J., & Schardt, M. (2007). Single tree detection in very high resolution remote sensing data. *Remote Sensing of Environment, 110*(4), 533–544.

Holme, K., Zhou, W., & Troy, A. (2008). An object-oriented approach for analyzing and characterizing urban landscape at the parcel level. *International Journal of Remote Sensing, 29*(11), 3119–3135.

Hung, C., Bryson, M., & Sukkarieh, S. (2012). Multi-class predictive template for tree crown detection. *ISPRS Journal of Photogrammetry and Remote Sensing, 68*, 170–183.

Jennings, S. B., Brown, N. D., & Sheil, D. (1999). Assessing forest canopies and understorey illumination: Canopy closure, canopy cover and other measures. *Forestry: An International Journal of Forest Research, 72*(1), 59–74.

Jensen, J. R. (1996). *Introductory Digital Image Processing: A Remote Sensing Perspective.* Upper Saddle River, NJ: Prentice Hall, pp. 225–232.

Koukal, T., Atzberger, C., & Immitzer, M. (2012). Tree species classification with random forest using very high spatial resolution 8-band WorldView-2 satellite data. *Remote Sensing, 4*(9), 2661–2693.

Lachowski, H. (1996). *Guidelines for the Use of Digital Imagery for Vegetation Mapping.* Collingdale: DIANE Publishing, pp. 1–19.

Madhura, J., Suganthi, S. R. L. (2015). A review of assessing the accuracy of classifications of remotely sensed data and of methods including remote sensing data in forest inventory. IIASA Interim Report. IIASA, Laxenburg, Austria.

McHale, T., Tang, Y., Chen, A., & Zhao, S. (2016). Carbon storage and sequestration of urban street trees in Beijing, China. *Frontiers in Ecology and Evolution, 4*(1), 53–59.

McPherson, G. K, Nowak, D. J., & Rowntree, R. A. (1994). Chicago's urban forest ecosystem: Results of the Chicago Urban Forest Climate Project. *US Department of Agriculture, Forest Service, Northeastern Forest Experiment Station, 201*(1), 55–186.

Plantier, T., Loureiro, M., Marques, P., & Caetano, M. (2006). Spectral analyses and classification of IKONOS images for forest cover characterization. *Center for Remote Sensing of Land Surfaces, 28*(5), 260–268.

Pouliot, D. A., King, D. J., Bell, F. W., Pitt, D. G. (2002). Automated tree crown detection and delineation in high-resolution digital camera imagery of coniferous forest regeneration. *Remote Sensing of Environment, 82*(2–3), 322–334.

Rango, Albert, Goslee, Sarah, Herrick, Jeff, Chopping, Mark, Havstad, Kris, Huenneke, Laura, Gibbens, Robert, Beck, Reldon, & McNeely, Robert. (2002). Remote sensing documentation of historic rangeland remediation treatments in southern New Mexico. *Journal of Arid Environments, 50*(4), 549–572.

Sarup, J. & Singhai, A. (2010). Image fusion techniques for accurate classification of remote sensing data. *International Journal of Geomantic and Geosciences, 2*(4), 0976–4380.

Shamshad, A., Wan Hussin, W. M. A., & Mohd Sanusi, S. A. (2004). Comparison of different data fusion approaches for surface features extraction using QuickBird images. *International Symposium on Geo-Informatics for Spatial Infrastructure Development in Earth and Allied Sciences 1*, 423–431.

Tyrväinen, L., Ojala, A., Korpela, K., Lanki, T., Tsunetsugu, Y., & Kagawa, T. (2014). The influence of urban green environments on stress relief measures: A field experiment. *Journal of Environmental Psychology, 38*, 1–9.

Vora, Priya D., & Chudasama, N. (2015). Different image fusion techniques and parameters: A review. *International Journal of Computer Science & Information Technology, 6*, 889–892.

20 Public Value Capture
An Opportunity to Improve the Economic Situation of African Municipalities

Andreas Hendricks

CONTENTS

20.1 INTRODUCTION

"There is nothing more important to the progress of our economies and societies than good regulations. By good regulation is meant the sort that serves to enhance the wellbeing of the community at large" (OECD 2015).

Public administration shapes economic prosperity, social cohesion, and sustainable growth. It molds the environment for creation of public value (European Commission 2016).

The shortage of financial resources is a global problem. Coming out of the economic and financial crisis, countries as well as municipalities have decreasing means to fulfill all their public commitments. This chapter provides options to solve this highly topical problem. An effective tax system is one of the key policies in this connection. Modernizing governance is a way to relieve economic and budgetary pressures, to design and deliver needed structural reforms, to remove existing barriers, and to foster innovation. Public value capture is essential to improve the refinancing of public infrastructure to keep the necessary budget for important duties such as education and health care. For this reason, it is one of the key factors of responsible land management and smart tools are needed for a successful implementation.

In Latin America, many value capture initiatives are also motivated by the mobilization of new and more flexible funds to finance special social programs. In the land policy realm, value capture has been associated with many constitutional and legislative reforms that redefine property rights, obligations (often embodying the social function of property), and the ability of public administrations to redistribute the benefits and costs of urbanization. These ideas contradict the pervasive and traditional mode of state intervention in Latin America, typified by the phrase "socialization of costs and privatization of benefits" (Smolka 2013).

Public value capture still attracts little interest on the African continent. However, it constitutes a type of financing with strong potential in cities with solid and regular space growth (ULCG Africa 2012).

The role of data is highly important for effective and reliable policies, as data provides a solid evidence base to draw upon for successful policy design. This implies gathering and interpreting data from an array of sources and viewpoints, and challenging preconceived ideas and current practices in the search for more effective policy solutions. Therefore, innovative financing is of growing importance and much can be learned from approaches in different countries. However, a "one-size-fits-all" approach should be avoided. The optimal tool box has to be adapted individually to country-specific circumstances.

20.2 THEORETICAL FRAMEWORK

Increasing property values have deep social, economic, and distributive-justice implications (Alterman 2012). Historically, this has raised fundamental questions. Is the economic value increase a private property or a social good? Do governments have the right to reap some (or all) of the increments in value? Are property owners responsible for the externalities of the development of their land and should they thus internalize the costs of mitigating the impact? Still, the issue has been hotly debated in relation to the concept of *public value capture,* which can be seen "as a method or a strategy to capture value increase to use it for specific purposes" (de Wolff 2007).

The ethical basis to address the distributive-justice problem of dividing value increase between private property owners and the governments is the concept of *unearned increment.* A general definition of this concept can be formulated as an increase in the value of property through no work or expenditure by the property owner (Morales Schechinger 2007).

What then constitutes work or expenditure by the property owner? Obviously, land values are determined by a number of factors, that is, as a result of both public and private investments and actions. In order to "sort things out," Hong and Brubaker (2010) divided the roots of increasing values into five main categories:

(1) the original productivity of the land, the value of current land-use
(2) changes in land-use regulations, extension of property rights
(3) public investments in infrastructure and social services
(4) private investments that increase land value
(5) population growth and economic development

In this chapter, we share the view of Ingram and Hong (2012) that a conceptual delineation of these five elements of land value and their ownership can facilitate the discussion of who should capture what. If we apply the five elements to a property development situation, we can construct Figure 20.1.

In the "predevelopment phase," property value is based on the current land-use, that is, farm, forest, or derelict urban land.

The "development phase" contains three principal measures and activities that influence property value.

- A change of land-use normally requires some form of permission from authorities, that is, extension of the property rights. This regulatory system differs between countries, but often it is based on land-use plan(s) and subsequent permits.
- Individual properties must be supplemented with public infrastructure. We can then make a distinction between "internal infrastructure" servicing only the development area (roads, water, and sewage, etc.) and "external infrastructure" for larger areas (main roads and parks, schools, etc.).
- For individual properties, it is the owners' responsibility to construct buildings and facilities with their own investments.

The value increase due to a specific government decision such as specific types of land-use regulatory decisions or the execution of public infrastructure can be also classified as betterment. This is a term used especially in Great Britain and its former colonies. Betterment can be further divided in the subtypes "development-right based betterment" and "infrastructure-based betterment." Infrastructure-based betterment levies are, historically, the earliest form of betterment capture (Alterman 2012).

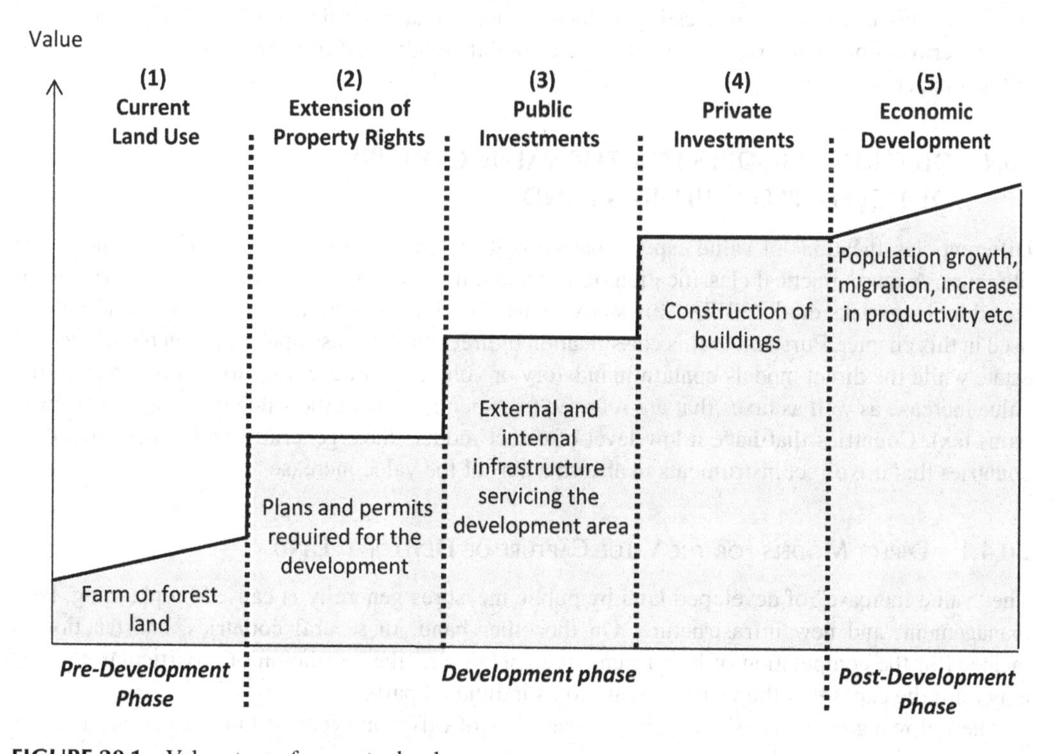

FIGURE 20.1 Value steps of property development.

When the property is completed, the "postdevelopment phase" begins. Value increase in this phase is determined by, for example, population growth, migration, and the increase of societal productivity.

In real life, these activities do not necessarily follow this order, as they might overlap or even precede each other in different sequences. For example, agricultural land already rezoned for housing (property owner's rights are already extended) might remain undeveloped for years and its economic value might increase further due to general economic factors while the owner is still farming it, perhaps waiting for larger value increases. Or land already built might profit from the construction of public infrastructure by the municipality and/or from general economic factors before the owner decides to sell his property.

20.3 METHODOLOGY

This chapter is mainly based on a literature analysis. Important literature concerning the previously mentioned problems has been summarized systematically and has been merged into a new generalizing presentation. Therefore, it is not only an additive or synoptic presentation of existing literature. The added value is given by the synthesis of different sources. For this reason, the available material has been analyzed with reference to quality and relevance. Latin American sources have been used because of existing similarities between Latin America and Africa concerning the urban development but more substantial experience concerning public value capture in Latin America.

Another important input has been given through interviews with experts. The COST action "Public Value Capture of Increasing Property Values" provides a network of European researchers representing over 30 countries. The author is Action Chair of this Action. It is funded by the European Cooperation in Science & Technology. It began in 2018 and is expected to be conducted over a four-year period. Furthermore, valuable input has been provided by colleagues from Latin America. This is especially interesting as those countries have similar problems to African countries in terms of population development, data availability, administrative problems, and the lack of public finance.

20.4 DIFFERENT MODELS FOR THE VALUE CAPTURE OF DEVELOPED BUILDING LAND

Different classifications of value capture based on its purpose or its outcome can be found in the literature. A more practical classification of direct and indirect models for value capture was introduced by Hendricks et al. (2017). This study focused on the subject of the instrument and will be used in this chapter. Pursuant to this classification indirect models base upon taxes on the whole real estate while the direct models contain mandatory or voluntary proceedings to absorb a part of the value increase as well as taxes that are referred to a specific part of the value increase (e.g., capital gains tax). Countries that have a low level of direct models have generally higher tax rates than countries that use direct instruments to absorb a part of the value increase.

20.4.1 Direct Models for the Value Capture of Developed Land

The "value increase" of developed land by public measures generally is caused by planning, land management, and new infrastructure. On the other hand, in several countries, construction is included in the computation of betterment levies. However, the legislation of a particular country may limit the capture of the value increase to its individual parts.

The following examples should give an overview of different types of models that are used in many countries worldwide.

20.4.1.1 Fees and Taxes

Many countries (e.g., Germany, France, and Sweden) use fees to recoup development costs (e.g., costs of infrastructure) or taxes to capture a specific part of the value increase (e.g., capital gains tax). Another option is the usage of linkage fees that have to be paid in return for the building right regardless of whether it can be shown that the need for increased housing is a direct result of a specific development. These funds are then used for public purposes, usually housing construction (Fainstein 2012).

20.4.1.2 Betterment Contributions

"A betterment contribution is a charge or fee imposed on owners of selected properties to defray the cost of a public improvement or service from which they specifically benefit" (Smolka 2013). This is widely used in the United States and Latin America. However, there is an enormous variance in the use of the instrument among and within countries. The application can be quite complicated and the estimation of the charge and its distribution among the beneficiaries of a project depend on important considerations.

In the first place, the question of which costs should be recovered has to be answered. Generally, there are at least included the direct costs, but several countries impose additional charges for items such as costs of feasibility studies, interest, and administration associated with public works. Most existing legislation limits the amount to be recovered to the lowest value of either the project cost or the land value increment. That is, if the project generates a larger increment than its cost, the latter prevails, whereas if the estimated increment is lower, then only this amount is to be recovered.

Another problem is the valuation of the overall land value increment and the individual benefit of the affected property owners. Not all benefits are necessarily reflected in land value increases. Infrastructure projects may also result in negative values due to generated noise and pollution. Different studies for the same project have generally varying results illustrating the difficulty of this kind of assessment. These technical imperfections in properly estimating how each property in the defined impacted area will benefit often leads to situations in which the charge may be higher than the net benefit for some properties while others receive a lower share of the cost than of the accrued benefit. The problem is strongly correlated with the challenge to define the impacted area and to identify all benefiting properties.

Possible criteria for the distribution of the charge are the size, frontage, and/or position of the property.

20.4.1.3 The "Real Estate Consortium" or "Conjoint Urbanization"

Public authorities and private landowners aspire to cooperate in the urban development of former rural areas. After development, landowners receive a plot that has the same value as the plot before development. Due to the increasing prices per square meter, a part of the developed area will remain with the municipality. This realization may be based on voluntary agreements or may be part of a mandatory proceeding (Lungo & Rolnik 1998).

20.4.1.4 Negotiated Development (Developer and Municipality)

The developer prepares the plan jointly with the municipality. Nevertheless, the municipality should take into account all private and public interests before the approval of the plan. The affected area can be (complete or partial) property of the developer or municipality. Generally, the developer is responsible for the construction of the infrastructure.

20.4.1.5 Flexible Building Rights

In Latin America, there are different models that imply exceptions to the general use regulations in favor of investors or property owners who paid a certain amount of money. The allowed exceptions and financial considerations are generally regulated in legal norms. In doing so, the core of the urban planning will be conserved in spite of more flexible building rights.

One possibility is the definition of a "basic floor space index." This index defines how much floor space generally may be constructed on a plot. On the other hand, the planner may fix a maximum floor space index and the payments of the property owners to reach a higher utilization of his plot. The municipality has to use the received amount of money for defined urban objectives (Smolka 2013).

Housing shortages and concomitant inadequate urban planning are further reasons for a deviation from general use regulations. In that case, constructions are oftentimes approved in deviation of the plan to advance housing projects in public interest. It is a selective exception in anticipation of the coming revision of the urban planning. In general, compensation has to be paid.

A third option is the transfer of building rights on another plot of the same owner. The main objective of this regulation is the preservation of a current use deserving protection (e.g., architectural important building or open space), which is lower than the permissible use. In this case, the owner of the real estate may transfer the difference between regular and substandard use on another plot taking into account the value ratio of both plots (Lungo 1998).

20.4.1.6 Urban Development or Redevelopment Measures

The new development of urban areas or the elimination of urban deficits are typical urban duties. In Germany, these problems are generally solved by mandatory proceedings. Urban redevelopment measures are those measures where an area is substantially improved or transformed with the purpose of eliminating urban deficits. The purpose of urban development measures is to develop local districts in a manner that is in keeping with their particular significance for urban development within the municipality. If the municipality does not buy the real estate, the affected landowners have to pay compensation corresponding to the value increase of their real estate.

In Latin America, urban development and redevelopment measures are generally realized through the cooperation between public authorities and private investors or property owners. The process may be initiated by the public or private partner. For example, in Sao Paulo the "Operation Urban Center" was realized to revitalize the city center through the creation of recreation and leisure time areas and cultural facilities. On the other hand, the measure "Faria Lima" was used for the development of a public transport network. The first step in this process is the urban and financial analysis of the project. The main duties of the municipality are the planning, the generation of the legal framework (e.g., the formal designation of the redevelopment area) and the coordination of the construction of infrastructure. On the other hand, the execution of all building activities is the duty of the private partners. In general, both get a part of the developed area as compensation for their activities. The corresponding agreement is controlled by a mixed commission occupied by representatives of public authorities and civil society (Lungo 1998).

20.4.1.7 Interim Acquisition

The buildup of land stocks oftentimes is used in municipalities to have available plots for public objectives within urban development. The acquisition of plots should be done long before urban development to avoid the anticipation of price increases by landowners. Otherwise, legal norms should state that in no case may the calculation of the buying price consider any influence or impact generated by planned public or private investments in the immediate area, nor the expected returns derived from uses established by urban land-use norms and regulations. Afterward, a part of the plots may be sold to maintain the stock balance by a so-called revolving land stock (Hendricks 2006). The administration of land stocks is frequently realized by external corporations. Furthermore, the urban development is frequently done by public–private partnerships. In this case, the benefits of the constructions remain with the private partner and the benefits of the increasing land value remain with the municipality (Morales Schechinger Chapter 5/2005).

Land banking is advantageous to control the use of the land, to prevent speculation, and through their ultimate sale or lease to capture for the community any increase in land value resulting from public or market actions. However, its effective implementation is oftentimes limited due to lack of resources, higher short-term priorities for scarce public funds, the local influence of strong private landowning interests, high inflation on land prices, and poor management practices (Smolka 2013).

20.4.1.8 Contract Models

The agreement of certain duties of the private partner in return to subsequent building rights is an alternative to interim acquisition. The most common duties are the provision of the needed area for the infrastructure or provision of plots for social, public, or ecological objectives. The affected land owners are sometimes integrated in a "real estate consortium." In this case, the land owners get back a plot after development that has the same value as the plot before development (Morales Schechinger Chapter 3/2005). On the other hand, financial compensations may be agreed for the generated value increase of the developed land due to public measures (Morales Schechinger Chapter 5/2005).

20.4.2 Value Capture of Developed Land by Taxes on the Whole Real Estate (Indirect Models)

On the one hand, the real estate tax rarely exceeds 1% of the gross domestic product in developing countries and countries in transition (in developed countries 1.5% in average). On the other hand, the income of this tax forms oftentimes a big part of the municipal budget (up to one-third). So it is of great importance for municipalities.

There is a big variety of methods of taxation. The system oftentimes is influenced by the colonial heritage in developing countries and countries in transition. On the other hand, their tax structure is frequently the consequence of deficient registers (Bahl 1998).

20.4.2.1 The Basis of Taxation

Basically, three different models can be distinguished. Land value may be used in addition to the value of the buildings. Alternatively, only the land value or the profitability of the real estate may be used.

Many countries (e.g., the United States, Canada, Germany, and most countries in Latin America) use the first option. Generally, professional valuation experts create maps of land values which have to be updated in a certain interval. Valuation of buildings is done using a simplified version of the cost approach method, multiplying the usable area of the buildings by price per square meter, which is listed in a valuation table. The price depends on both the use and the quality of a certain building and has to be estimated by a valuation expert.

However, many countries (e.g., Denmark, Jamaica, and parts of Australia and New Zealand) use only the land value as basis for the calculation of real estate taxes. The main objective of this approach is a higher efficiency of the land-use in urban areas. The biggest problem is the fact that the land value oftentimes is quite small in comparison to the value of the real estate in the whole (especially in the case of commercial properties, e.g., hotels or offices). This may lead to unjust taxation.

Another important argument in this discussion focus on the source of the value increase. If the basis just includes the land value, the taxation charges only the unearned increment of the price increase. If the basis includes construction, the taxation charges also the value increase due to investments of the property owner (Morales Schechinger 5/2005).

There are two reasons causing concern considering the profit of the real estate as basis of taxation. On the one hand, the valuation of unused plots is a problem in this system and on the other hand there is oftentimes a lack of information concerning the income of commercial or industrial properties. In that case, the valuation has to be based on theoretical assumptions. For these reasons, this approach is used less often than the others (Morales Schechinger 2007).

20.4.2.2 Tax Rates

In principle, the options to determine a single tax rate or different tax rates exist.

20.4.2.2.1 Progression in Stages

Generally, countries use a system of different rates distinguishing between plot and constructions, use of the plot and/or buildings (e.g., living or commercial use) or built or unbuilt plots (Hendricks 2015).

20.4.2.2.2 Continuous Progression

The progression in stages causes leaps in the taxation. For this reason two owners of real estate of more or less the same value may have different financial burdens, if they are in different tax brackets. The continuous progression is an approach to avoid this problem.

One option for this kind of progression is the usage of additions within a bracket, which rise continuously and adapt the maximum value in a bracket to the starting value in the following bracket.

An alternative is the definition of different rates for different "value sections," that is, for example, a tax rate of 0.6% for "the first" €25000, 0.7 % for "the second" 25000, and so on.

20.4.2.2.3 Extraordinary Tax for Unbuilt Plots

The definition of higher tax rates for unbuilt plots is quite common in many countries. One reason for this regulation is the pressure on the landowner to start building activities on the plot. If this does not work, the higher tax income of the municipality can be at least used to refinance the higher costs of technical infrastructure which are caused by the fact that the municipality has to develop new building areas while parts of the developed urban areas are unused. However, in rapidly growing areas it is unlikely that the tax on undeveloped land will be so high as to wholly discourage speculation. Thus, high taxation of undeveloped land, unless it is based on 100% of increased value, will not ensure contiguous development (Fainstein 2012).

20.5 OBSTACLES

20.5.1 Legal Requirements

First of all, legal requirements must be clarified. The constitution must define private property and the social function of property to allow the absorption of at least a part of the surplus value by law. Furthermore, legal norms for mandatory or voluntary proceedings for this absorption must exist and these regulations must be known and accepted in the local governments (Hendricks 2013). Local officials often allege that their hands are tied and they avoid taking action even when they actually are permitted to apply many value-capture instruments. This situation resonates where principles (sometimes in explicit value capture parlance) established administratively or by law are essentially ignored in practice, or at best are implemented partially or selectively in a few jurisdictions (Smolka 2013).

On the other hand, the concept of land tenure must be accepted by the population. The system of land tenure in the majority of countries in sub-Saharan Africa is currently incompatible with the establishment of the mechanisms mentioned earlier. In urban areas in particular, a parallel system has developed that results in the existence of a market on which "titles" are exchanged (at best "residence permit"), a market structuring the development and organization of informal districts accommodating nearly the entire growth of cities (ULCG 2012).

20.5.2 Deficient Planning/Unregulated Population Development

Unregulated processes of urbanization are another problem. For example, in Brazil the percentage of urban population increased between 1960 and 2000 from 44.7 to 81.2%. Due to this process the municipalities had to handle an influx of population of around 100 million. These movements are a very important reason for the big part of so-called informal development in Latin America. A big part of these people buys plots without technical infrastructure and building rights. Afterward, the municipalities have to solve this problem. Oftentimes they do it before political selections. For this reason, this can be called an "exchange" of infrastructure/building permission against votes. The consequences are a chaotic urban development and inflated costs of infrastructure, which has to be constructed between the existing buildings (Hendricks 2015).

Africa is facing similar problems. While in 2012 only 40% of the population lived in urban areas (in contrast to Europe, Latin America, and North America where nearly 80% did so), currently the urban population increases more quickly than that of any other area in the world. By the end of 2020, out of the 30 cities that will witness the fastest growth worldwide, 24 will be on the African continent. In 2030, the urban population of sub-Saharan Africa should double compared to 2010 and reach nearly 600 million people. African cities are already confronted with enormous problems as more than half of their inhabitants live in overpopulated slums. Another negative result is speculation. In many African big cities, the best investment of the last 10 years has been to acquire a land plot in order to resell it—without developing it—10 years later. In certain countries, the same land can have its market value multiplied by 10, or even more, without changing its use value (ULCG 2012).

20.5.3 Tax Collection

The administration must fulfill especially three tasks to guarantee an adequate tax collection. In the first place updated and complete registers are needed to investigate the land owners or other liable persons to tax. Furthermore, the local governments need updated data for real estate valuation and an effective system of tax collection must be installed.

20.5.3.1 Registers

The compilation of an updated and complete list of all liable persons to tax including all the needed data of the subject to tax is one of the biggest problems especially in developing countries and countries in transition. While most of the countries have a cadaster, it is oftentimes neither updated nor complete. The share of registered parcels is oftentimes smaller than 40% (Lungo 1998). Additional land registers are in even worse condition, if the formal proof of property is given by an official document in property of the owner and not by the land register. If the value of the buildings is used in addition to the land value as basis of taxation, there is the additional problem that the "informal constructions" are not included in the registers. Furthermore, the quality of the registered data is oftentimes not good enough for the real estate valuation. For this reason, many countries base their taxation only on the land value despite the problems mentioned.

20.5.3.2 Valuation

Real estate valuation is another essential problem. In the first place the taxable property has to be defined and its value has to be determined. Many countries use the market value of the real estate as basis of taxation. The problem is that a lot of highly skilled valuation experts are needed to keep the needed data in current state. Due to an insufficient number of experts, many local governments implement an updating rate of five years or longer. On the other hand, real estate markets in developing countries and countries in transition are oftentimes characterized by high fluctuations. This leads to unjust taxation at the end of this five-year period and a big skip in taxation when the valuation is updated. Furthermore, a collection of relevant data for real estate valuation is often missing.

20.5.3.3 Process of Tax Collection

The process of tax collection is deficient in many countries for two reasons. On the one hand, the probability to get caught in case of tax evasion is quite low. There often is a lack of qualified staff or adequate registers to control the payments. On the other hand, if someone gets caught, the punishment is too low. The penalty is frequently a more symbolic act or the execution is not consequent enough (Bahl 1998). Another problem might be too low taxation. In consequence, many taxpayers do not consider necessary to pay this small amount of money and the local authorities do not want to spend money to prosecute the claim. Finally, there is frequently established a culture of not paying and not collecting (Morales Schechinger 2007).

20.5.4 Corruption

Corruption is a problem for the successful use of direct and indirect models. In case of indirect models it hinders the efficient tax collection. The problem for the direct models is the uncontrolled development without building permit or with unjustified permit against payment. As long as developers are able to develop land in this way there is no incentive for any agreement to cover public costs of development. On the other hand, the "inner logic" of corruption and negotiations for development agreements is quite similar ("building right against payment"), that is, the developer is willing to pay for the building permit. The municipality has just to force the developer to cooperate with public authorities instead of criminal actors.

20.6 CONCLUSION

Public value capture is one of the key factors of responsible land management. However, changing from the prevailing complacency toward property development, whereby individual landowners capitalize unearned income from public investments into a new regime in which private benefit are balanced with social costs, involves a painstaking cultural shift that may take a long time and is expected to face significant resistance. Politicians commonly dislike public value capture because the voters do so and the administration because of the resulting efforts of new tools. For this reason, the final part of this chapter is dedicated to recommendations for a successful implementation of smart tools of public value capture.

First of all, it is important to increase the knowledge about the complex nature of varied value capture approaches and to promote greater understanding among public officials and citizens about how value capture tools can be used to benefit their communities (Smolka 2013).

Conducting research and documenting and disseminating implementation experiences are essential to assist public officials and decision makers in understanding the operating mode and opportunities of public value capture. It is advisable to shift the debate on value capture from ideological and social justice rhetoric to a more technical and practical context and to relate it to fundamental principles of economic theory. Furthermore, documentations are helpful how value capture has fostered investments in urban infrastructure and services and improved land-use development. The willingness to accept tools of public value capture increases, if these tools are perceived to be associated directly with the solution of a locally recognized problem. It should be also emphasized that value capture policies can reduce speculation and corruption because land value increments are less volatile. The main objective is the creation of a win-win situation for private developers and the general public.

After improving the understanding of public value capture in general, there are several key factors for a successful implementation. Management skills are required to deal with many complex factors and diverse stakeholders including fiscal, planning, and judicial entities as well as local government leaders. The administrative continuity has to be ensured in the whole implementation process. Trial-and-error is part of the process of refining and institutionalizing policy tools and there is no one-size-fits-all solution. Cadaster and land register and valuation data has to be created, maintained, and updated.

A stringent planning system is an essential requirement for the use of direct models of public value capture. The municipality must penalize any violation of this planning to get the control of urban development. As long as developers are able to buy land on the "pirate market" and to develop it without building permit there is no incentive for any agreement to cover public costs of development. For the same reason, corruption has to be reduced.

The public control of building rights and land uses should be preferred instead of public ownership of land, if the public authorities do not have the financial means for buying the land.

A key factor for the success of indirect models is the improvement of tax collection. An efficient control system has to be established as well as an adequate punishment of tax evasion. Furthermore, the structure of the tax system should be kept as simple as possible and exemptions should be minimized.

REFERENCES

Alterman, R. 2012. Land use regulations and property values: The "windfalls capture" idea revisited. In *The Oxford Handbook on Urban Economics and Planning*, ed. N. Brooks, K. Donangy, G.J. Knapp, 755–786. Oxford: Oxford University Press.

Bahl, R. 1998. El impuesto al suelo frente al impuesto a la propiedad en países en vías de desarrollo y en transición. http://200.41.82.27/cite/media/2016/02/Bahl-R_ND_El-impuesto-al-suelo-frente-al-impuesto-a-la-propiedad-en-paises-en-vias-de-desarrollo-y-en-transicion.pdf (accessed: March 19,2019).

de Wolff, H. 2007. The new Dutch land development act as a tool for value capturing. *International Conference in Sustainable Urban Areas*, 25–28 June in Rotterdam.

European Commission. 2016. *European Semester Thematic Factsheet: Quality of Public Administration.* Brussels: European Commission.

Fainstein, S. 2012. Land value capture and justice. In *Value Capture and Land Policies.* Cambridge, MA: Lincoln Institute of Land Policy.

Hendricks, A. 2006. Einsatz von städtebaulichen Verträgen nach § 11 BauGB bei der Baulandbereitstellung—Eine interdisziplinäre Theoretische Analyse und Ableitung eines integrierten Handlungskonzeptes für die Praxis. PhD diss., University of Darmstadt.

Hendricks, A. 2013. Urban redevelopment east: A Programme to handle the problems in shrinking cities in Germany. In *Land Management: Potential, Problems and Stumbling Blocks Landmanagement*, 257.

Hendricks, A. 2015. Different models for the absorption of the surplus value of developed land to refinance the costs of urban development. In *Challenges for Governance Structures in Urban and Regional Development. Proceedings of the 2nd and 3rd International and Interdisciplinary Symposium of the European Academy of Land Use and Development, Ankara and Dresden*, ed. E. Hepperle, R. Dixon-Gough, V. Maliene, R. Mansberger, J. Paulsson, A. Pödör and K.-F.Schreiber, 161–172. Zürich: vdf-Hochschulverlag.

Hendricks, A. et al. 2017. Public value capture of increasing property values—What are "unearned increments"? In *Land Ownership and Land Use Development. Proceedings of the 4th and 5th International and Interdisciplinary Symposium of the European Academy of Land Use and Development, Krakow and Oslo*, ed. E. Hepperle, R. Dixon-Gough, R. Mansberger, J. Paulsson, J. Hernik and T. Kalbro, 257–282. Zürich: vdf-Hochschulverlag.

Hong, Y.-H., and Brubaker, D. 2010. Integrating the proposed property tax with the public leasehold system. In *China's Local Public Finance in Transition*, ed. J. Y. Man and Y.-H. Hong. Cambridge, MA: Lincoln Institute of Land Policy.

Ingram, G.K., and Hong, Y. 2012. Land value capture: Types and outcomes. In *Value Capture and Land Policies.* Cambridge, MA: Lincoln Institute of Land Policy.

Lungo, M., and Rolnik, R. 1998. Gestión Estratégica de la Tierra Urbana. Programa Salvadoreño de Investigación sobre Desarrollo y Medio Ambiente. https://prisma.org.sv/gestion-estrategica-de-la-tierra-urbana (accessed: March 19,2019).

Morales Schechinger, C. Chapter 3/2005. Notas sobre la regulación del mercado de suelo y sus instrumentos. http:// http://www.vecinalesdecomodoro.org/wp-content/uploads/2015/08/Notas_sobre_regulaci%C3%B3n_del_suelo. pdf (accessed March 19, 2019).

Morales Schechinger, C. Chapter 5/2005. Notas sobre la movilización de plusvalías para financiar a la ciudad. Texto preparado para el curso de Educación a Distancia Financiamiento de las Ciudades Latinoamericanas con suelo urbano. Lincoln Institute of Land Policy.

Morales Schechinger, C. 2007. Políticas para el financiamiento público en el impuesto a la propiedad inmobiliaria. In: *Catastro multifinalitario aplicado a la definición de políticas de suelo urbano.* https://www

.lincolninst.edu/publications/books/catastro-multifinalitario-aplicado-la-definicion-politicas-suelo-urbano (accessed March 19, 2019).

OECD. 2015. *Regulatory Policy Oulook 2015*. Paris: OECD Publishing.

Smolka, M. 2013. Implementing value capture in Latin America—policies and tools for urban development. https://www.lincolninst.edu/sites/default/files/pubfiles/implementing-value-capture-in-latin-america-full_1.pdf (accessed March 19, 2019).

ULCG Africa. 2012. Land value capture: A method to finance urban investments in Africa? Report of the Africities summit from December 4th to 8th 2012. *EULCG Africa, Dakkar.* https://www.uclg.org/sites/default/files/exe_cglu_-_uk_web_1.pdf (accessed March 19, 2019).

21 Increasing Tenure Security to Enhance Sustainability by Using Smart Land Management Tools in Kenya

Tobias Bendzko

CONTENTS

21.1 INTRODUCTION

Land is a limited resource that suffers from many demands and is exposed to external factors which can affect the usability of land. Due to its nature it is difficult to increase land except in rare cases where soil is recovered from the ocean or additional layers/levels are added. The external factors that influence the availability or scarcity of land are for example weather and climate change or population growth. Last indicator for the scarcity of land can be mostly found in the global south and comes along with huge challenges toward land management.

According to official numbers from the World Bank, Kenya has grown more than five times in population from 1960 (8.1 million) to 2017 (49.7 million), which puts an enormous pressure on land and its utilization as agricultural land. An increased demand of agricultural products can not only be provided by new seeds, improved agricultural techniques or fertilization. The recent past has shown that the overuse of land especially in Kenya has resulted in large deforestation, water scarcity, soil degradation, soil erosion and a severe decrease on biodiversity.

This phenomenon can be witnessed in many countries of the global south and leads to a large set of problems. In combination with external threats and scarcity of water or nutrition in agricultural areas it threatens the existence of local small-scale farmers and food security of the people depending on the grown crops and puts its country in a dangerous imbalance. In order to balance that imbalance several multinational and international organizations are trying to bring up solutions by handing out guidelines on agriculture or giving out regulations like the Paris climate convention which was ratified from Kenya in 2016 (Ministry of Environment and Forestry 2019).

Despite all of these attempts to improve the situation a persisting development toward a larger unbalance can be witnessed. Therefore a sustainable and smart approach on land management is urgently needed to secure the livelihood of local farmers and future generations. This chapter will put a spotlight into a very important agricultural area in Kenya and analyses its current situation to develop possible smart measurements for improvement of tenure security inside that study area. Although the research area seems to be very specific, it shares its problems with most other areas of agricultural land-use within subsistence farmers inside a customary tenure system setting.

21.2 THEORETICAL PERSPECTIVE

Tenure security is a very complex topic that is influenced by many different factors. For this chapter a short introduction toward tenure security will be made to understand the factors influencing tenure security and its effects on land-related decisions.

Therefore, tenure security will be divided into three dimensions of security all interacting and creating a triangle of security in which the subsistence farmers are making decisions on the land and its use. The triangle can be seen in Figure 21.1: "Rules of tenure define how property rights to land are to be allocated within societies. They define how access is granted to rights to use, control, and transfer land, as well as associated responsibilities and restraints" (FAO 2002).

21.2.1 Legal Tenure Security

Legal tenure security is given by the institutional framework that is represented by the individual government. The government sets the legal framework for tenure security and land rights in particular.

21.2.2 Personal/Received Tenure Security

Perceived security is mainly developed through personal experiences the persons were having and influenced by the happenings in the past and it is describing the personal relationship toward the

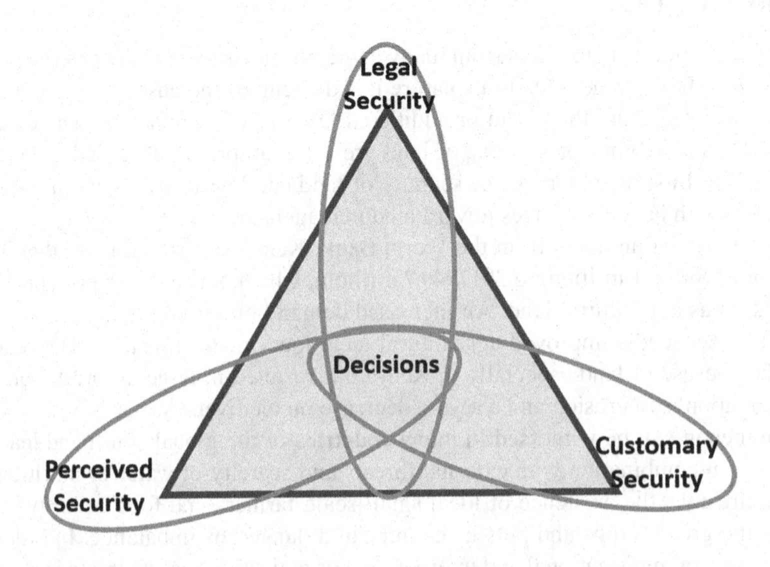

FIGURE 21.1 Dimensions of tenure security. (From Author 2019.)

own land. It can be influenced by media and stories from others and usually represents the most important factor for the individual over all security.

21.2.3 Customary Tenure Security

In countries with an existing customary land system the community and the customary tenure sets a third party of this triangle on tenure security. Customary tenure security can enforce social justice inside a tribe or cultural group on tenure where the legal framework is failing. Depending on the country and the legal land rights the customary land rights can be included and recognized by the formal land rights but customary land rights usually are adjusted by each customary group and can take part in providing tenure security to a large extend.

In order to get to the central decisions the individual persons will have to balance the three dimensions of security and add individual weight to them. To result into a sustainable decision all three dimensions have to be analyzed for their influencing indicators. Therefore, a set of decisions has to be analyzed within a multicriteria decision analysis to result into a multicriteria analysis (MCA). The influencing indicators that lead to the decisions inside the dimension triangle are analyzed and described within this chapter and smart solutions to overcome these insecurities are discussed then in the findings to give possible smart solutions in order to improve tenure security.

21.3 METHODOLOGY

For this research a comprehensive assessment of land management data inside a specific region in Kenya was conducted. The selected region for the case study is lying inside the coastal area of Kenya and marks one of the most important biodiversity hotspots and one of the so called water towers of Kenya. The research site is located in the Taita Taveta County inside the Taita Hills which are at an elevation level of over 1,650 meters over sea level (Pellikka et al. 2018).

Due to its unique location and intensive agricultural importance to the entire region and beyond this area has been part of several international research projects on soil erosion, land degradation or effects of sacred forest remnants on urban areas (Boitt, Ndegwa, and Pellikka 2014; Erdogan, Pellikka, and Clark 2011; Pellikka et al. 2018). For entire Kenya the Taita Hills act as one of five water towers of the country which play an important role for an entire region to supply water for agriculture (Kenya Forestry and Research Institute 2018). According to Dean of the Agricultural Campus of the Taita Taveta University Dr. M. Maghenda this source of water is endangered due to changes in this region indicated by humans. As some of the conducted research shows major external factors on the availability of land have been issued in this region and some research projects were dealing with ways of improvement of the persistent tenure system and customary tenure security. These factors that were leading toward soil degradation and the loss of more than 90% of its pristine forests inside the Taita Hills are rooted inside tenure insecurity in the region. Therefore, a comprehensive collection on the tenure situation were part of the overarching research in which framework this data collection took place in July until August 2018. Supported and funded by the DAAD several researchers collected data inside the research area to outline impacts of the remaining pristine forests on biodiversity connecting the people of Taita Taveta (Habel 2019).

In order to identify different indicators and solutions to enhance tenure security first of all a comprehensive assessment of the existing legal tenure framework had to be accumulated. In a second step a multicriteria analysis (MCA) was used to identify solutions which can be supported by smart systems (Department for Communities and Local Government 2009); (Wang, et al. 2009). The collected quantitative data was acquired by standardized questionnaires toward the local community to identify their tenure situation and personal opinions. For a better holistic view all conducted questionnaires were followed by mapping of the individual parcels by the researchers using handheld GNSS devices with an accuracy better than 10 meters and sufficient enough for this purpose. Additionally key informant interviews were conducted to fill knowledge gaps with expert

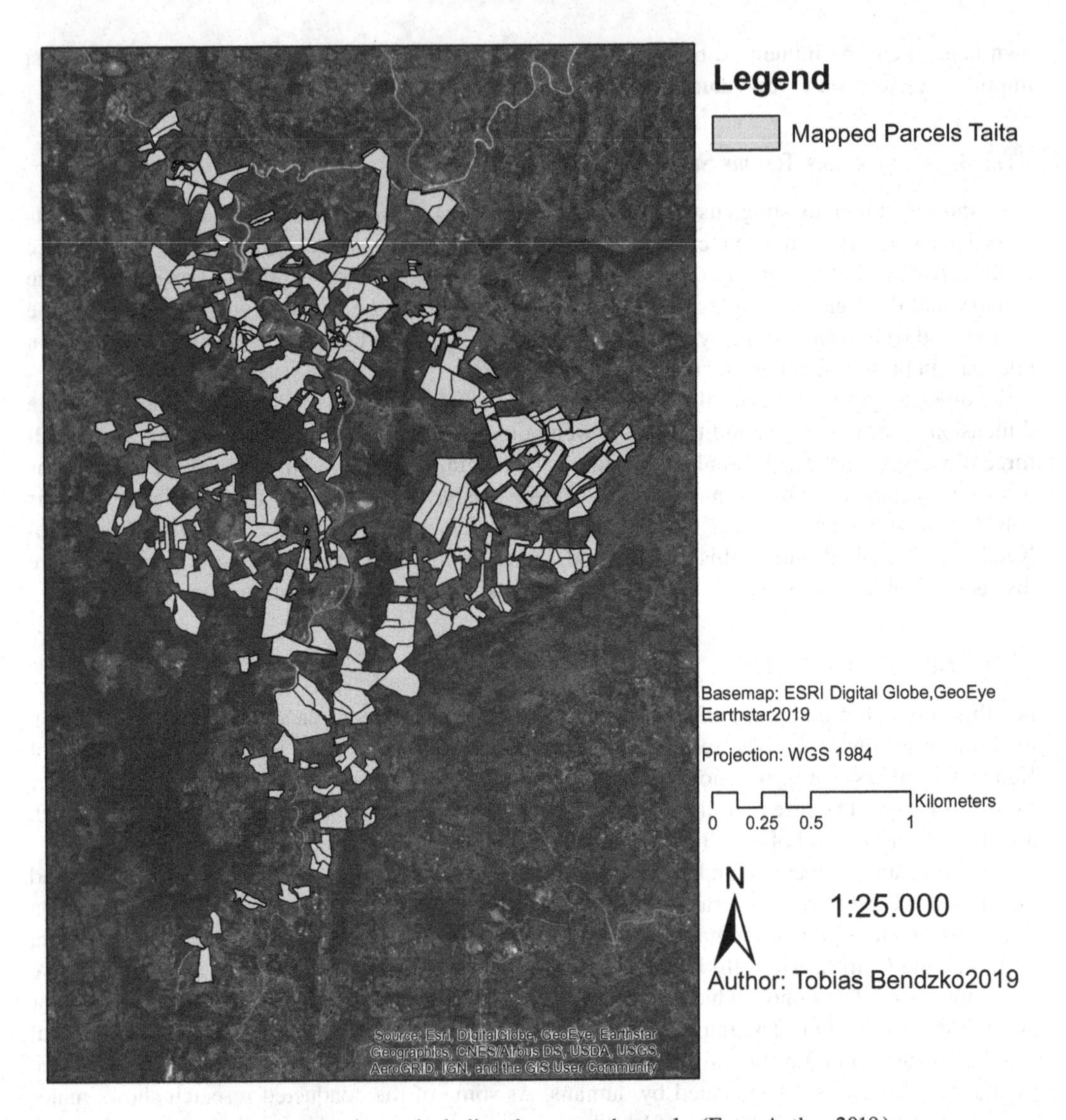

FIGURE 21.2 Map of the research area including the mapped parcels. (From Author 2019.)

interviews on qualitative basis. Last but not least, cartographic methods were applied by the author to identify patterns in their appearance and compare the digitized cadaster map with the collected mapped parcels.

The mapped parcels of individual farmers can be found in Figure 21.2. In total, 337 parcels were mapped in combination with standardized interviews on land tenure. The second set of 301 questionnaires were conducted randomly and independent from individual farmers to identify customary values and cultural identity toward their environment as it can be seen in following map:

21.4 RESULTS

21.4.1 Legal Assessment

In order to set the legal framework for the tenure security and sustainability on land-use it is important to first have a closer look at the laws Kenya provides on land in general and on environmental protection that are influencing land-use.

Kenya has a special role in sub-Saharan Africa with its new constitution that was enacted in 2010 and which recognizes customary land as itself. But already a year before the new constitution was launched, the Ministry of Land and physical planning released in 2009 the National Land Policy (Ministry of Lands and Physical Planning 2009). Its main goal was to provide a national land-use policy with regard to an comprehensive "legal, administrative, institutional and technological framework for [an] optimal utilization and productivity of land-related resources in a sustainable and desirable manner at national, county and community level" (Ministry of Lands and Physical Planning 2009). The legal principle of the acts are based on the premise of "philosophy of economic productivity, social responsibility, environmental sustainability and cultural conservation. Key principles informing it include efficiency, access to land-use information, equity, elimination of discrimination and public benefit sharing" (Ministry of Lands and Physical Planning 2017).

The Land Act from 2009 (Ministry of Lands and Physical Planning 2009) is opting to provide an overall framework and to include all critical key issues toward:

> land administration, access to land, land-use planning, restitution of historical injustices, environmental degradation, conflicts, unplanned proliferation of informal urban settlements, outdated legal framework, institutional framework and information management. It also addresses constitutional issues, such as compulsory acquisition and development control as well as tenure. (Ministry of Lands and Physical Planning 2009)

The new constitution in Kenya provides three ownership types: public, private, and community land (The Constitution of Kenya, Chapter 5, Article 61). Public land means land, which is kept and operated by the government (The Constitution of Kenya, Chapter 5, Article 62); (Bruhn 2019). Private land is attributable to a specific or legal person (Sirbira, Voß and Malaku 2011); (The Constitution of Kenya, Chapter 5, Article 64). The majority of Kenya's land is community land, which means that land is adjudicated to communities and is "acquired, possessed and transferred under community based regimes" (Willy 2018); (Parliament of Kenya 2010); (The Constitution of Kenya, Chapter 5, Article 63). At almost 70%, it is the most prevailing type of land in entire Kenya and it is mostly situated in rural areas, where land has for the most part not been officially registered (Bruhn 2019); (Wayumba 2015); (Sirbira, Voß and Malaku 2011). According to the 2010 constitution, land rights and property rights for women are now protected and recognized under national law inside the land act and the Kenyan constitution (Musangi 2017).

With the establishment of the Land Registration Act (2012), the National Land Commission Act (2012) and the Land Act (2012), which was later complemented and revised by the Land Law (Amendment) Act in 2016, the government consolidated the complex structure of land laws. The Land Act and the Land Registration Act repealed most of the individual acts and built a coherent land system which is consistent with today's constitution (Bruhn 2019).

With the new constitution (2010) and the Community Land Act (2016) communal land rights were officially recognized by law. The Customary Land Hold, Freehold and Leasehold land rights are the three existing tenure types for individuals and come up with different obligations and liabilities (Government of Kenya 2016); (Wayumba 2015).

Also in 2016 Kenya ratified the Paris Agreement on Climate Change and the declaration of the Climate Change Act and committed itself toward an active environmental conservation and sustainable development, which has been enforced by putting inside the national law and regulations (Ministry of Environment and Forestry 2019).

Despite of all these intentions on environmental protection, food security in Kenya is still a difficult topic with many related issues and a close relationship toward land tenure. The main indicator toward land-related conflicts, among others, is still a too low food production with a large population growth inside Kenya (Olielo 2013). According to the FAO the main task in fighting food insecurity is sustainable approaches in agriculture which comprises conservation aspirations regarding the ecology and focuses on socioeconomic factors. Land-use and land tenure are thus the key factors and act as a vital link between nature conservation and sustainable agriculture and therefore were analyzed in particular for this research.

In total there are five major laws that have to be considered within the limits of this research as the legal framework and that are seen in Table 21.1.

21.4.2 Situation in the Study

From the interviewed farmers the following family situation can be derived:

The average size per household inside the research area counts 4.39 persons with a standard deviation of 2.21 and a range between single households up to 16 persons per household. Slightly more than 85% of all interviewed households included children, which is contrary to the county-wide development of a stable toward shrinking population according to municipality numbers. The data received from the questionnaires suggest an average number of more than four per family (Figure 21.3).

The financial structure of all interviewed households (compare Figure 21.4) shows that most of the households have less than 5,000 KES available for the month, which has to be compensated by agricultural activities to ensure subsistence farming. A total of 79.5% of all interviewed people were engaged in farming as their main source of income.

TABLE 21.1
Overview of the Considered Legal Frameworks for Tenure Security

ELEMENT	LAW	YEAR	CONTENTS (only in parts)	AUTHORITY
Land	Land Act	(2016) (2012)	Comprehensive approach and overall framework for a sustainable land-use including customary tenure regimes and gender equality in land rights	Ministry of Lands and Physical Planning
Forest	Conservation Act for Forest Forest Act	(2016) (2012)	Provision of laws for a sustainable management and development of all forest resources including its rational utilization for socioeconomic development; Limitations and regulations on the type of trees which have to be protected and which can be utilized (KFS 2016)	Kenya Forest Service (KFS)
Water	Water Act	(2002)	"provide for the management, conservation, use and control of water resources and for the acquisition and regulation of rights to use water; to provide for the regulation and management of water supply and sewerage services" (Water Resources Management Authorities 2002)	Water Resources Management Authorities
Air	Environmental Management and Coordination Regulations	(2013)	Preventing air pollution and ensuring an healthy and clean ambient air quality for all people	National Environment Management Authority
Soil	Agriculture and Food Authority Act	(2013) (1955)	Uphold and maintain a stable agriculture, which protects the soil quality and its fertility under the utilization of a sustainable land management	Agriculture and Food Authority

(Based on A. Bruhn 2019, modified by Author)

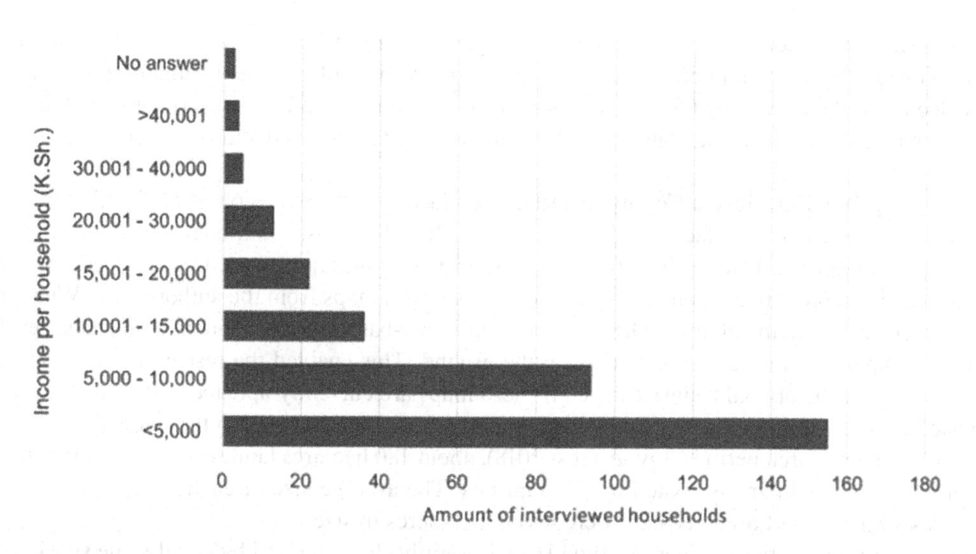

FIGURE 21.3 Monthly income per household in KES (n = 331). (From research group including Author 2019.)

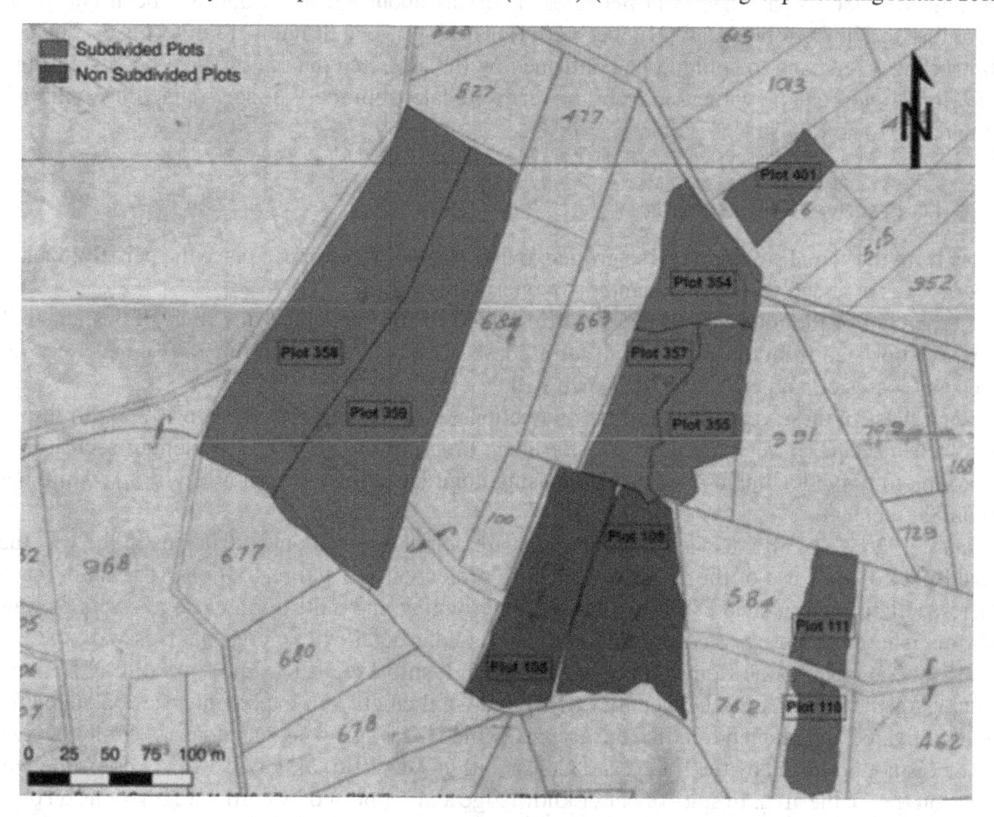

FIGURE 21.4 Subdivision of plots without updating the paper-based cadaster (base map). (From Author 2018.)

Ownership or family ownership can be found in 93.7% of all interviewed cases inside the case study area. Only 6.3% are leaseholders of the land in between one to five years and with an average of 2.75 years remaining during the data collection. Of those owning the land either individually or as shared property just 54.5% can proof their title by official documents. Although 32.6% of the interviewees stated they were born on the land, only 2.9% of all respondents named a family grave on the property as valid proof of their ownership over all tenure types.

According to official numbers of the county government in Taita Taveta County, 37% of the land is owned by the community and 35% of them are owning a land title document in the form of title deeds. Of the remaining 65% of all community owned land, 55.2% have no registered title and 5.8% are held under leasehold contracts, which are not offering deeds (County Government of Taita Taveta 2018).

The people of Taita Taveta County identified two factors for subdivision of land and these were economic and sociocultural factors (Museleku, et al. 2018). The most common reason for subdivision of land is inheritance. Due to the customary system male inheritances are treaded predominantly (Soini 2005). In order to compare, several selected cadastral maps from the authorities in Wundanyi were received and digitized in ArcGIS to convert the paper-based cadaster into a digital cadaster that can be compared with the mapped areas on the ground. This enabled the research to compare the situation inside the official cadaster to see if these maps are currently updated and when they were updated for the last time. The results of this comparison will be discussed in the findings.

Within the research period (July–August 2018), about 180 hectares land was mapped cumulating from 347 individual parcels owned by 331 families. The average size of each household was 0.54 hectares, and 79.4% of all household were less than 2 acres in size.

Agricultural activities all over the Taita Hills is mainly characterized by small-scale subsistence farming. In 1986 the average land per household was about 2.0 acres and has been cut to half according to Himberg by today (Himberg 2011). These very small numbers in terms of land size, Himberg was referring to couldn't be confirmed by the collected data in the field but a splitting of the original plots can be witnessed from the received data from the Wundanyi mapping authorities (Wundanyi Land Registry).

21.4.3 Perceived Tenure Security

Apart from the actual situation on the ground and the legal framework in place the perceived tenure security was assessed within the quantitative questionnaires (n = 331).

1.8 percent of the interviewed people were forced to sell their land due to evictions and 1.5% were expropriated from the government for at least one time. The remaining 97.7% never encountered any losses of land except on their own will.

Regardless of these very low numbers of evictions or expropriations, close to 10% of all respondents felt insecure or very insecure. Mostly named reason for this perceived tenure security was either the missing document or unofficial leasing contracts which were not legally fixed and only on oral agreement.

In total, 26.9% of all interviewed households are complaining about restrictions in the way they would like to make use of the land due to nature protection or legal restrictions. The main reason for restrictions were named as restrictions on tree logging and cultivation close to rivers or water streams. Therefore these two constraints were discussed in detail with representatives and experts of local associations like the Kenyan Forestry Research Institute (KEFRI), the Agricultural Research Campus of the Taita Taveta University (TTU) and local chiefs. All expert interviewed stated that most cultivated and planted tree like eucalyptus, which is fast-growing and consumes a lot of water but that is not protected by the Forest and Conservation Act (2016) although it has become the most common tree in the area. In spite of better knowledge and supported by cutting regulations on original tree species, original tree species are rarely planted and only remaining in very low percentages of pristine forests inside the research area. In the expert interviews on cultivation in proximity to rivers or other water sources, the experts from TTU and the local chiefs were taking adversary opinions. Whereas the local chiefs were defending the farmer's opinions why it is important to move closer to the rivers, the researchers insisted on the important needed buffer zones to avoid fertilizers and pesticides to enter the water and prevent too much reduction of the existing waterbody. Both sides had the same opinion when they argued that too many regulations are in place which makes it hard for farmers to comply with the laws.

21.4.4 Customary Perception

In the final part of the results, customary perception on land and connected land-use is described from the data collected and analyzed from the interviews and questionnaires distributed in the case study area.

According to 73.1% of all respondents, the government is in charge when it comes to land management and regulation. In 16.9% of all cases the community was seen as responsible party, 4.8% saw the village elders and 3.9% saw others like local chiefs in authority to supervise land and its management.

Regarding the transfer of ownership the opinion of the interviewee's shows with 67% that the majority is aware that ownership can be transferred but nearly all of them considered only transfer within their family as far as possible. A total of 29.9% revealed that transfer of ownership is not possible. This large amount of people believing their land can be only shared within their family can be explained by the 75.7% of the respondents that were sharing their land with family members. Only 2.3% mentioned they are using shared land with the community.

Lukorito has analyzed in his research the forms of inheriting land inside the area of Taita Hills and came to the conclusion that the most critical aspect is the fragmentation of land plots due to inheritance between generations. He also found that land usually is only sold or subdivided if there are financial plights due to health issues or financial needs for educational purposes (Lukorito 2018).

The expert interviews have shown that most people feel insecure with their land and because of this insecurity and missing customary knowledge from their ancestors they tend to overuse the soil too much and neglect farming techniques like terrace farming in the research area. This leads to terraces which are in very poor condition and that cannot be upkeep because of missing customary knowledge from older generation.

In terms of land rights toward women, 22.4% of all interviewees neglected that women are allowed to own land as it is written in the Kenyan constitution, instead keeping to their customary rules and completely deny ownership to women. A total of 0.6% of the respondents were not able to answer the question, while 77% stated that they were aware that women were allowed to own land.

In the research area, 68.6% of the people can remember being taught about the importance on protection of nature in school and 31.4% stated they never have been thought the values of nature conservation at all. A total of 41.7% of all respondents assumed that nature and the environment is not affected by human actions like farming or water extraction. A detailed insight of the results can be seen in Table 21.2, with the largest percentages highlighted in bold type.

Another striking result from the analysis of the questionnaires (n = 301) conducted by the researchers is that most people distinguish nature between plants and animals. According to their customary education and knowledge, only plants are worth of protection while animals are not

TABLE 21.2
Importance of Nature Conservation toward Different Agricultural Aspects (n = 331)

	Not Important	Less Important	Neutral	Important	Very Important	Don't Know
Soil	13.9%	7.6%	9.1%	31.1%	**36.0%**	2.4%
Water	0.9%	6.3%	5.3%	37.8%	**48.9%**	0.9%
Wood (timber)	16.0%	19.0%	5.7%	**30.2%**	26.6%	2.4%
Wood (charcoal)	**49.8%**	13.6%	8.2%	4.5%	3.3%	20.5%
Climatic conditions	0.0%	1.8%	7.3%	33.2%	**55.9%**	1.8%
Animal grazing	25.1%	**30.2%**	12.4%	22.4%	7.6%	2.4%

(Research group including Author 2019)

included and can be killed even toward extinction. The reason for this perception is that animals are mostly perceived as disturbing factors for their agriculture and therefore should be hunted or killed.

21.5 FINDINGS

With Kenya's fast-growing population and increasing demands, the Taita Hills as one of the five water towers is becoming more important for its provision of water to the lowlands and its provision of agricultural products. Inside the research area farming is the main source of income and the reduction of water accessibility, soil erosion and availability of land has made it difficult for subsistence farmers inside the area to sustain their financial needs. According to the World Bank poverty classification, 88.2% of the farmers inside the research areas would be classified as poor and beyond the line of US$1.90 per day and person (World Bank Data Lab 2019)

A decreasing parcel size and the loss of traditional agricultural knowledge are identified as the main two sources for this decreasing agricultural income. One of the main outputs of a related research on this data was the:

> lower income potential per household on agricultural products has led to poor farming techniques and to the dilapidation of agricultural plots. Monetary resources for seeds, farming tools and fertilizers are missing which resulted in an increase of partly fallowed and less maintained plots. (Bruhn 2019)

According to the expert interviews and the data collected from the questionnaires the main reason for people to move away from the region are possibilities for the local youth with shrinking possible income and plot sizes for agriculture. This accelerates the previous mentioned situation and leads to more young people moving away to larger cities and an overall decreasing population for this area. In order to tackle this problem, smart solutions to increase the tenure security and income possibilities has to be found.

The analysis of the retrieved registry index maps or paper-based cadaster maps has shown that most maps were not very accurate with regard to the accuracy including roads or public infrastructures. As shown in Figure 21.4, the georeferenced paper-based maps are showing roads and paths at wrong locations. In the process of geo-referencing the paper-based maps several adjustments had to be made to make the collected ground truth points fitting to the cadaster map. The maps were created in the year 1986 and the base map were never updated since then, only a relative low number of subdivisions were noted on the maps and covering only very few cases. Through the analysis of the selected parts in comparison to the mapped parcels, it is becoming very obvious that most titles in place are still referring to an old status in the official cadaster of the mapping authorities.

All official affairs toward land are handled by the regional registry office in Wundanyi and quite costly in their services and require long waiting periods to eight weeks on average, which people cannot afford according to the local land registrar S. Manyarkiy (Lukorito 2018) (Manyarkiy 2018):

> A common practice is to keep the former owner as a registered authority without an amendment. Unregistered property rights result in the problem that these people are just an unofficial owner of the land, who are then not backed by the land law in its entirety. (Manyarkiy 2018)

An update and digitization of the existing land registry could lead toward an improved accuracy and correct location of the individual and recent plots but it would have to be done using alternative approaches for low cost data collection like the so called Social Tenure Domain Model (STDM) (Enemark et al. 2010). STDM was introduced by the Global Land Tenure Network (GLTN) as a tool and concept that provides standardized and easy to accomplish smart methods for data recordation in order to create a pro-poor cadaster (UN Habitat 2017)

Despite improving the legal tenure situation by handing out legal binding documents it will also improve the perceived tenure situation at the same time. However the interviews have shown that most people are not aware of their actual rights and need further education on land rights and also woman's tenure rights. Also a large gap has been shown with 28.9% of all respondents being unaware who is actually in charge of land management. A possible solution to remedy this missing knowledge is an education program for local farmers and their children taking place in schools or community halls to tackle these identified gaps and enlighten the farmers toward a more sustainable land-use and better understanding of their rights provided by the Kenyan constitution and its connected laws.

In order to have a greater impact it is important to create these trainings in a manner that all farmers are able to understand the content and provide the information in their local dialects. Experiences in the field have shown that a smart approach via smartphone apps or educational videos might fail since infrastructure for these technologies are not in place and town hall meetings would be given more attention to ensure that the knowledge is spread within the community.

Also important to share is knowledge regarding the environmental protection laws. Already inside the briefly introduced nature protection laws the topics are presented in a complicated language which is difficult for the farmers to understand for most farmers. In addition to the national regulations the international conventions for example under the Paris Agreement that are imposing additional regulations on the farmers have resulted into uncertainty on what is allowed and what is not permitted under the rules. As the first step the imposed regulations and laws by the government have to be analyzed and restructured to avoid contradictory rules like different sized buffer zones along water bodies or unwanted promotion of eucalyptus tree species for tree cutting. The second step is the codification of the laws in a simpler manner so that they can be understood by the small-scale farmers and ensuring they are able to survive with their agricultural businesses. The third and final step would be to train the farmers themselves to protect the nature and educate them also on traditional knowledge but conceptualizing the knowledge within latest insights from science.

From these results, it can be seen that a majority of the local farmers are convinced that it is only possible to transfer ownership within the family and only to male members even though the law provides and permits different practices. To close this educational gap and to avoid further subdivision, additional training would be needed that can be supported by establishing a transparent market for vacant or unused land. This legal land market would have to be hosted in one of the official institutions like for example in the mapping authority for each sub-district. Task for this land market is to ensure a fair and comparative price for land or leases on agricultural land and to avoid fallow land that can be put into better use and on which the established infrastructures like terraces would be maintained. This market should be supported by three-dimensional GIS to ensure an accurate display of the plot size and distinguish its precise location within the district. With this support, an accurate price for the land can be calculated and a possible local market for small-scale farmers can be established on which those with small land sizes could increase their agricultural land.

In order to increase the learning effects and to optimize the planted crops remote sensing technology can be applied to receive multispectral satellite imagery. Applying this state of the art technologies, for example possible erosion can be calculated using the Revised Universal Soil Loss Equation (RUSLE) and to start sensitizing of local farmers on the effects of excessive corn farming and give reasons to maintain or rebuild the terrace farming. The existing research station inside the Taita Hills and the Agricultural Campus of TTU can be used especially as a perfect distributor of that knowledge and even apply these technologies for a smarter land-use and adaptations against soil erosion and soil degradation.

According to different authors, a large amount of its original woodland has vanished and the pristine forest has been reduced to only 1% (Piiroinen et al. 2015). Due to the promotion of eucalyptus trees and an increased demand of agricultural areas has drastic impacts on the future of the ecosystem and biodiversity hotspot of the Taita Hills. In order to keep the remaining forest spots untouched and ensure an extension of pristine forest types drone flights can be put in place to ensure a better conservation and to prevent tree logging inside protected areas. Therefore larger areas which are hard to pass can be assessed with areal images received from a drone in order to map changes and monitor illegal activities inside the protected forests.

21.6 CONCLUSION

The multicriteria decision analysis of the different factors influencing the three dimensions of tenure security has shown several issues related to land management. All comprised measurements entail both hands-on solutions and the applica of smart technologies to receive a sustainable change inside land management. Although the individual measurement seems to be very specific to the region they can be applied in a similar way inside different settings with customary tenure regimes. But the research has shown that not only smart solutions have to be followed up in order to achieve a change in the mind-set of the subsistence farmers because in regions with a very low access rate to internet some methods cannot be applied like the usage of apps to inform local people.

However, digitizing the cadaster and establishing a local transparent market and a one stop shop for a fast update of tenure can help the subsistence farmers. Supporting the nature conservation via drone supervision of the borders and using drone imagery for helping local farmers on their agricultural techniques can add not only value to nature conservation but also toward the agricultural yield of the individual farmers. As a further independent step also satellite data can be utilized to predict soil erosion and dry periods within the research area and help the farmers to choose and select their crops in a sustainable manner. The previous mentioned measurements all need to be supported by educational campaigns to educate the local people on several issues ranging from land rights to customary knowledge.

All of the measurements discussed in this chapter can be achieved with little financial support but require proper implementation and supervision. Therefore further research is needed to identify possible ways to translate written regulations and recommendations for the common people and subsistence farmers to make them understand how a sustainable behavior can maximize their overall benefit and conserve their precious land for future generations.

REFERENCES

Boitt, M., C. Ndegwa, and P. Pellikka. "Using Hyperspectral Data to Identify Crops in a Cultivated Agricultural Landscape—A Case Study of Taita Hills, Kenya." *Journal of Earth Science and Climatic Change* Nov 10, 5(9), 2014: 232.

Bruhn, A. *STDM Application in the Region of Kidaya-Ngerenyi, Kenya - Conciliation of Sustainable Agriculture and Land Tenure.* Munich: Technical University of Munich, 2019.

County Government of Taita Taveta. *County Integrated Development Plan 2018–2022.* Wundanyi: Taita Taveta County, 2018.

Department for Communities and Local Government. *Multi-Criteria Analysis: A Manual.* London: Communities and Local Government Publications, 2009.

Enemark, S., I. Williamson, J. Wallace, and A. Rajabifard. "Land Administration for Sustainable Development: Facing the Challenges—Building the Capacity." *FIG Conference,* Sydney: FIG, 2010.

Erdogan, H. E., P. K. Pellikka, and B. Clark. "Modelling the Impact of Land-Cover Change on Potential Soil Loss in the Taita Hills, Kenya, between 1987 and 2003 Using Remote-Sensing and Geospatial Data." *International Journal of Remote Sensing* 32(21), 2011: 5919–5945.

FAO. Land Tenure and Rural Development, 2002. Available at: http://www.fao.org/3/Y4307E/Y4307E00.htm.

Government of Kenya. *Community Land Act.* Nairobi: Government of Kenya Press, 2016.

Habel, J. "Biodiversitynetwork Kenya." Apr 2019. https://biodiversitynetworkkenya.wordpress.com/ (accessed April 20, 2019).

Himberg, N. *Traditionally Protected Forestś Role within Transforming Natural Resource Management Regimes in Taita Hills, Kenya.* Helsinki: Helsinki University Print, 2011.

Kenya Forestry and Research Institute. *Water Towers.* 2018. http://kefriwatertowers.org/ (accessed April 15, 2019).

KFS. *Forest Conservation and Management Act.* Nairobi: Kenya Forest Service, 2016.

Lukorito, J. "The Process of Subdividing from 1st. November 2018." *Daily Nation*, 2018.
Ministry of Environment and Forestry. *Ministry of Environment and Forestry*, 2019. http://www.environment.go.ke/?p=3001 (accessed April 5, 2019).
Ministry of Lands and Physical Planning. *National Land Policy*. Nairobi: Ministry of Lands, Kenya, 2009.
Ministry of Lands and Physical Planning. *Sessional Paper, No. 1*. Nairobi: Ministry of Lands and Physical Planning, 2017.
Musangi, P. "Women Land and Property Rights in Kenya. Responsible Land Governance: Towards an Evidence Based Approach." *World Bank Conference*. Washington, DC: World Bank Conference, 2017.
Museleku, E., P. M. Syagga, M. Kimani, and W. Mwangi. "Implications of Agricultural Land Subdivision on Productivity: Case Study of Kajiado County." *International Journal of Innovative Research and Knowledge*, 3(6), 2018: 195–212.
Olielo, T. "Food Security Problems in Various Income Groups of Kenya." *African Journal of Food, Agriculture, Nutrition and Development*, Sep 2013.
Parliament of Kenya. *Constitution of Kenya*. Nairobi: Parliament of Kenya, 2010.
Pellikka, P. K., V. Heikinheimo, J. Hietanen, E. Schäfter, M. Siljander, and J. Heiskanen. "Impact of Land Cover Change on Aboveground Carbon Stocks in Afromontane Landscape in Kenya." *Applied Geography*, May 2018: 178–189.
Piiroinen, R., J. Heiskanen, P. Hurskainen, J. Hietanen, and P. Pellikka. *Mapping Land Cover in the Taita Hills, SE Kenya; Using Airborne Laser Scanning and Imaging Spectroscopy Data Fusion*. Berlin: University of Helsinki, 2015.
Sirbira, D. N., W. Voß, and G. C. Malaku. "The Kenyan Cadastre and Modern Land Administration." *ZfV Zeitschrift für Geodäsie, Geoinformation und Landmanagement*, May 2011.
Soini, E. *Livelihood Capital, Strategies and Outcomes in the Taita Hills of Kenya*. Nairobi: World Agroforestry Centre, 2005.
UN Habitat. *Social Tenure Domain Model*, 2017. http://stdm.gltn.net (accessed August 12, 2019).
Wang, J., Y. Jing, C. Zhang, and J. Zhao. "Review on Multi-Criteria Decision Analysis Aid in Sustainable Energy Decision-Making." *Renewable and Sustainable Energy Reviews*, Dec 2009: 2263–2278.
Water Resources Management Authorities. *Water Act*. Nairobi: Water Resources Management Authorities, 2002.
Wayumba, G. *A Review on Special Land Tenure Issues in Kenya*. Nairobi: FIG, 2015.
Willy, L. A. *The Community Land Act in Kenya Opportunities and Challenges for Communities*. Nairobi: MDPI, 2018.
World Bank Data Lab. *World Bank Data*, 2019. https://worlddata.marketpro.io/ (accessed April 16, 2019).

Section VI

Future of Land Management

22 Advancing Responsible and Smart Land Management

Walter Timo de Vries, John Bugri, and Fathima Mandhu

CONTENTS

22.1 INTRODUCTION

Advancing responsible and smart land management relies on number of factors. On the one hand is the need for a better understanding of the effects that contemporary external drivers have on existing socio-legal-economic relations to land. On the other hand, on the advances in land management sciences and professional practices, manifested through newly derived and tested concepts, cause-effect relations, frameworks, methods, and instruments are also important considerations. Contemporary drivers are significant if these are persuasive and disruptive. Behavioral sciences call a driver persuasive if it is sufficiently attractive and credible, associated with power and status, has relevant meaning to the recipient, and leads ultimately to a change in attitude (Guyer et al. 2019). Technologies are persuasive if they come without coercion, manipulation, or deception (van Delden, de Vries, and Heylen 2019) and yet change socioeconomic relations and perceptions and expectations. In a similar way, disruptive drivers and changes are often associated with innovations which displace and replace existing socio-organizational structures and workflows, interpersonal and interinstitutional relations, utilization of technologies, and societal situations. From a technological point of view, Gartner's yearly technological trend reports define a strategic technology trend as one with *substantial disruptive potential that is beginning to break out of an emerging state into broader impact and use, or which are rapidly growing trends with a high degree of volatility reaching tipping points over the next five years.** Hence, disruption does not necessarily reflect a radical and instant change, but a gradual yet finite change, whereby something is actually becoming obsolete and is being replaced. New form of land management has to deal with such changes and consequently may need to undergo persuasive and disruptive changes itself.

* https://www.gartner.com/en/newsroom/press-releases/2018-10-15-gartner-identifies-the-top-10-strategic-technology-trends-for-2019.

22.2 EVIDENCE OF PERSUASIVE AND DISRUPTIVE DEVELOPMENTS AFFECTING LAND MANAGEMENT

Which developments are so persuasive and disruptive that they will foster significant transitions and transformations in responsible and smart land management? Zevenbergen, de Vries, and Bennett (2018) identify population growth, urbanization, large-scale economic investments, climate change adaptation, individualization, conflicts and disasters as key drivers for persuasive and disruptive changes for people-to-land relations. For Africa specifically (Shackleton et al. 2019) point to a complex of interrelated drivers (biophysical, socioeconomic, and political) which have affected landscapes and livelihood. They argue that despite considerable improvements in road and communication infrastructure, enhanced water supply and other services and new opportunities for entrepreneurship in Africa, the negative effects of the changes in the ecosystems due to climate changes are outweighing the improvements in agricultural production and manufacturing, food processing, employment of alternative energies and smarter technologies, for example. Hence, climate change is therefore one of the crucial persuasive and disruptive changes. It is leading to decreasing availability of productive land, depletion of fertile soils and desertification, shortages in water, uncertainties in rainfall, increased frequency of storm surges, biodiversity loss, and increased CO_2 levels. All of these changes have an effect on land tenure and the location where people can still make use of land.

Blockchain, artificial intelligence, machine learning, big data science, and robotics are indeed examples of persuasive and disruptive technologies. These technologies are increasingly becoming anthropomorphic, in the sense that they can simulate, imitate, and predict human behavior and decisions. They can increasingly also be employed for replacing people, surveillance and intrusion in human lives, and sustaining power. What is, however, lacking in these technologies is an intrinsic system of values and responsibility. The technologies are completely dependent on their makers rather than on negotiated and democratically established social agreements on what is considered right and fair. Hence, to produce new opportunities to manage land in a smart and responsible manner the focus should be on adopting the 10 strategic technological trends. This intelligent digital mess is about connections between hardware and software. The use of robots and drones as autonomous things that can promote collaborative intelligence and working together as the first strategy.

The various chapters in this book provide further evidence of the specific nature of these persuasive disruptions in the African context.

The first disruption can be seen in an emerging distrust in data quality, or otherwise put, the quality of available data is no longer taken for granted. Chapter 2 has shown that valuation data exhibit a large degree of variation, even though they are assembled by professional valuers. This has two possible implications that are paradoxically contradictory. One could advocate for better standards and uniformity in data capturing and data assembling processes. This would theoretically lead to lower levels of variation and higher degrees of uniformity. The discourses on data modeling and spatial data infrastructures have been advocating such processes. Reversely, if data are uniform there is no possible way to check the quality with alternative processes. Hence, more "redundancy" of data is needed to ensure proper mechanisms of quality checks and validations. There are, however, options to overcome this paradox. In the case of Uganda, de Vries and Nyemera (2010) and de Vries and Lance (2011) argue that additional redundant data remain relevant as long as no data are available, as is the case for most land rights related data, or as long as quality procedures are insufficiently transparent or documented. However, if these two conditions are fulfilled, variations in data quality are likely to become smaller.

Chapters 3, 8, and 13 specifically address issues of urban relocation of tenants and guided and unguided urban sprawl. These developments can also be considered disruptive, especially when they lead to grassroots voiced social-spatial policies. Existing institutional policies build on conventional ideologies of economic growth and centrally guided land policies still tend to stimulate the continuation and increase of sociospatial inequities and injustices, and do not provide sufficient attention for local and contextual knowledge, experiences, and realities. In contrast, there is a growing movement of locally mobilized stakeholders who are increasingly more effective in voicing their key demands for more spatial justice and more customized solutions for land-related problems that are mostly felt at the local level. Such voices are present in both profoundly urbanized areas where competing interests for adequate space is most contested and in cripplingly rural areas where access to services and sustainable livelihood is most at stake. The disruption can again be found in the underlying philosophy of land policies, which is becoming more pragmatist and radical at the same time. They are pragmatists because stakeholders prefer immediate action with immediately visible and measurable results in outcomes; radical, because the conventional procedures and protocols to implement administrative, regulatory, and organizational changes may be completely bypassed.

Another element that is persuasive in the African land tenure context is the social tenure domain model and the associated land administration domain model. Chapters 9, 14, 18, and 21 address this model in different ways and at different stages and contexts of implementation. A focus on land tenure, instead of a focus on freehold land rights as a basis provides a completely different perspective and a new frame. Instead of reporting that more than 70% of all land is not captured by a system of land rights or land titles, one could start by framing that all land where people have an interest is probably captured by some form of tenure. The core problem is then not per se to convert that from informal to formal tenure but to capture the actual tenure in its own right. The social tenure domain model has actually this purpose. This is a fundamental change to the existing discourse of titling. If such a paradigm is accepted, it would address the point that ownership of land rights in the freehold context is alien to Africa and would also imply more tenure responsive land policies (Chigbu et al. 2019).

A last disruption, which is reported in Chapters 17 and 20, is that of public value capture. As Africa is developing, its infrastructures need to expand as well. Public value capture would be one of the ways to finance such developments, assuming that the benefits can be equally shared among the public. In itself this provides a radical change to current financing mechanisms, such as taxation. Given that processes of direct land taxation and cost redistribution through public investment in most African countries is far from adequate, public value capture provides a realistic alternative.

22.3 MAKING LAND MANAGEMENT MORE RESPONSIBLE AND SMARTER

Is land management science and practice sufficiently advancing to cope with newly arising challenges? On the one hand, it should improve the living conditions of people at household level, and it should also make the institutions more responsible and enhancing the creation of shared responsibilities. In search for new ways of land management, de Vries (2018) uses two dimensions to formulate new research directions which link smart information to responsible decisions and behavior. The information dimension describes to what extent the information regarding land is complete, accurate and guaranteed, or incomplete, inaccurate, and not guaranteed. The decision and behavioral dimension reflect the extent to which social rules and legal norms of different stakeholders coincide or are contentious. Using these two dimensions, one can derive four specific research and development directions for responsible and smart land management. Figure 22.1 describes the key challenges for these two dimensions.

When the information is complete and all of the rules are respected, the main task is to increase the efficiency of enforcing the rules and to maintain a set of rules that is still relevant to society in the face of ever-changing social preferences. In this case, there is a risk that the information will lead to unwanted surveillance and that a focus on administrative efficiency may lead to a technocratic bureaucracy that gradually ignores human values and human dignity (Georgiadou, Lungo, and Richter 2014). If the information is complete and supports a system of social rules but the legal framework is ignored and circumvented, there is a problem with the legal norms. The challenge in this case is to adapt the legal framework to societal needs or to invest in enforcement and compliance. Both options lead to significant transaction costs (Williamson 1998). Therefore, a key question is which option should be chosen. When information is incomplete and rules are

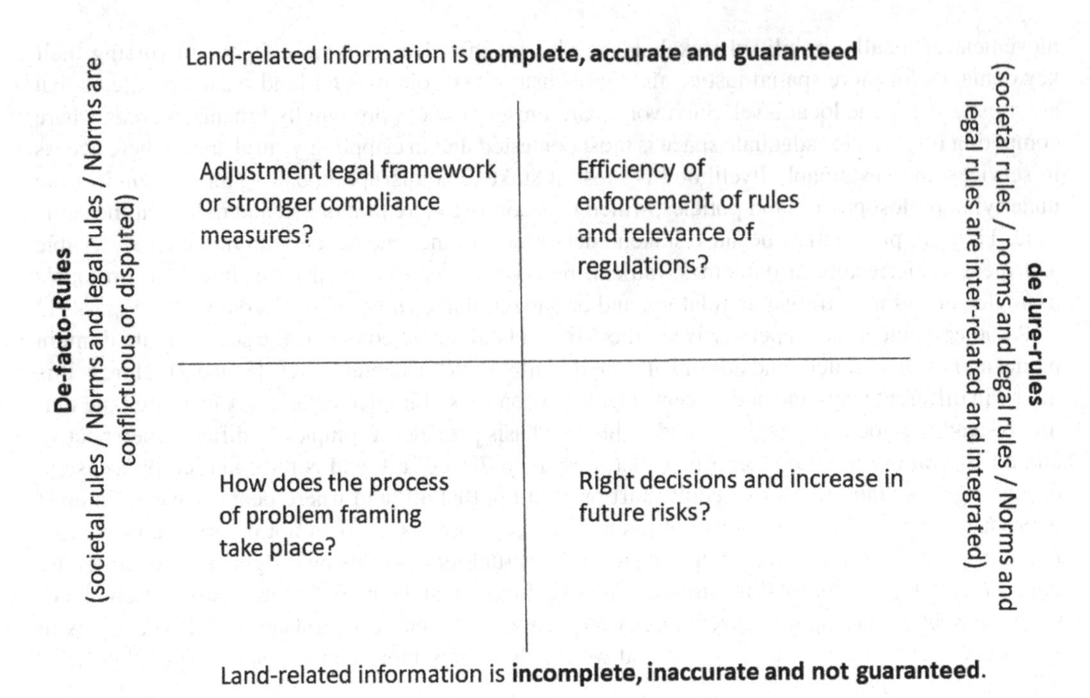

FIGURE 22.1 Research framework for responsible and smart land management.

enforced without any challenge, the biggest challenge is whether decisions are really made for the right reasons, increasing the risks in the future. An important question then is dealing with risk and uncertainty. When information and rules are controversial, there are unstructured governance issues (Hoppe 2011). The challenge of unstructured governance issues lies in the political domain; in other words, how does the process of problem framing take place?

For land management interventions specifically, one can derive the following challenges in each of the quadrants (numbering starting bottom left, continuing upper left, upper right, and completing the circle bottom right).

Quadrant 1 combines research challenges to address systems that exhibit a low quality of land information, and a high degree of effective informal rules. In such circumstances it is necessary to rethink existing paradigms and discourses. Two specific elements have been addressed in this book. The first is the inclusion of the concept of human recognition in the discourse of land tenure. By including human recognition, a bridge is made to perceived individual rights and existing individual behavior. Currently little information is available about the interpersonal dynamics within couples and groups of people that make up either public or communal land claims or interests. By evaluating human recognition this quandary is addressed. Secondly, there is a need to reexamine the relation between public and social values, which we could label as an "epistemic continuum of economic and social values in and for land management." This provides a new manner to look for and specify the ground rules for land management intervention. One of the missing links in connecting the different norms and value systems existing in land management is the difference in epistemologies (de Vries and Voß 2018). When comparing the epistemology of economic and social values, one needs to look at how each of the respective value systems gain meaning in and from the other value system. Value has meaning to stakeholders by how stakeholders engage with the manifestation of the values (de Vries and Miscione 2010). This engagement is usually in the form of social contracts emerging from consistent and mutually agreed social interactions. These social contracts, which also emerge in the discussion on the relevance of blockchains for example, gain meaning through a basic assumption that one can predict and measure the effects of choices in transactions. Professional actors rely on such objective models for their decisions, and also act upon these models even though some of the benefits may be qualified as intangible. One could, however, also argue that there exists a certain epistemic continuum of values, which is connected through the common ground of optimization—as both economic and social value systems tend to favor some sort of optimal state. This state is reached through favoring behavior that reinforces certain shared values and disfavoring actions that contradict the main logic of the respective value systems. de Vries and Voß (2018) provide a table of this continuum of land management values, connected to respective disciplines. For Africa specifically, this implies revising the value systems in use for land management interventions. As indicated in the previous section, public value capture provides one of the alternatives, if it sufficiently coincides with new systems of value capturing, which derive more consistent values and that connect better to public investments with public benefits.

Quadrant 2 assembles research challenges whereby there is a high quality, and possibly also a high volume, of land information, yet the society still tends to function on the basis of many informal rules. In these circumstances the legal framework is out of sync with the monitoring and/or enforcement framework. There exists, for example, much data on the direction, size, and volume of urban sprawl, but the formal means to control and direct this development are completely ineffective. Instead, the developments are guided by unknown or hidden mechanisms out of the public eye of administrators or citizens. A core question in these cases is how to align information collection better to policies and vice versa.

Quadrant 3 brings together research challenges whereby there is both a high quality of land information, combined with effective formal rules. In such a scenario there are fundamentally two problems: effective maintenance and relevance. The first challenge is largely an operational management problem, that is, how to execute agreed and formalized workflows in the most effective manner. The second is a probe of regular reflection whether existing solution strategies are still relevant and appropriate. Perhaps good functioning titling solutions fall into this category. Is land titling in Africa still solving and addressing the problem for which titling was started?

Quadrant 4 assembles the research challenges in situation where there is a low quality of land information, but an effective apparatus of formal rules. In other words, there is an effective administration and system of land governance, but the information sources are incomplete or ineffective. The effect of this situation is that while the average quality of land governance may be acceptable, but at some specific locations or in some specific cases there is a major gap in information provision or in information transparency. In such cases a possible solution strategy is to make better use of connected monitoring systems, and open, big and linked data.

22.4 MAKING LAND POLICIES MORE RESPONSIBLE AND SMARTER: REFLECTION OF THE FOUR QUADRANTS TO THE LAND POLICY FRAMEWORK IN AFRICA

The Framework and Guidelines on Land Policy (African Union [AU] African Development Bank [ADB], Economic Commission for Africa [ECA] 2010) provides a clear overview of the historical, political, economic, and social background of the land question in Africa. To show how smart and responsible these overreaching principles are a reflective connection has been drawn between the four quadrants with the selected parts of some of the guidelines. Two main thematic areas that cut across all the issues discussed in the various chapters of this book is the strategic as well as the social and cultural context of land as a valuable resource in Africa. In terms of constructing a social and religious identity, land remains an important factor. The link across generations that the family lineages and communities share and control constitutes land resources. Land therefore is part of the spirituality of society. The policy indicates the appropriate strategies for land policy development. In terms of information and data the best practice should include wide public consultation in the process, followed by engagement with civil society organization and communication for land policy development. In terms of development of a smart and responsible land policy the initial stage of stating the problem and formulating the correct strategy to provide a solution should be the basis of land policy. This process of problem forming in quadrant 1 is reflected in the African Land Policy framework and guidelines (AU,ADB & ECA 2010) under the development and application for tacking principles that provide for measures direct or indirect, either positive or negative effects on land policy on beneficiaries and natural resources. For quadrant 2, one of the necessary steps for effective land policy implementation includes identifying components of the policy that must be legislated and the preparation of instruments, structures and procedures for the management of those components. In addition, the domestication of relevant regional and international commitments into the land policy. On efficiency of enforcement of rules and relevance of regulations under quadrant 3 the guidelines on land policy state that the primary role of land in development must focus on the state of land administration systems. The two aspects are land rights delivery and efficiency and efficacy of the laws, structures and institutions for land governance. Finally, under quadrant 4 the right decisions and increase in future risks for Africa is to put in place adequate policies to ensure that the risks associated with global ecosystem, demand for energy supplies and rapid increase in foreign direct investment, are avoided or effectively managed. In terms of overall reflection, the guidelines lack the elements of responsible and smart land governance.

22.5 MAKING LAND MANAGEMENT MORE RESPONSIBLE

If responsible land management incorporates the eight Rs of de Vries and Chigbu (2017), a core question remains when and under which conditions this can be achieved in Africa. Does this require new types of ideologies and new norms guiding new types land policies? Is it achieved by fundamentally changing the manner in which fiscal gaps of governments are being addressed? Is it just a matter of focusing on effects of climate change? The answer is yes and no to all of these questions. There is simply no priority list and no single solution to these sets of problems. Making land management more responsible is a continuously evolving matter that requires regular reflections and updates given changing circumstances. Titling was once considered a major condition for economic development and public value capture currently seems an effective instrument for spatial development. However, these instruments do not generate the same solutions and benefits across all countries. The issue of responsibility is still at stake. Therefore, there is still a need to better understand how mutual systems of trust, accountability, legality, and legitimacy are established in the land domain. Such requires not only a thorough investigation of the functioning of existing and new types of legislations but also further developments in adopting and refining the continuum of land rights. The continuum of land rights concepts should also incorporate new concerns about surveillance possibilities that may arise by increasing the amount and usage of linked and big data. The effectiveness of land rights may be eroded if the information of the land rights is misused for other purposes than to secure the right.

Last but not least, new forms of land interventions may need to be sought. Experience has shown the effective use of land consolidation programs in the past, but the instrument may need to be revised and improved. Given the concerns related to the exchange and tradeoffs of allocating rights and redistributing rights, especially when they lead to more social and economic injustice and inequality, one may need to develop tools of social land consolidation, whereby land rights are not redistributed based on economic values but on social values. Part of this may be to include some older ideas of having public benefits act as social stakeholders (Grear 2012, Starik 1995, Stone 1972). The rights of nature and also the rights to the city discourses turn over the legal foundations that (spatial) governance and land management is rooted in the primacy in the rules of law and not in the rules of men and women, local communities, or the rules that benefit nature. Imagine if nature was considered one of the stakeholders with an equal right of voting as people or governments: how would the allocation of land rights look like? One could also approach it from the implications of excludability which are connected to private ownership rights. If a constitution recognizes that everybody can have an ownership or property right, every individual ownership right also immediately comes with a social responsibility. One cannot simply do whatever one wants to do with private ownership. The social responsibility forces an individual right holder to do the right thing (socially/publically) with private rights. Allocating and acknowledging a private right to the environment, as a kind of public restriction, also reflects a public responsibility that can override the excludability which is typical of private rights.

22.6 MAKING LAND MANAGEMENT SMARTER

Can we make land management smarter? In addition to fit-for-purpose approaches, developments in utilization of blockchain technologies , voluntary geographic information and open cadasters combined with big and linked data, revised versions of LADM, 3D, 4D, and 5D cadasters, new simulation tools, artificial intelligence (AI), neural networks are still ongoing (Wagner and de Vries 2019, de Vries, Bennett, and Zevenbergen 2014). Spatial data remain crucial and should not be just for collecting data but also to make the data useful and accessible in multiple forms and shapes and in all sorts of application fields. Taking into account both the time dimension and the third dimension of land data, would allow better information descriptions and management of subsoil and of resources and infrastructures. In addition, it could help to support managing the space above soil and to secure rights in spaces which are currently still unused. When such developments are accomplished even in the current Africa context, technology can certainly be used to build scenarios for responsible use of space and to find more usability for new and old spaces. Examples of application areas include investigating the reasons and causes for the existence of vacant land and depleted land, monitoring of land-use and land ceiling, assessing which land is used ineffectively or inappropriately, determining which land requires which energy sources, and determining the relations between land tenure and outbreaks of diseases of people, plants, and animals.

22.7 FURTHERING RESEARCH OF LAND MANAGEMENT IN AN AFRICAN CONTEXT

Land management in the African context needs cross–state land information systems and approaches that do not stop at boundaries. Monitoring and assessment systems work best if they are connected and rely on open, big and linked data, coupled with enhanced quality descriptions and transparency. Titling of land in a continent where the majority of land parcels are not registered remains relevant, but there has to be a Plan B. Reports of some titling projects are not very promising in terms of seeing a major improvement in tenure security or in uptake of loans and credits. In other words, having better titling systems in place without enhancements in accompanying and interrelated institutional structures and associated information infrastructures will not create better development opportunities, neither will it lead to a more secure and equitable distribution of land. The crucial question still remains whether advances in technology can really address poverty and social injustices. Or in other words, is development and access to technology enough, or does it need to be accompanied by crucial disruptive changes in behavior, expectations, values, and institutions? Will there be truly new types of information repositories, for example, neocadasters (de Vries, Bennett, and Zevenbergen 2014), which reflect new institutional structures that do not only reflect existing status quos but also affect behavior and legitimacy in different manners? This remains one of the key advancements to be made.

REFERENCES

Chigbu, Uchendu Eugene, Pierre Damien Ntihinyurwa, Walter Timo de Vries, and Edith Ishimwe Ngenzi. 2019. "Why tenure responsive land-use planning matters: Insights for land use consolidation for food security in Rwanda." *International Journal of Environmental Research and Public Health* 16(8):1354. doi: 10.3390/ijerph16081354.

de Vries, Walter Timo. 2018. "Suche nach neuen Grenzen des Landmanagements." *zfv* 143(6):373–383. doi: 10.12902/zfv-0228-2018.

de Vries, W. T., R. M. Bennett, and J. A. Zevenbergen. 2014. "Neo-cadastres: Innovative solution for land users without state based land rights, or just reflections of institutional isomorphism?" *Survey Review* 47(342):220–229. doi: 10.1179/1752270614Y.0000000103.

de Vries, Walter T., and Uchendu Eugene Chigbu. 2017. "Responsible land management—Concept and application in a territorial rural context." *fub. Flächenmanagement und Bodenordnung* 79(2 April):65–73.

de Vries, W. T., and K. T. Lance. 2011. "SDI reality in Uganda: Coordinating between redundancy and efficiency." In: *Spatial Data Infrastructures in Context: North and South*, edited by Zorica Nedovic-Budic, Joep Crompvoets, and Yola Georgiadou, 103–119. Boca Raton, FL: CRC Press.

de Vries, W. T., and G. Miscione. 2010. "Relationality in geo-information value: Price as product of socio-technical networks." *International Journal of Spatial Data Infrastructures Research: IJSDIR* 5: 77–95.

de Vries, W. T., and B. W. Nyemera. 2010. "Double or nothing: Is redundancy of spatial data a burden or a need in the public sector of Uganda." *Information Research* 15(2010):4, 15 p.

de Vries, Walter Timo, and Winrich Voß. 2018. "Economic versus social values in land and property management: Two sides of the same coin?" *Raumforschung und Raumordnung | Spatial Research and Planning* 76(5):381–394. doi: 10.1007/s13147-018-0557-9.

Georgiadou, Y., J. H. Lungo, and C. Richter. 2014. "Citizen sensors or extreme publics? Transparency and accountability interventions on the mobile geoweb." *International Journal of Digital Earth* 7(7):516–533. doi: 10.1080/17538947.2013.782073.

Grear, Anna. 2012. *Should Trees Have Standing? 40 Years On.* Cheltenham, UK: Edward Elgar.

Guyer, Joshua J., Pablo Briñol, Richard E. Petty, and Javier Horcajo. 2019. "Nonverbal behavior of persuasive sources: A multiple process analysis." *Journal of Nonverbal Behavior* 43:1–29.

Hoppe, R. 2011. "Governance of problems: Puzzling, powering and participation." Portland, OR: Policy Press. http://www.adobe.com/products/digital-editions/download.html.

Shackleton, Sheona, Vanessa Masterson, Paul Hebinck, Chinwe Ifejika Speranza, Dian Spear, and Maria Tengö. 2019. "Editorial for special issue: 'Livelihood and landscape change in Africa: Future trajectories for improved well-being under a changing climate'." *Land* 8(8):114.

Starik, Mark. 1995. "Should trees have managerial standing? Toward stakeholder status for non-human nature." *Journal of Business Ethics* 14(3):207–217.

Stone, Christopher D. 1972. "Should trees have standing—toward legal rights for natural objects." *Southern California Law Review* 45:450.

van Delden, Robby, Roelof A. J. de Vries, and Dirk K. J. Heylen. 2019. "Questioning our attitudes and feelings towards persuasive technology." In: *Persuasive Technology: Development of Persuasive and Behavior Change Support Systems*, Oinas-Kukkonen, H., Win, K. T., Karapanos, E., Karppinen, P., & Kyza, E. (Eds.). Cham, Switzerland: Springer.

Wagner, Magdalena, and Walter Timo de Vries. 2019. "Comparative review of methods supporting decision-making in urban development and land management." *Land* 8(8):123.

Williamson, O. E. 1998. "Transaction cost economics: How it works, where it is headed." *De Economist—Quarterly Review of The Royal Netherlands Economic Association* 146(1):23–58.

Zevenbergen, Jaap, Walter de Vries, and Rohan Bennett. 2018. "Dynamics in responsible land administration: Change at five levels." *XXVI FIG Congress 2018: Embracing Our Smart World Where the Continents Connect: Enhancing the Geospatial Maturity of Societies.*

Index

Printed in the United States
by Baker & Taylor Publisher Services